Neptune

The planet, rings and satellites

Springer
London
Berlin
Heidelberg
New York
Barcelona
Hong Kong
Milan
Paris
Santa Clara
Singapore
Tokyo

Ellis D. Miner and Randii R. Wessen

Neptune

The planet, rings and satellites

Springer

Published in association with
Praxis Publishing
Chichester, UK

Ellis D. Miner
Cassini Science Adviser
NASA Jet Propulsion Laboratory
California Institute of Technology
Pasadena
California
USA

Randii R. Wessen
Telecommunications and Mission Systems Manager
NASA Jet Propulsion Laboratory
California Institute of Technology
Pasadena
California
USA

SPRINGER–PRAXIS BOOKS IN ASTRONOMY AND SPACE SCIENCES
SUBJECT *ADVISORY EDITOR*: John Mason B.Sc., Ph.D.

ISBN 1-85233-216-6 Springer-Verlag Berlin Heidelberg New York

British Library Cataloguing in Publication Data
  Miner, Ellis D.
    Neptune: the planet, rings and satellites. –
      (Springer-Praxis books in astronomy and space sciences)
    1. Neptune (Planet)
    I. Title   II. Wessen, Randii R.
    523.4′81
    ISBN 1-85233-216-6

Library of Congress Cataloging-in-Publication Data
  Miner, Ellis D., 1937–
    Neptune: the planet, rings and satellites / Ellis D. Miner and Randii R. Wessen.
      p. cm.—(Springer-Praxis books in astronomy and space sciences)
    Includes bibliographical references.
    ISBN 1-85233-216-6 (acid-free paper)
    1. Neptune (Planet)   I. Wessen, Randii R., 1958–   II. Title.   III. Series.
    QB691 .M56 2001
    523.4′81—dc21   2001020837

Project copy editor: Alex Whyte
Cover design: Jim Wilkie
Typesetting: Originator, Gt Yarmouth, Norfolk, UK

Printed on acid-free paper supplied by Precision Publishing Papers Ltd, UK

To the dedicated men and women of
the space program, both human and robotic exploration
and both in the United States and elsewhere, for the
amazing accomplishments of the past half century

Contents

List of figures

See also colour section between pages 134 and 135

List of tables

Authors' preface

It has now been more than a decade since the authors worked together as part of the science effort within the Voyager Project at the Jet Propulsion Laboratory. Each of us was involved in robotic exploration missions to the planets prior to the Voyager Project, and each has been involved in other missions and endeavors since our Voyager experience. We are, nevertheless, both in full agreement that no other experience is likely to come close to matching the excitement of anticipation and discovery that accompanied the Voyager Mission – and that was especially true for the Neptune encounter of Voyager 2 in 1989.

This book follows the general format and content of *Uranus: the Planet, Rings and Satellites*, written as a part of the Wiley-Praxis Series in Astronomy and Astrophysics, and no attempt has been made to avoid repetition of materials and information contained in the 1998 edition of that book. On the contrary, each book has been written as a standalone project. It is also our belief that the inspiring story of the continuing discoveries associated with spacecraft and telescopic exploration of the Solar System in which we live cannot be told too often. We have been fortunate to be able to witness much of that discovery and we desire to share the excitement and wonder of those days, months, and years.

A professional text, entitled *Neptune and Triton*, was published in 1995 by The University of Arizona Press. It contains 23 chapters spanning more than 1,200 pages. Its editor, Dale Cruikshank, and most of its contributors, are our friends and colleagues from Voyager days. Much of the description of the findings and conclusions about Neptune and its system from the Voyager Mission that we include in this book are based on the results chronicled in *Neptune and Triton*. In contrast with that text, however, we attempt here to present a more coherent account in a language and style more accessible and understandable to the 'non-expert'. We have attempted to include the most salient conclusions, without getting into esoteric details of interest primarily to those whose lives and careers are intimately tied to the scientific investigation of the giant planets of the outer Solar System.

With the advent of new tools with advanced capabilities, such as the Hubble Space Telescope and large Earth-based telescopes (such as the Keck telescope atop Mauna Kea on the island of Hawaii) equipped with adaptive optics, observations from Earth are beginning to rival those of spacefaring robots like Voyagers 1 and 2. We attempt to include in this book relevant results from such investigations, especially as they alter or expand upon the Voyager results. It is likely, in a field where new discoveries are announced on an almost daily basis, that some of the material in this book will already have been superseded by new observational or theoretical studies. Such is the nature of science in general, and Solar System studies in particular, and the authors welcome constructive criticism on the depth, breadth, and accuracy of this book's contents.

For esthetic reasons, and to show respect for the investigators, some of the illustrations that appear in the text have also been reproduced in the color section between pages 134 and 135.

Ellis D. Miner
Randii R. Wessen
Pasadena, California, December 2000

Acknowledgments

The authors gratefully acknowledge Caltech's Jet Propulsion Laboratory who provided the equipment, part of our time, and most of the figures contained in this book. Funding for this work is part of JPL's contract with the National Aeronautics and Space Administration in the United States to conduct exploration of the Solar System. We also acknowledge the editor and authors of the professional text, *Neptune and Triton*, for their permission to use the findings and figures contained in that book in Chapters 8 through 12 of this book.

List of acronyms

Acronyms are regularly used in space program work, and have become an invaluable part of NASA vocabulary. There is, however, a tendency to use three-letter acronyms which, at times, entails dropping or adding a letter from the relevant acronym. Where acronyms have been used in this book we have tended to use those that were adopted during the Voyager Mission, rather than invent new (perhaps more sensible) ones.

AACS	Attitude and Articulation Control Subsystem
AO	Adaptive Optics
AP	Advanced Planning
APS	Assistant Project Scientist
ASD	Advanced Software Development Group
ASSET	Automated Science/Sequence Encounter Timeline
BLF	Best Lock Frequency
BML	Backup Mission Load
CCD	Charge-Coupled Device
CCS	Computer Command Subsystem
CDT	Capability Demonstration Test
Co-I	Co-investigator
COMSIM	Command Simulation
CRS	Cosmic-Ray Subsystem
CSIRO	Commonwealth Scientific and Industrial Research Organization
CST	Canopus Star Tracker
DDOR	Delta Differential One-way Ranging
DMT	Data Management Team
DMWF	Desired Memory Word File
DSN	Deep Space Network
DSSCAN	Digital Storage Subsystem Cancellation
DTR	Digital Tape Recorder
ER	Experiment Representatives
ESF	Event Sequence File

EVTSDR   Event System Data Record
FDS   Flight Data Subsystem
FE   Far Encounter
FEO   Flight Engineering Office
FOH   Flash-Off Heater
FOO   Flight Operations Office.
FPA   Failure (or Fault) Protection Algorithms
FSO   Flight Science Office
FT   Final Timeline
GCMD   Ground Command
GDSEO   Ground Data System Engineering Office
GDT   Gyro Drift Turns
GSDT   General Science Data Team
HET   High Energy Telescope
HFM   High Field Magnetometer
HGA   High Gain Antenna
HST   Hubble Space Telescope
IDC   Image Data Compression
IMC   Image Motion Compensation
IRIS   Infrared Interferometer Spectrometer and Radiometer
ISAS   Institute of Astronautical Science
ISDT   Imaging Science Data Team
ISS   Imaging Science Subsystem
IT   Integrated Timeline
JATO   Jet-Assisted Take-Off
JPL   Jet Propulsion Laboratory
KBO   Kuiper Belt Object
LASP   Laboratory for Atmospheric and Space Science
LECP   Low Energy Charged Particle
LEMPA   Low Energy Magnetospheric Particle Analyzer
LEPT   Low Energy Particle Telescope
LET   Low Energy Telescope
LEU   Late Ephemeris Update
LMS   Low Field Magnetometer
LSU   Late Stored Update
MAG   Magnetometer
MCCC   Mission Control and Computing Center
MCT   Mission Control Team
MIMC   Maneuverless Image Motion Compensation
MIRIS   Modified IRIS
MJS   Mariner Jupiter–Saturn mission
MPO   Mission Planning Office
NA   Narrow Angle (camera)
NASA   National Aeronautics and Space Administration
NAV   Navigation Team

NE   Near Encounter
NET   Near Encounter Test
NIMC   Nodding Image Motion Compensation
NOCT   Network Operations Control Team
NRAO   National Radio Astronomy Observatory
NSWG   Neptune Science Working Group
OB   Observatory Phase
OPNAV   Optical Navigation
OPSGEN   Observation Pointing Sequence Generator
PE   Post Encounter
PI   Principal Investigator
PKE   Pluto–Kuiper Express
PLS   Plasma Spectrometer Subsystem
POR   Power-On Reset
PPS   Photopolarimeter Subsystem
PRA   Planetary Radio Astronomy
PS   Project Scientist
PWS   Plasma Wave Subsystem
RSS   Radio Science Subsystem
RSST   Radio Science Support Team
RTG   Radioisotope Thermoelectric Generators
SCT   Spacecraft Team
SEQ   Sequence Team
SEQGEN   Sequence Generator
SIS   Science Investigation Support Team
SOE   Sequence of Events
SRF   Sequence Request File
SS   Sun Sensor
SSG   Science Steering Group
TCM   Trajectory Correction Maneuver
TET   The Electron Telescope
TL   Team Leader
TLC   Tracking Loop Capacitor
TMT   Torque Margin Test
TOL   Time Ordered Listing
UP   (Final) Uplink Product
UVS   Ultraviolet Spectrometer
VCB   Voyager Change Board
VIM   Voyager Interstellar Mission
VLA   Very Large Array (radio telescopes)
VUIM   Voyager Uranus/Interstellar Mission
WA   Wide Angle (camera)

1

The discovery of Neptune

1.1 THE DISCOVERY OF URANUS

The story of Neptune begins with the discovery of Uranus, for the discovery of that planet showed astronomers that there were planets to be found in the Solar System other than those discovered by the ancients. It also gave astronomers tantalizing clues that an as-yet-unseen planet, further away from the Sun, was still waiting to be discovered.

In the small English town of Bath, William Herschel (1738–1822; Figure 1.1) was gaining a growing reputation as a fine musician. However, his passion was not only for music but also for astronomy. He developed this love for the stars from his father, who told him countless stories about the night sky. Those experiences as a young boy led Herschel to design and build his own telescopes as an adult. With practice and great patience, Herschel was able to advance the art of grinding and polishing telescope mirrors, and eventually his telescopes became far superior to anything to be found at the Royal Observatory. Once he was satisfied with his first telescope, Herschel decided to point it skyward to chart the distribution of stars in the entire universe!

On March 13, 1781, Herschel spotted something near the star H Geminorum that appeared as a 'Nebulous Star or perhaps a Comet' [1]. After many nights of observations, he concluded that the object moved against the background stars. Based on the object's motion Herschel assumed that he had observed a comet. He was unaware that on that night, from his backyard (Figure 1.2) at 19 New King Street, he had discovered a seventh planet and had virtually doubled the radial extent of our Solar System.

Since Herschel believed he had only discovered a comet, he continued with his work and never felt that the discovery was worth mentioning to his fellow astronomers. However, one day Herschel was having a discussion with Nevile Maskelyne (1732–1811), the fourth Astronomer Royal, and briefly mentioned his comet discovery. Maskelyne quickly realized that this object could be a comet but might

Figure 1.1. Sir William Herschel. (P-29487A)

also be a new planet. With coaxing from Maskelyne, Herschel decided to present his finding to the Royal Society and, on April 26, 1781, presented his discovery titled an 'Account of a Comet'. This title did very little to help convince his fellow astronomers that the object he had found was anything but a comet.

Once Herschel made his presentation, astronomers began the task of trying to determine the object's orbital elements. However, observations of the object did not converge on a solution. This strange behavior, coupled with the fact that the object lacked a coma or a tail, made astronomers think twice about its nature. Additional observations showed that the object gave off a steady light, moved slowly in latitude, and traveled along the zodiac. Based on these facts, Johann Lexel (1740–84), an astronomer at Saint Petersburg, realized that Herschel's comet might really be a planet. The object was indeed found to be a planet and was given the name Uranus.

1.2   THE PERTURBATIONS OF URANUS

Years went by and astronomers around Europe tried in vain to calculate accurate orbital elements for Uranus. Unfortunately, the great distance of the object meant

Figure 1.2. Rear of Herschel's house. (P-29488B)

that it had an orbital period that had to be much longer than Saturn's 29-year period. This made determination of an orbit very difficult, because Uranus would move very slowly against the background star field. No significant progress was made until Johann Bode (1747–1826) realized that the solution might be found in observations made prior to the object being reclassified as a planet. He deduced that there might be many earlier observations in which Uranus was assumed to be a star. If one could find them and include them with the new observations (those made after Uranus's discovery), a more accurate calculation of the planet's orbital elements could be made.

Bode was joined by fellow astronomer Placidus Fixlmillner of Kremsmünster (1721–91), and together they found observations from 1660, 1756, 1781 and 1783 that were possible prediscovery observations of Uranus. From this data a new orbit of the planet was calculated and its orbital elements determined. This solution was then proposed as the correct orbital elements for Uranus. By 1788, however, astronomers had found that errors in the predicted position of Uranus were starting to grow unacceptably large. It appeared that for long periods of time only the old or new observations, but not both, could be used to predict the planet's motion. For years, astronomers tried to reconcile the problem of Uranus's motion but failed. Jean Baptiste Delambre (1749–1822) made a particularly noteworthy innovation by adding the gravitational influences of Jupiter and Saturn into his calculations. His resulting tables did a much better job of matching the observed position of Uranus, but like others before him, he failed to predict Uranus's long-term motion accurately.

Alex Bouvard (1767–1843), using 17 observations of Uranus spanning the prior four decades, could not find the one solution that satisfied all the observations. Bouvard stated with respect to the old and new observations, '... I leave to the future the task of discovering whether the difficulty of reconciling the two systems results from the inaccuracy of the ancient observations, or whether it depends on some extraneous and unknown influence which may have acted on the planet' [2].

In 1832, at a meeting of the British Association for the Advancement of Science, George Biddell Airy (1801–92), later Sir George B. Airy, reported that the tables made 11 years earlier by Bouvard were off by more than half a minute of arc. These errors were not acceptable, and the astronomical community increasingly demanded an explanation.

One of the first to suggest the possibility of additional planets was Alexis Claude Clairaut (1713–65). Speaking about the motion of Halley's comet in November 1758, he said, '... that a body which travels into regions so remote, and is invisible for such a long period, might be subject to totally unknown forces, such as the action of either comets, or even of some planets too far distant from the Sun ever to be perceived' [3]. This was pure speculation, but some force acting upon Uranus had to be found to explain its strange motion. Nevertheless, many astronomers of the day, including Airy, believed that the problem with Uranus's motion lay more in its great distance from the Sun or perhaps in Newton's Laws of Motion rather than in perturbations from an as-yet-undiscovered planet.

In 1841, Johann Heinrich von Mädler (1794–1874), Director of the Dorpat Observatory, published his *Populäre Astronomie* in which he stated, '... we may even express the hope that analysis will at some future time realize in this, her highest triumph, a discovery (of a planet) made with the mind's eye in regions where sight herself was unable to penetrate' [4]. By 1838, with many astronomers alluding to the probable existence of other planets, a trans-Uranian world was by far the most popular explanation for the irregular motion of Uranus.

1.3   URBAIN JEAN JOSEPH LE VERRIER

On March 11, 1811, in the small town of Saint-Lô, Normandy, Urbain Jean Joseph Le Verrier was born (Figure 1.3). As a small boy, Le Verrier showed a great interest in science. By the time he finished his education in the town's local school, it was obvious that he had great mathematical abilities. In 1830, Le Verrier graduated from Caen at the head of his class. In that same year he took the entrance exam for École

Figure 1.3. Urbain Jean Joseph Le Verrier. (P-33662)

Polytechnique in Paris, but did not make the grade. This was a bitter disappointment and was one that Le Verrier would remember all his life.

However, his passion for mathematics was not so easily deterred. To help to satisfy that craving, his father sold their home to raise money to send Le Verrier to the Collége de Saint Louis. During his first year at college Le Verrier won the Prix de Mathématiques Spéciales. Though Le Verrier had an exceptional ability for mathematics and science, he did not have a preference for any particular field of science.

Upon graduating from the university, Le Verrier began work with Joseph Louis Gay-Lussac (1778–1850) at the Administration of Tobacco. His work with Gay-Lussac was in chemistry, which did not quench his thirst for advanced mathematics. Accordingly, in his spare time he would attempt to solve very long and tedious mathematical problems in astronomy. On September 10, 1839, Le Verrier submitted his first paper to the Académie des Sciences concerning planetary orbits. The paper was well received by the astronomical community.

In 1840, Le Verrier published his second paper, in which he developed a new and innovative approach for solving equations of planetary motion. With this technique he could solve such problems by using a simple substitution rather than having to resort to the usual long hours of difficult calculation.

After this work, and with growing credentials in astronomy, Le Verrier turned his attention to short period comets. He continued to work with comets until the summer of 1845, when François Arago (1786–1853), the Director of the Paris Observatory and Secretary of the Academy of Sciences, suggested that he should study the problem of Uranus's motion. At Arago's suggestion Le Verrier happily turned his attention to the strange behavior of Uranus.

1.4   JOHN COUCH ADAMS

On June 5, 1819, in a small farmhouse at Lidcot in the parish of Laneast, Cornwall, John Couch Adams was born (Figure 1.4). John was one of seven children born to

Figure 1.4. John Couch Adams. (P-33662)

Thomas and Tabitha Adams. When John was eight years old he was sent to school in a farmhouse, but after only two years, his teacher could not keep pace with Adams's abilities. It was obvious to all in this small parish that Adams was a gifted mathematician, and if he were to continue his education he would have to attend a better school.

To this end, Adams was sent to a school headed by his cousin, the Reverend John Couch Grylls. The school was able to provide Adams with a solid education, but one that was weak in mathematics. To compensate for that shortcoming, Adams, in his free time, would go to the Devonport Mechanics Institute's library to continue his studies, and it was there that his passion for astronomy grew. In 1834, Adams was awarded with J.F.W. Herschel's book of Astronomy because of his great academic ability and his love of the stars. Adams always looked back at this bestowal as one of the major motivating events in his life.

Adams continued to learn, reading anything he could to better his understanding of astronomy. By 1837 he had taught himself conic sections, differential calculus, theory of equations, theory of numbers, and mechanics. There were very few students in his school who could keep up with his mathematical abilities, including the teachers. As with Le Verrier, it became obvious that if Adams were to improve his skills he would have to be sent to a higher school.

In October 1839 Adams applied for, and was accepted at, St John's College of Cambridge University. Unfortunately, the school was very expensive and Adams knew that his parents would have difficulty paying for his education. So, upon entering Cambridge, Adams took and won a sizarship. This award allowed him to earn a scholarship, the money from which was used to offset some of his education expenses.

Unlike Le Verrier, Adams knew he had a love for astronomy. On June 26, 1841, while Adams was in a Cambridge bookshop, he came across a copy of George Airy's report on the question of Uranus's motion. This report for the British Association for the Advancement of Science described the problems in using Bouvard's tables for determining the position of Uranus.

Adams took out a small piece of paper and wrote (Figure 1.5), '1841. July 3. Formed a design, in the beginning of this week, of investigating, as soon as possible after taking my degree, the irregularities in the motion of Uranus, wh. are yet unaccounted for; in order to find whether they may be attributed to the action of an undiscovered planet beyond it; and if possible thence to determine the elements of its orbit, &c approximately, wh wd. probably lead to its discovery.'

During his next holiday to Lidcot in 1843, Adams began the work to determine the motion of Uranus. In that year, at the age of 24, he had developed an approximate solution which proved to him that he could explain the motion if there were a distant planet whose gravitational influence was perturbing Uranus. He recognized that, in order to solve the problem more completely, he was going to have to do a lot more work.

To solve the problem, Adams divided it into four parts. First he had to calculate the forces on Uranus for any given time. He then had to separate these forces into known and unknown components. Next he would have to relate the unknown forces

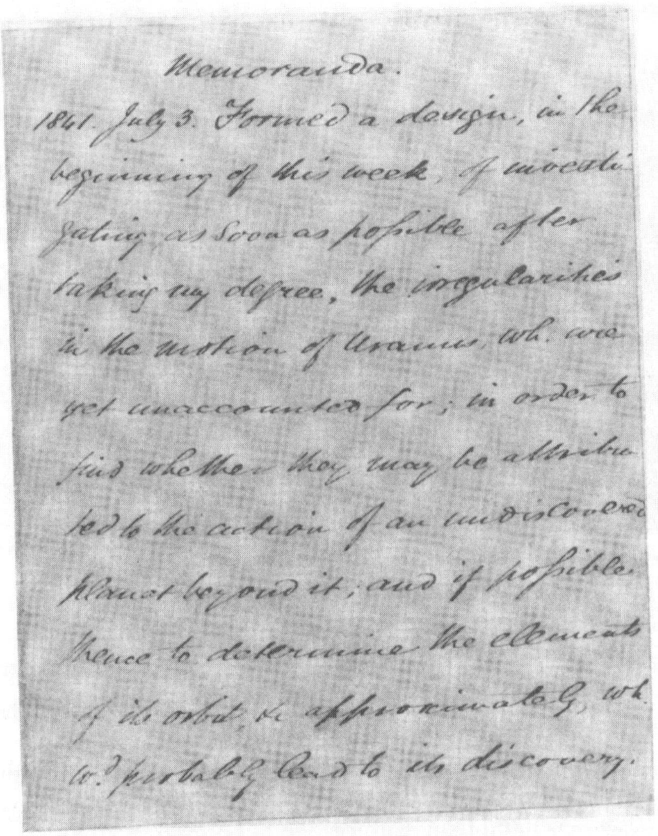

Figure 1.5. John Couch Adams's memorandum.

to the mass and orbital elements of a hypothetical planet. He would then have to solve the equations for an answer.

Adams first assumed that an orbital radius for the hypothetical planet could be determined from the empirical law proposed by Daniel Titius (1729–96) and Johann Bode (Table 1.1). This law stated that the orbital radius of the planets, in astronomical units (one astronomical unit [AU] is equal to the mean Earth–Sun distance) could be determined from the equation

$$r = 0.4 + 0.3 \times 2^n$$

where r was the Sun–planet distance (AU) and n was a simple integer. This law gave Adams an orbital radius of 38.8 AU for the hypothetical planet corresponding to $n = 7$. He then presupposed that since most of the other planets had orbits with small eccentricities, the hypothetical planet could be assumed to have a circular orbit.

Unfortunately, his college work prevented him from tackling this problem im-

Table 1.1. Titius–Bode's law of planetary orbital radii

Planet	n	Observed	Predicted
Mercury	$-\infty$	0.39	0.40
Venus	0	0.72	0.70
Earth	1	1.00	1.00
Mars	2	1.60	1.52
Ceres (Asteroid)	3	2.77	2.80
Jupiter	4	5.20	5.20
Saturn	5	9.56	10.0
Uranus	6	19.2	19.6
Neptune	7	30.1	38.8

mediately and he tried to devote as much time as he could spare to the problem between work assignments. When he finally reached a solution, he considered the problem again, this time removing the assumption that the hypothetical planet had to have a circular orbit. Adams solved this case in September 1845.

Excited by his success in actually reaching an answer, Adams brought the orbital elements and the mass for his hypothetical planet to James Challis (1803–82). Challis, the Director of the Cambridge Observatory, was the person who succeeded Airy as the Plumian Professor of Astronomy. As Challis doubted the ability of mathematics to predict the location of a planet, he suggested that Adams should send his results to the Astronomer Royal. In later years Challis admitted that he discounted Adams's predictions, because it was '... so novel a thing to undertake observations in reliance on merely theoretical deduction' [5].

1.5   THE SEARCH

As the Director of the Cambridge Observatory, and based on Adams's demonstrated mathematical abilities, Challis should have undertaken the search for the trans-Uranian world immediately. At the very least he should have counseled Adams to publish his work. Unfortunately, Challis only suggested that Adams should mail his results to the Astronomer Royal. Taking Challis's advice, Adams decided to go further and hand deliver his results to the Astronomer Royal while on his way home for a month-long vacation.

When he arrived at the Airy home on September 30, 1845, Adams was disappointed to learn that Airy was not there. His meeting with Airy would have to wait. On October 21, 1845, on his way back to Cambridge, Adams tried again but this time found only Mrs Airy at home. Mrs Airy stated that the Astronomer Royal was not at home but would be back shortly. Adams left his card and decided to return some time later. On Adams's next visit Airy was at home, but was having his usual early dinner and, once again, Adams was unable to see the Astronomer Royal. It seems that one of Airy's peculiar habits was not to be disturbed while he was eating, and his butler not only faithfully carried out these instructions but never told the Astronomer Royal of Adams's visit.

Adams left his paper at the Royal Observatory but was irritated by the entire affair. It was later discovered that Mrs Airy had not delivered Adams's card to her husband. This could be explained by the fact that she probably had a lot on her mind as, one week later, she gave birth to the Astronomer Royal's son Osmund [6].

On November 5, 1845, Sir Airy finally did write to Adams to thank him for his paper. However, the letter made it seem as if the Astronomer Royal was unaware that Adams had derived the orbital elements for the hypothetical planet rather than just assuming them. In his letter, Airy asked if Adams's calculations could explain why Uranus was further from the Sun than current tables predicted. Adams was so disgusted by this irrelevant question that it took him more than a year to respond. Airy had succeeded in discouraging Adams from pursuing further counsel.

Meanwhile, on November 10, 1845, Le Verrier had finished his analysis and presented his work, 'Premier Mémoire sur la Théorie d'Uranus' to the French Academy of Science. His paper described the results of his analysis on Uranus's motion. Half a year later, Le Verrier submitted his second paper, 'Recherches sur les Mouvements d'Uranus' on June 1, 1846. His analysis revealed that the hypothetical planet should have a mean longitude of 325 degrees on January 1, 1847. He was sure that the error in his calculated position was less than 10 degrees.

Le Verrier's analysis was detailed and thorough. Unfortunately, his mathematical analysis did not inspire a single French astronomer to take up the search. It seems that the amount of work required to find a planet, which may not even exist, was enough to discourage most astronomers.

On July 9, 1846, Airy was in the town of Ely talking with George Peacock, his old professor at Cambridge. Eventually the conversation turned to the problem of Uranus's motion. When Airy told the professor of the work done by both Adams and Le Verrier, he was amazed that Airy had not requested an immediate search and urged Airy to act at once. Three days later Airy wrote to Challis, asking him to begin a search for the hypothetical planet.

On July 18, 1846, Challis agreed to take up the search, but as Airy refused to believe that a planet could be found mathematically he did not suggest that Le Verrier's or Adams's data should be used to look for the planet. Instead, he proposed a detailed search around the predicted location. This search was made even more difficult by Challis who decided to record all stars down to 11th magnitude. Adams's work had predicted that the hypothetical planet would have a disk, and should be no dimmer than a star of 9th magnitude.

Challis estimated that the search he was about to begin would require at least 3,000 stars to be recorded and take roughly 300 hours (approximately 8 continuous weeks) of telescope time with good weather!

Finally on August 1, 1846, Challis began his search, but the moon was not on Challis's side. It was waxing and as England was having its typical poor weather the search had to be delayed until visibility improved. Eleven days later Challis resumed his search, and on the fourth night of good visibility – two weeks into the search – he recorded the position of the new planet. Unfortunately, he did not realize that he had

Figure 1.6. Johann Gottfried Galle.

seen anything other than a star. Meantime, on September 2, 1846, Adams mailed Airy his sixth solution, which removed the need to use Bode's law for the mean radius of the hypothetical planet.

In France, Le Verrier was giving up hope of ever being able to convince any of his countrymen to take up the search, and by September 1846 began to look elsewhere for someone who was willing to search for the trans-Uranian world. Once committed to looking elsewhere, Le Verrier remembered that he had once received a paper from Johann Gottfried Galle (1812–1910; Figure 1.6), an assistant at the Berlin Observatory, and used this paper as an excuse to resume correspondence.

Le Verrier's letter reached Galle on September 23, 1846. Upon receipt, Galle was intrigued by Le Verrier's unusual request for time on the observatory's telescope (Figure 1.7). However, as an assistant, he had to obtain permission and immediately approached Johann Franz Encke (1791–1865), the Director of the Berlin Observatory. Encke was not as enthused by the request as his assistant had been and tried to convince him of the extremely small probability of success. However, Galle

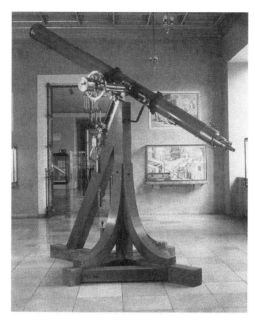

Figure 1.7. Galle's telescope.

was persistent and was finally given permission to begin the search. Unbeknown to Galle and Encke, a young astronomer named Heinrich Louis d'Arrest (1822–75) had overheard their conversation and later pleaded with Galle to let him help.

On that same night, September 23, 1846, Galle and d'Arrest pointed the 9-inch Fraunhofer Refractor at the location specified by Le Verrier (RA 22 h 46 min; DEC −13 deg 24 min). At first glance, Galle did not see the 3 seconds of arc disk that had been predicted by Le Verrier. D'Arrest suggested that better star maps be used to look for an object that was not charted.

D'Arrest consulted the files to look for star maps of the area that Le Verrier had predicted for the new planet. To his pleasant surprise he found just what he was looking for – a new map of Aquarius. This map (Figure 1.8), printed near the end of 1845 and not yet distributed, was known as the Hora XXI of the Berlin Academy's *Star Atlas* and was drawn by Dr Bremiker.

Galle went back to the telescope and d'Arrest positioned himself over the new map. Together they resumed the search: Galle called out star positions and d'Arrest searched for them. Within minutes Galle called out a star that was not on the map! Both astronomers were dumbfounded. Could this be the new planet? Not sure of what they found, d'Arrest brought Encke to the telescope. It was about midnight, and they waited for a few hours to see if the object moved against the stars. Unable to tell, they had to wait another night to be certain.

The next evening revealed to them the truth about its nature. The object had moved against the background stars. They had indeed discovered the eighth planet in

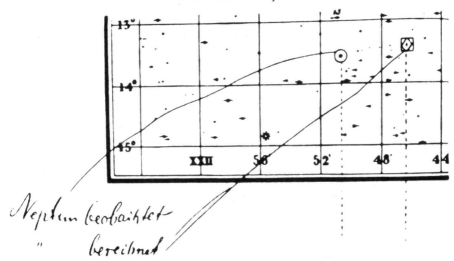

**NEPTUNE PREDICTED (RIGHT) AND
NEPTUNE OBSERVED (LEFT)
SEPTEMBER 23, 1846**

PORTION OF HORA XII OF BERLIN ACADEMY'S STAR ATLAS BY CARL BREMIKER
(USED BY PERMISSION: ARCHENHOLD-STERNWARTE BERLIN-TREPTOW, BERLIN, GDR)

Figure 1.8. Star map with predicted and observed positions for Neptune. (D2001-0724-cl.jpg)

our Solar System. The planet was found at RA 22 h 53 min 25.84 s with a magnitude of 8, and was within one degree of the location predicted by both Adams and Le Verrier!

1.6   CO-DISCOVERERS

News of the discovery reached Airy on September 29 in a letter written by Encke. The discovery would almost certainly be followed by a storm of questions, such as: 'Why was Mr Adams kept silent for four months?' and 'If Airy knew of Adams's and Le Verrier's results, why did he not mention them earlier?'

On the same night that Airy received his letter from Encke, Challis noticed a star that appeared to have a disk. This star was one of hundreds he had observed that night, and being a proficient astronomer he noted it in his logbook before moving on to the next source. Challis never returned to that star for further observation – a decision that would come back to haunt him.

Two days later, when Challis learned of Galle's discovery, he referred to his logbook and looked at his notes for that region of the sky. To his horror, he had not only seen the planet, but had seen it twice! As a matter of fact, many astronomers had seen the planet without realizing its true nature. The records show that Galileo

Galilei (1564–1642) had probably seen Neptune on December 28, 1612, and again on January 28, 1613, 234 years before Galle! [7].

In astronomy, the discoverer of a planet is given the honor of naming the new world. Galle choose the name Janus, after the Roman god of gates, doors, and boundaries and because Janus was the forefather of Saturn. However, Le Verrier did not like the name. He thought that the name Janus, and its association with boundaries, would imply the edge of the Solar System and instead chose the name Neptune, which had been suggested by the Bureau of Standards.

By the middle of October, Le Verrier was informed that an Englishman was disputing his claim as the discoverer of Neptune. He was incensed. How could someone else take credit for his mathematical discovery when there was no prior mention of it in the professional journals or at conferences?

Meanwhile, all across the world astronomers began observing Neptune to learn its secrets. On October 1, 1846, John Herschel (1792–1871) sent a letter to William Lassel (1799–1880) to see if he would search for any satellites that might be found around the new planet. Nine days later Lassel discovered Triton, Neptune's largest satellite.

As time went on, more and more of the story of Neptune's discovery became public. On October 17, 1846, Challis wrote a detailed description of his role in the search. This description highlighted Adams's work, as well as all the observations Challis had made between July and August. Then national pride took over. England and France each claimed that one of its citizens was responsible for the discovery of Neptune.

Who was the rightful discoverer of Neptune? Was it Adams, who may have produced the answer earlier, but who never published his results? Or was it Le Verrier, who might not have been the first to calculate the right solution, but whose answer was used to guide Galle's telescope? By the end of October both Airy and Challis were under great pressure to explain their lack of action.

On November 13 at the Royal Astronomical Society meeting, Adams, Airy, and Challis presented their roles in the search for Neptune. Both Airy and Challis had allowed a historic chance for England to slip through their fingers. An Englishman may not be given his rightful credit to the discovery of a new world. And initially this came to pass. Le Verrier was given the title of 'Discoverer of Neptune'.

Almost a year later, Adams and Le Verrier first met at the June 1847 meeting of the British Association for the Advancement of Science. When they met, there were no hostilities or resentment. They both recognized the magnitude of their work and their respective roles in the planet's discovery and became life-long friends who were eventually known as the co-discoverers of Neptune.

Today, a few historians of astronomy find tantalizing clues that Adams's solutions were not converging but varied by more than 35 degrees of longitude. It was perhaps this fact, and not his opinion of Adams, that had kept Airy initially silent about Adams's claim to be the discoverer of Neptune. These historical sleuths think that Airy, Challis, and Herschel put forward Adams's claim to give the discovery to an Englishman. We may never know for sure [8].

NOTES AND REFERENCES

1. Royal Astronomical Society Manuscript, Herschel, W. 2/1-2, 23; also quoted in Alexander, A. F. O'D. (1965) *The Planet Uranus*. American Elsevier Publishing Company, Inc., p. 26.
2. Bouvard, A. (1821) *Tables astronomiques, publiées par le Bureau des Longitudes de France contentant les Tables de Jupiter, de Saturne et d'Uranus construites d'après la théorie de la Mécanique céleste*, p. xiv.
3. Clairaut, A. -C. (1759) *Journal des sçavans*. Paris, p. 86.
4. von Mädler, J. H. (1841) *Populäre Astronomie*, p. 345.
5. Airy, G. B. (1846) Account of some circumstances historically connected with the discovery of the planet exterior to Uranus. *Monthly Notices of the Royal Astronomical Society*, **7**, 145.
6. Airy, G. B. (1896) *Autobiography*, edited by Wifrid Airy. Cambridge, p. 172.
7. Drake, S. and Kowal, C. T. (1980) Galileo's sighting of Neptune. *Scientific American*, **243**, 74–81.
8. Rawlins, D. (1999) Recovery of the RGO Neptune Papers: Safe and Sounded, DIO Vol. 9, No.1

BIBLIOGRAPHY

Cruikshank, D. P. (ed.) (1995) *Neptune and Triton*. University of Arizona Press, Tucson.

Grosser, M. (1962) *The Discovery of Neptune*. Cambridge, Massachusetts.

Littmann, M. (1988) *Planets Beyond: Discovering the Outer Solar System*. Canada.

Miner, E. D. (1998) *Uranus: the Planet, Rings and Satellites* (2nd edn.). Wiley–Praxis, Chichester, UK.

Moore, P. (1988) *The Planet Neptune*. Wiley–Praxis, Chichester, UK.

2

Neptune's position in the Solar System

2.1 LAST OF THE GIANT PLANETS

Neptune is a cold, dark world. With a mean solar distance of 4,504 million km (30.11 astronomical units (AU)), light levels at Neptune are more than 900 times dimmer than they are at Earth. Noon on Neptune would appear no brighter than what a human would experience at dusk on Earth. Even the Sun would take on an appearance more like a star than the bright disk seen from Earth. The planet's great distance from Earth gives it a mean opposition magnitude of 7.8, which is below the capability of the unaided human eye to discern. However, with a pair of binoculars, Neptune has a clearly defined disk and the appearance that one would expect of a planet.

The unique mathematical discovery of Neptune opened up new opportunities. Planets could be found with a piece of paper and a pencil. Newton's Laws of Gravitation, perturbation theory, and differential calculus could be used to predict very precisely the location of a hidden 'wanderer'. These wanderers fall mainly into two groups: terrestrial and giant planets. Of these, Neptune is the last of the giant planets.

2.2 TERRESTRIAL PLANETS AND GIANT PLANETS

The terrestrial planets (Mercury, Venus, Earth and Mars) are relatively high-density, rocky objects with surfaces that reveal clues to their early histories. They are surrounded by thin gaseous envelopes, devoid of ring systems, and may have moons.

The giant planets are much more massive than their terrestrial cousins. Although they have molten rocky cores that may be roughly the size of a terrestrial planet, they are surrounded by an immense envelope composed predominantly of hydrogen and helium. Although much more massive than the terrestrial planets, these outer planets are much less dense, due to the large amount of lighter elements contained within them.

Table 2.1. Physical characteristics of the terrestrial planets

	Mercury	Venus	Earth	Mars
Mean distance from Sun (AU)	0.3871	0.7233	1.0000	1.5237
Orbital period (tropical years)	0.2408	0.6152	1.0000	1.8807
Inclination to Earth's orbit (deg)	7.00	3.39	0.00	1.85
Equatorial diameter (Earth = 1)	0.382	0.949	1.000	0.532
Volume (Earth = 1)	0.056	0.857	1.000	0.151
Mass (Earth = 1)	0.0553	0.8150	1.0000	0.1074
Mean density (g/cm^3)	5.43	5.20	5.52	3.91
Rotational period (h)	1,407.5	5,832.5	23.9	24.6
Tilt equator to orbit (deg)	0.1	177.4	23.4	25.2
Average surface magnetic field (gauss)	0.003	<0.0001	0.31	<0.0003
Magnetic source strength (Earth = 1)	0.0007	<0.0004	1.0000	<0.0002
Magnetic dipole tilt (deg)	+14	N/A	+10.8	N/A
Magnetic dipole offset (planetary radii)	Unknown	N/A	0.0725	N/A
Solid surface?	Yes	Yes	Yes	Yes
Ring system?	No	No	No	No
Number of known satellites	0	0	1	2

The giant planets are sometimes referred to as gas giants; however, the extreme temperatures and pressures inside these worlds make their interiors more liquid than gaseous. On the other hand, at these extreme temperatures and pressures, there is very little difference between liquids and gases.

If a space traveler were unfortunate enough to enter the atmosphere of one of the giant planets, he or she would not find a single solid surface! Instead, as he or she descended into the planet, our traveler would find that the temperature, pressure, and density would all continue to increase smoothly, with no sharp transitions. Assuming that he or she was adequately protected from the temperature, pressure, and radiation, our traveler would eventually 'float' at that level in the atmosphere where the surrounding density and his or her own density were equal. Some of the physical characteristics of the planets in our Solar System are described in Tables 2.1, 2.2, and 2.3.

Figure 2.1. Pluto and Charon from the Hubble Space Telescope.

Table 2.2. Physical characteristics of the giant planets

	Jupiter	Saturn	Uranus	Neptune
Mean distance from Sun (AU)	5.2026	9.5549	19.2184	30.1100
Orbital period (tropical years)	11.8565	29.4235	83.7474	163.723
Inclination to Earth's orbit (deg)	1.30	2.49	0.77	1.77
Equatorial diameter (Earth = 1)	11.209	9.449	4.007	3.883
Volume (Earth = 1)	1321	764	63	57
Mass (Earth = 1)	317.7	95.2	14.5	17.1
Mean density (g/cm³)	1.33	0.69	1.32	1.64
Rotational period (h)	9.925	10.656	17.240	16.110
Tilt equator to orbit (deg)	3.1	26.7	97.9	29.6
Average surface magnetic field (gauss)	4.28	0.22	0.23	0.14
Magnetic source strength (Earth = 1)	20,000	600	50	25
Magnetic dipole tilt (deg)	−9.6	0.0	−58.6	−46.9
Magnetic dipole offset (planetary radii)	0.10	0.05	0.33	0.55
Solid surface?	No	No	No	No
Ring system?	Yes	Yes	Yes	Yes
Number of known satellites*	28	30	21	8

* Number of confirmed satellites. These numbers may grow with improved observing technologies.

By any definition, Pluto is a maverick. It has a solid surface and its own moon, Charon. However, Pluto is devoid of any appreciable atmosphere, has a relatively low density, and has a highly elliptical orbit (eccentricity of 0.25). Its classification has always been an enigma. Indeed, its closest twin is probably Neptune's largest moon, Triton.

Many hypotheses have been generated to try to explain the presence of Pluto, one of which proposes that Pluto and Charon are escaped satellites from Neptune (tiny Pluto's orbit spans the orbital distances of Neptune), but as no reasonable scenario has been proposed to explain how Pluto could reach its present orbit from Neptune, this hypothesis now has few proponents. Most classify Pluto as an outer planet because of its location in the Solar System; some feel it is appropriately classified as a terrestrial planet because of its solid surface. The uniqueness of Pluto defies its being grouped with any other Solar System planets.

One way to solve this conundrum is to suggest that Pluto is not a planet at all. Pluto and Charon could be the largest known Kuiper Belt objects. There are hundreds of known Kuiper Belt objects which reside in a region bounded by Neptune's orbit and the most distant portions of Pluto's orbit. The logic goes as follows: In 1950, Jan Oort (1900–92) published a paper proposing an origin for comets [1]. He proposed that a spherical shell of comets surrounded the Solar System between 50,000 AU and 150,000 AU from the Sun. It is estimated that this region, known as the Oort Cloud, could contain 190 billion comets! Unfortunately, present-day instruments cannot detect comets at these great distances from the Sun. It is nevertheless a fact that a spherical shell of comets does not fit well with the observational data, which suggests that comets are generally close to the ecliptic

Table 2.3. Physical characteristics of Pluto

	Pluto
Mean distance from Sun (AU)	39.5447
Orbital period (tropical years)	248.02
Inclination to Earth's orbit (deg)	17.14
Equatorial diameter (Earth $= 1$)	0.180
Volume (Earth $= 1$)	0.006
Mass (Earth $= 1$)	0.002
Mean density (g/cm^3)	(2.0)
Rotational period (h)	153.3
Tilt equator to orbit (deg)	119.6
Average surface magnetic field (gauss)	Unknown
Magnetic source strength (Earth $= 1$)	Unknown
Magnetic dipole tilt (deg)	Unknown
Magnetic dipole offset (planetary radii)	Unknown
Solid surface?	Yes
Ring system?	No
Number of known satellites	1

and have prograde orbits. One year after Oort's paper was published, Gerard Kuiper (1905–73) suggested that comets came from the region of space approximately 35 to 50 AU distant from the Sun. If this region were disk shaped, in the ecliptic, and had prograde motion, it could produce comets that match the observed data.

Kuiper proposed that the Oort Cloud dropped comets into this disk region, from which comets then entered the inner Solar System. This model is now generally accepted because (a) it seems to fit the observed data, (b) many objects have been observed in this proposed disk region (now known as the Kuiper Belt), and (c) the IRAS spacecraft may have observed the same structure around the star Vega in 1983.

By the end of the year 2000, more than 300 Kuiper Belt Objects (KBOs) had been identified. It is possible that Pluto is the largest member of the Kuiper Belt, and is orbited by another KBO, Charon. At present, however, the American Astronomical Society's Division for Planetary Sciences (the world's largest organization of planetary scientists) is adamant that Pluto should continue to be identified as a planet, and not as a KBO.

2.3   THE ORIGINS OF NEPTUNE

Humans have never observed the birth of a solar system, nor the development of a planet. It is therefore a challenging task to deduce how Neptune or any other planet was formed. However, planetary theoreticians continue to learn more and more about Earth's place in the Solar System, and their understanding of the laws of

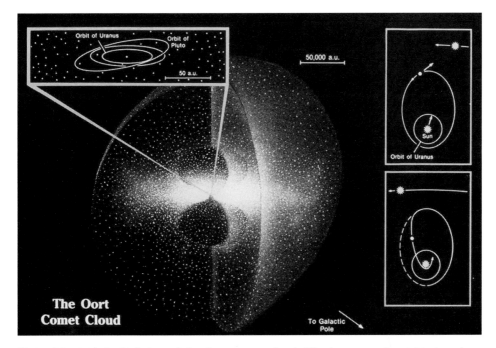

Figure 2.2. Artist's depiction of the Oort comet cloud. The insets on the right show how pristine comets from the Oort cloud can be gravitationally deflected to the inner solar system by a passing star.

nature is growing dramatically. They assume (as we must) that the laws of physics that describe our world operate similarly in all corners of the universe. Without such an assumption, their progress in studying planetary formation (or any other theoretically based study) would be stymied.

Planetary scientists have a rich array of observations to assist them in their task. In addition to the data collected from investigation of conditions in our Solar System, discovery and characterization of planets around other stars has become possible over the past decade. By late 2000, evidence for more than 50 such extra-solar planets existed, though none had been seen directly. The techniques these planet hunters use generally involve the laws of gravitation and the consequent effects of the unseen planet on the parent star, as evidenced either by tiny wobbles in the apparent position of the star, or by cyclic changes in the star's velocity toward or away from Earth, as deduced from Doppler shifts of spectral features in the observed light from the star.

Each of these techniques requires observations over long periods of time, usually on the order of months to years. If a planet is orbiting a star, the gravitational pull from the unseen planet tugs on the star just as the gravitational pull from the star keeps the planet in its orbit. The tugging from the planet causes small radial velocity changes in the direction of the planet and the small change in velocity appears as a

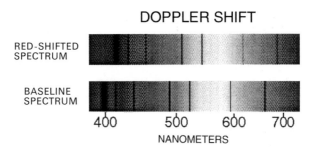

Figure 2.3. The red-shifted spectrum (top) is of a star moving away from the Sun.

slight increase or decrease in the wavelength of spectral features in the light coming from the star as seen from the Earth. This wavelength shift was first explained by Christian J. Doppler (1803–53) and hence bears his name.

Transverse shifts also require long periods of repeated careful observations of the precise position of a nearby star relative to the background stars. This technique does not work well for stars more than a few tens of light years from Earth, because the apparent motions become too small to discern. It also requires that the planets be large enough relative to the parent star to induce significant wobble in the stellar positions.

The Doppler technique also has its limitations and works best for planets that have relatively short orbital periods and relatively large masses. The smaller the planet's mass, the smaller the Doppler shift, and the harder it is to measure the change in frequency. The short orbital period is required to allow the astronomers to validate the shift by observing multiple 'years'. Obviously if a planet in another star system had an orbital period like Neptune's (163.7 years), no astronomer would be able to observe it for one complete orbit.

To complicate matters further, astronomers must assume the orientation of the new solar system's ecliptic. An orientation edge-on to Earth would provide the most accurate masses for these new worlds. Any other orientation would imply that the masses of these worlds are really much greater. It is possible that a number of these new worlds may not be planets at all. They may really be Brown Dwarfs, failed stars that were unable to sustain nuclear fusion in their cores.

Once the dataset is large enough and diverse enough, astronomers may begin to see patterns developing that might begin to explain the formation of solar systems and of terrestrial and giant planets. Currently there are two models that seem to best describe the origin of outer planets: accretional models [2] and giant protoplanet models [3].

2.3.1   Accretional models

These models assume that solar systems form in regions of the universe that contain relatively large amounts of dust and gas which come from debris ejected from others

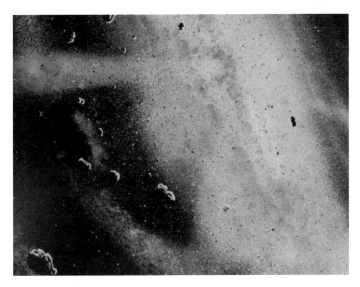

Figure 2.4. Artist's depiction of an accretion disk.

stars. The stars are the instruments that process hydrogen and helium into the heavier elements that are found in planets and in life itself.

A local instability in a particular dust and gas cloud can cause a collapse towards a single point. As more and more material 'falls' toward the center of this instability, the material's potential energy is converted into kinetic energy. The increase of kinetic energy manifests itself as an increases in temperature, density, and pressure. The material that does not fall toward the center begins to orbit the protostar, and to conserve angular momentum, the dust and gas form a disk of orbiting material around the protostar.

The temperature and pressure at any point in the surrounding disk determine whether the material is in a gaseous or a solid state. During this process, additional local instabilities may form, producing locally collapsing regions, which will be either additional stars or planets, depending on the total mass of the dust and gas involved in the local collapse. If temperatures rise above 10 to 15 million degrees Celcius, fusion begins and a star is born. Lower masses and temperatures result in a proto-planet.

Each protoplanet begins as a small collection of solid material. This mass captures its own gaseous envelope from the surrounding disk of material. The new star is usually very active at this point and begins to dissipate the surrounding disk of material. Planets that are relatively close to the star lose most of their low-mass gaseous envelopes as the stellar wind blows the material away from the central star. These worlds become terrestrial planets. Planets that are further from the star experience colder temperatures and a less dense solar wind. Gases from the surrounding nebula begin to form water, ammonia, and methane ices. Planets at sufficiently large distances from the star become ice-rich and retain their hydrogen and helium envelopes. These worlds become giant planets.

Some weaknesses of accretional models are: they predict very long time durations for planet formation; they do not predict the observed (slight) compositional differences between the giant planet atmospheres; and they do not predict major differences in giant planet masses (for example, why Jupiter and Saturn are so much larger than Uranus and Neptune). As future accretional models become more and more sophisticated, they may either account for these observed differences or predict results contrary to those observed. Obviously incorrect predictions would lead to rejection of the hypotheses.

2.3.2   Giant protoplanet models

These models assume that all planets start with the same composition – that is, that all protoplanetary compositions are equivalent to the composition of their central star. It is then assumed that different 'evolutionary' processes begin to alter their compositions. These processes might be the seizure of material from another developing star during a close encounter [4] or internal collisions and momentum transfers [5]. Alternatively, the contracting central star, as it begins to stabilize, may leave rings of material which interact differently with each protoplanet, resulting in non-uniform compositions [6].

The primary weakness of protoplanet models is that they do not adequately describe these evolutionary processes. Some theoreticians invoke exotic mechanisms to overcome these deficiencies, but the supporting observational data is absent.

2.3.3   Occum's Razor

Accretional models seem to require fewer *ad hoc* assumptions to explain protoplanetary evolution from small rocky cores into the observed giant planets. Given a choice between accretional and protoplanetary models, Occum's Razor would lead planetary scientists to pick the former. Occum's Razor implies that if two theories both adequately predict the outcome, the simpler theory (with fewer *ad hoc* assumptions) is more likely to be correct.

NOTES AND REFERENCES

1. Oort, J. (1950) The structure of the cloud of comets surrounding the Solar System, and a hypothesis concerning its origin. *Bulletin of the Astronomical Institutes of the Netherlands*, Vol. XI, No. 408.
2. Lewis, J. S. (1984) The origin and evolution of Uranus and Neptune. In Bergstralh, J. T. (ed.) *Uranus and Neptune*. NASA Conference Publication 2330, National Aeronautics and Space Administration, Scientific and Technical Information Branch, pp. 3–24.
3. Podolak, M. (1982) The origin of Uranus: Compositional considerations. In Hunt, G. (ed.) *Uranus and the Outer Planets*. Cambridge University Press, pp. 93–109.

4. Woolfson, M. M. (1978) The capture theory and the evolution of the solar system. In Dermott, S. F. (ed.) *The Origin of the Solar System.* John Wiley & Sons, New York, pp. 179–217.
5. McCrea, W. H. (1978) The formation of the Solar System: A protoplanet theory. In Dermott, S. F. (ed.) *The Origin of the Solar System.* John Wiley & Sons, New York, pp. 75–110.
6. Cameron, A. G. W. (1978) Physics of the primative solar nebula and of giant gaseous protoplanets. In Gehrels, T. (ed.) *Protostars and Protoplanets.* University of Arizona, Tucson, pp. 453–87.

BIBLIOGRAPHY

Cruikshank, D. P. (ed.) (1995) *Neptune and Triton.* University of Arizona Press, Tucson.
Littmann, M. (1988) *Planets Beyond: Discovering the Outer Solar System.* Canada.
Miner, E. D. (1998) *Uranus: the Planet, Rings and Satellites* (2nd edn). Wiley–Praxis, Chichester, UK.
Moore, P. (1988) *The Planet Neptune.* Wiley–Praxis, Chichester, UK.
Yeomans, D. K. (1991) *Comets: A Chronological History of Observation, Science, Myth, and Folklore.* John Wiley & Sons, Inc.

3

Speculation about Neptune's rings

3.1 DISTANT STARS AS PROBES FOR THE RINGS

With the discovery of rings around Uranus, the possibility of finding rings around Neptune increased. Saturn had rings, Jupiter had rings (discovered by Voyager 1 in March 1979), and now Uranus had rings (discovered by James L. Elliot *et al.* and Robert L. Mills *et al.* during the stellar occultation of SAO 158687 on March 10, 1977). The search for rings about Neptune was on.

Two techniques could be used for this pursuit: (a) direct observations using back-scattered light and (b) stellar occultations. Back-scattered light observations would only be possible if the proposed ring system was composed of ring particles that were substantially larger than the wavelength of visible light (0.3 to 0.6 µm). If the particles were about the same size or smaller, the photons would scatter beyond the planet rather than reflect back toward the Sun (forward-scattered light, Figure 3.1).

Stellar occultation observations require the planet to cross in front of a known, stable star. The existence of rings can be determined if the star brightness flickers prior to occultation and then again after the star emerges from behind the planet. A stable star of known output is essential if the data is to be interpreted correctly. If the occulted star's luminosity is variable, astronomers would not be able to tell if the change in brightness was due to characteristics of the star or to material found around the planet.

Occultation observations were much more difficult with Neptune than they were for Uranus. One reason is that, although Uranus and Neptune are both about the same size, Neptune is one and a half times more distant. The greater distance implies that Neptune has a smaller apparent disk, and a smaller disk blocks fewer stars, and, because of this, Neptune stellar occultations occur only a few times each year. Another reason why these observations are less likely to discover rings around Neptune concerns the axial tilt of Uranus compared to Neptune. If we assume that rings are equatorial, Uranus's axial tilt of $98°$ increases the chance that its rings will appear opened, as viewed from Earth. An open ring system subtends a

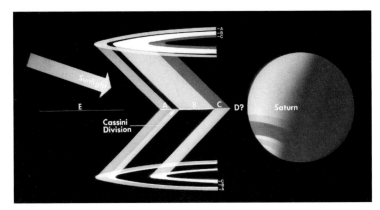

Figure 3.1. This drawing of Saturn's rings shows the rings in back-scattered light (top) as compared to forward-scattered light (bottom). (260–868)

larger amount of sky, thus increasing the chances that the rings will occult a star. Uranus's rings are fully open to Earth observers for approximately half a Uranian year (42 Earth years) while Neptune's rings (if they existed and were in its equatorial region) would almost always be edge-on to Earth-based astronomers.

The prospect of discovering rings about Neptune was more than enough to overcome the difficulty of the challenge. D. A. Allen decided to try infrared observations of Neptune based on his success with Uranus. His choice of infrared wavelengths was driven by his desire to decrease the relative brightness difference between the planet and its proposed rings. Reducing the relative brightness increases the probability of detecting rings by decreasing the dynamic range between the relatively bright disk and the purported dim rings. Unfortunately, his Neptune observations were not successful.

> The results were disappointing. The planet is too small to see clearly, and shows no hint of a ring ... and to complicate the issue, the *K* window (2.0 to 2.4 microns) does not always find Neptune dimmed by methane absorption ... [1]

Theoreticians also took up the search. A. R. Dobrovolskis determined mathematically that, in general, Triton's gravitational pull would 'warp' a ring system and prevent such a system from forming, but in spite of this he found regions in which dust particles in orbit about Neptune could be stable. These fixed regions were found between 0–15 and 165–180°, and at 90° [2] and the idea of polar rings took some investigators by surprise.

3.2 OBSERVATIONS BY THE TEAM AT VILLANOVA

In 1968 a team of astronomers from Villanova University were studying the occultation of a 7th magnitude star by Neptune to determine the planet's diameter. The idea

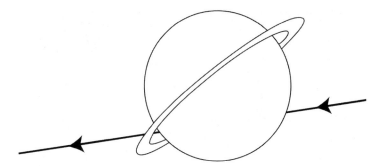

Figure 3.2. This diagram of Neptune and the proposed ring system depicts how the 1968 stellar occultation, recorded by the team at Villanova University, could miss the rings during immersion but be occulted during emersion.

that Neptune might have rings, and that stellar occultations might be the way to find them, had not yet been considered. The observation was a success and, like most other observations that serve their purpose, was written up and enshrined in papers documenting the results. Thirteen years later, on May 10, 1981, another stellar occultation was observed, this time by many astronomers. Results from that event prompted the Villanova team to re-examine the data from 1968, and they found that, a well-defined decrease in star brightness, lasting for 2.5 minutes, occurred about 3 minutes after emerging from Neptune (Figure 3.2). If the dimming was caused by a ring system, then the dip corresponded to a distance in the plane of the equator of 29,800 to 36,125 km from the planet's center [3]. The result was interesting but not strong enough to make a conclusive discovery because the fading could be from an experimental error, a new satellite or a short-lived ring.

The detection of a brightness 'dip' confirmed to astronomers that stellar occultations can be a powerful observing technique – a technique that could be used, among other things, as probes to search for a ring system about Neptune.

3.3   MORE APPARENT RING OCCULTATIONS

Astronomers did not have long to wait to try their technique to look for a ring system. On May 24, 1981, Neptune just grazed a well-behaved star, and James L. Elliot, from the team that was very successful with ring observations of Uranus, observed the encounter. During the event they found a brief drop in brightness, and although his analysis indicated that they had not necessarily found a ring, it was possibly a new satellite with a diameter of 180 km [4]. Whatever caused the fading, the event would have to be observed again in order to confirm the discovery. Eight years later, during the Neptune encounter, Voyager 2 confirmed the Elliot team's tentative conclusion. They had indeed found a satellite (eventually to be named Larissa).

The power of this technique, combined with the interesting Neptune results, led astronomers to await anxiously for the next event – which was more than two years away. When it did occur, astronomers were prepared for it. W. B. Hubbard and colleagues observed the event from six locations during the June 15, 1983 occultation. The event allowed them to search as close as 0.03 radii of the planet's surface for rings. Hubbard and his team had no success [5]; and Elliot's team also had negative results:

> ... no evidence for equatorial rings between 25,300 and 200,000 km ... These results rule out a Neptunian system similar to that of Saturn or Uranus, but not a system of low optical depth similar to the Jovian rings. [6]

The stellar occultation events were producing inconsistent results. Some teams thought they had detected rings, some thought they had found new satellites, and some had no detections at all. Only further events could hope to resolve the situation. On September 12, 1983, the Neptune occultation of MKE 31 was observed by Pandey, Mahra, and Mohan [7], who recorded a dimming of the star's output that corresponded to distances of 1.16, 1.20, 1.22, and 1.28 radii from the center of Neptune. Unfortunately, these changes in brightness were not seen by any other team, and although their data indicated that Neptune had a ring system between 28,700 and 31,700 km, was the data accurate?

After the first two attempts in early 1984 and a growing list of observed occultations (see Table 3.1), astronomers had to admit that the data did not support a ring about Neptune. Even unusual ring systems like ring arcs or rings, which vary in thickness with azimuth, could not be found [2] and further events had to be observed.

The next occultation, which occurred on July 22, 1984, was of the star SAO 186001, and this time Manfroid, Gutierrez, Häfner, and Vega saw a fading [8]. They estimated that the data would fit the discovery a new satellite about 10 to 15 km across at a distance of 75,000 km from the center of Neptune. Fortunately, a second team, led by Hubbard, at the Cerro-Tololo Interamerican Observatory, Chile, also recorded the event. His team was 100 km south of Gutierrez's team and also saw the dimming. The event was attributed to a partially transparent arc of material 67,000 km from the center of the plane, and the term 'ring arc' was used for the first time in Hubbard's paper to describe the observed change in brightness [9].

The idea of a stable arc of material in orbit about a planet was very difficult to believe. What would keep the material confined to an arc? Could astronomers be witnessing the birth of a ring? Or could it be something else ... even stranger? Astronomer Jack Lissauer [10] calculated that rotational shear forces would spread the arc material into a uniform ring in less than three years!

The subsequent event, which occurred on June 7, 1985 (Figure 3.3), involved a binary star system. During the occultation a single sharp fading was observed. Astronomers trying to describe the event could explain the data if there was a ring of material at 62,600 km (or 63,760 km depending on which star of the pair was occulted), but only one of the two stars was observed. The model also assumed that the ring material was in the planet's equatorial plane. Eighteen days later another occultation occurred but this time no indication of ring material was found [11].

Table 3.1. Neptune stellar occultations observed from 1968 to 1989 (from Cruikshank [12])

Date of occultation	Event seen	Corresponds to a 'ring' location at (km)	Observed from
1968 Apr 7	Possible occultation	29,800–36,125	MSO and Japan (2)
1980 Aug 21	Possible occultation		MSO
1981 May 10	None	N/A	AAO, IRTF, MSO, UH
1981 May 24	None	N/A	CTIO
1981 May 24	Satellite discovery	73,500	CAT, ML
1983 Jun 15	None	N/A	AAO, CFH, IRTF, KAO, UH, Australia (2)
1983 Sep 12	Possible occultation	28,700–31,700	UPSO
1984 Apr 18	Possible occultation	55,200	PAL
1984 May 11	None	N/A	PAL
1984 Jul 22	Confirmed event	67,000–75,000	CTIO, ESO
1985 Jun 7	Possible occultation	62,600 or 75,000	SAAO
1985 Jun 25	None	N/A	SAAO
1985 Jul 30	None	N/A	PAL, IRTF
1985 Aug 10	None	N/A	IRTF
1985 Aug 20	Confirmed event	62,900–63,000	CFH, CTIO, ESO, IRTF, LOW, MWO
1986 Apr 23	None	N/A	CFH, IRTF, UH
1986 May 4	None	N/A	KPNO, ESO
1986 Jul 27	Confirmed event	42,300	IRTF, PAL, UH
1986 Aug 23	None	N/A	IRTF, UH, UKIRT
1987 May 23	None	N/A	IRTF, PAL
1987 Jun 22	Triton occultation?	245,800	ESO
1987 Jul 9	Confirmed event	35,200	KPNO, MMT, PAL, UKIRT
1987 Aug 29	Possible occultation		IRTF
1988 May 26	None	N/A	IRTF
1988 Jul 9	Possible occultation	63,170 or 63,120	MMT, PAL
1988 Aug 2	None	N/A	ESO
1988 Aug 25	None	N/A	CFH, IRTF
1988 Sep 12	None	N/A	OHP, OPMT
1989 Jul 8	None	N/A	ESO, OPMT

AAO	Anglo-Australian Observatory	MMT	Multiple Mirror Telescope
CAT	Catalina Station	MSO	Mt Stromlo Observatory
CFH	Canada–France–Hawaii Telescope	MWO	Mount Wilson
CTIO	Cerro-Tololo Interamerican Observatory	OHP	Observatoire de Haute Provence
ESO	European Southern Observatory	OPMT	Observatoire du Pic du Midi
IRTF	Infrared Telescope Facility	PAL	Palomar Observatory
KAO	Kuiper Airborne Observatory	SAAO	South Africa Astronomical Observatory
KPNO	Kitt Peak National Observatory	UH	University of Hawaii 88″
LOW	Lowell Observatory	UKIRT	United Kingdom Infrared Telescope
ML	Mt Lemmon	UPSO	Uttar Pradesh State Observatory

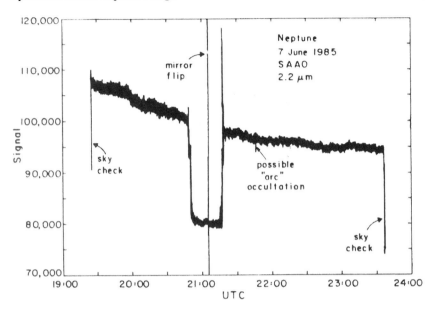

Figure 3.3. Possible arc occultation feature seen during the June 7, 1985 stellar occultation. Notice that even under the best visibility, rings or rings arcs are difficult to observe against the background noise. (From Covault *et al.* [11].)

Theoreticians were busy trying to combine all the data into a model that would describe a majority of the observations, but no single model would fit all the findings. The truth about Neptune's ring system would probably be contained in some combination of the following hypothesizes, if it existed at all:

- Two shepherding satellites, each between 100 to 200 km diameter, could confine the ring (they would be too small to be seen from Earth) [10].
- The ring system was either incomplete or at least highly azimuthally variable [13].
- Newly discovered satellites might produce some of the observations.
- Incomplete rings could consist of a series of short arcs of width 100 km that center on co-rotation resonances of a single satellite in an inclined orbit [14].

Further analysis of the data prompted Philip D. Nicholson to conclude that only three of the many stellar occultation events appeared to be real. Based on this conclusion he proposed that Neptune had at least three distinct ring arcs, the outermost of which was at 72,500 km from the center of Neptune. As for the inner two arcs, the data could not be resolved into a single location.

Another three occultations were observed in 1985, with a further four between 1986 and 1987. Of these only about half produced data that could be interpreted as having some sort of stellar dimming. As the Voyager 2 spacecraft continued to close in on Neptune, astronomers continued to observe the planet. In 1988, five more opportunities occurred, with only one showing signs of an occultation. In retrospect, the situation was summed up most eloquently in 1984 by Elliot and Kerr, who said, 'If Neptune has rings, they will almost certainly not be discovered from the ground' [15].

NOTES AND REFERENCES

1. Allen, D. A. (1984) An infra-red astronomer looks at cloud-covered planets. In P. Moore (ed.) *Yearbook of Astronomy* (1985), p. 138.
2. Dobrovolskis, A. R. (1980) Where are the rings of Neptune? *Icarus*, **43**, 222–6.
3. Guinan, E. F., Harris, C. C. and Maloney, F. (1973) Evidence for a ring system of Neptune. *Bulletin of the American Astronomical Society*, **14**, 658 (abstract).
4. Elliot, J. L. and Kerr, R. (1984) *Rings*. Cambridge, pp. 179–86.
5. Hubbard, W. B., Frecker, J. E., Gehrels, J. A., Gehrels, T., Hunten, D. M., Lebofsky, L. A., Smith, B. A., Tholen, D. J., Vilas, F., Zellner, B., Avery, H. P., Mottram, T., Murphy, T., Varnes, B., Carter, B., Nielsen, A., Page, A. A., Fu, H. H., Wu, H. H., Kennedy, H. D., Waterworth, M. D. and Reitsema, H. J. (1985) Results from observations of the 15 June 1983 occultation by the Neptune system. *The Astronomical Journal*, **90**, 655–67.
6. Elliot, J. L., Baron, R. L., Dunham, E. W., French, R. G., Meech, K. J., Mink, D. J., Allen, D. A., Ashley, M. C. B., Freeman, K. C., Erickson, E. F., Goguen, J. and Hammel, H. B. (1985) The 1983 June 15 occultation by Neptune. I. Limits on a possible ring system. *The Astronomical Journal*, **90**, 2615–23.
7. Pandey, A. K., Mahra, H. S. and Mohan, W. (1984) The occultation of MKE 31 on September 12, 1983. *Earth, Moon, and Planets*, **31**, 217–19.
8. Manfroid, J., Gutierrez, F., Häfner, R. and Vega, R. (1984) *Appulse of SAO 186001 to Neptune*. IAU Circular 3962, Central Bureau for Astronomical Telescopes.
9. Hubbard, W. B., Brahic, A., Sicardy, B., Elicer, L. R., Roques, F. and Vilas, F. (1986) Occultation detection of a Neptunian ring-like arc. *Nature*, **319**, 636–40.
10. Lissauer, J. J. (1985) Shepherding model for Neptune's arc ring. *Nature*, **318**, 544–5.
11. Covault, C. E., Glass, I. S., French, R. G. and Elliot, J. L. (1986) The 7 and 25 June 1985 Neptune occultations: Constraints on the putative Neptune 'arc'. *Icarus*, **67**, 126–33.
12. Cruikshank, D. P. (ed.) (1995) *Neptune and Triton*. The University of Arizona Press, pp. 734–5.
13. Sicardy, B., Roques, F., Brahic, A., Bouchet, P., Maillard, J. P. and Perrier, C. (1986) More dark matter around Uranus and Neptune? *Nature*, **320**, 729–31.
14. Goldreich, P., Tremaine, S. and Borderies, N. (1986) Towards a theory for Neptune's arc rings. *The Astronomical Journal*, **92**, 490–4.
15. Elliot, J. L. and Kerr, R. (1984) *Rings*. Cambridge, pp. 179–86.

BIBLIOGRAPHY

Bergstralh, J. T. (ed.) (1984) *Uranus and Neptune*. NASA Conference Publication 2330, National Aeronautics and Space Administration, Scientific and Technical Information Branch, Washington, DC, pp. 27–373, 575–88.

Littmann, M. (1988) *Planets Beyond: Discovering the Outer Solar System*. Canada.

Miner, E. D. (1998) *Uranus: the Planet, Rings and Satellites* (2nd edn). Wiley–Praxis, Chichester, UK.

Moore, P. (1988) *The Planet Neptune*. Wiley–Praxis, Chichester, UK.

4

Other pre-Voyager Neptune observations

4.1 THE INTERIOR OF NEPTUNE

Studies of the interior structure of any planet are necessarily complicated by a paucity of relevant data. Even for the Earth, direct observations probe only a tiny fraction of the total radius of our planet, and much of our knowledge of inner Earth is obtained by building mathematical models and comparing the predictions of such models with conditions at the Earth's surface.

Boundary conditions against which scientists may check possible models include several observable characteristics. The total mass of Earth (or of the remote planets) can be determined with fair precision by comparing the orbital periods of the satellites with the physical dimensions of those orbits. The size and shape of the planet, measured from its vicinity or from remote telescopic observations, can then be used to determine its volume and hence its overall bulk density (mass per unit volume). If satellite (or ring) orbits are elliptical or inclined to the planet's equator, the apsidal precession [1] rates of the elliptical orbits or the nodal regression [2] rates of the inclined orbits also provide clues about internal mass distribution. Reflected sunlight, infrared thermal radiation, and radiowaves given off by the planet contain information about the chemical composition of at least the outer layers of the atmosphere. Under certain conditions, such data can also reveal variations in composition with altitude and the atmospheric temperatures and pressures at those altitudes.

Armed with knowledge of a planet's distance from the Sun and of the Sun's brightness at Earth's orbit, one may calculate the amount of sunlight that reaches any planet in the Solar System. The difference between the amount of sunlight incident upon the planet and the amount of sunlight it reflects back into space represents the solar energy absorbed by the planet. That solar energy heats the planet to a predictable temperature level. If the observed temperature of the planet (from infrared and radiowave data) is higher than that attributable to solar heating, the excess must come from heat sources within the planet, providing an additional constraint on interior models.

Any magnetic field around a planet generally arises from processes occurring in that planet's interior. Although it is often difficult to measure the strength and nature of a magnetic field from a remote observing site, one can make crude estimates of those characteristics by observing the product of interactions within the field. One such interaction results in ultraviolet and visible auroral glows, as electrons streaming outward from the Sun are channeled by the planet's magnetic field into the atmosphere near the magnetic poles. Another interaction typical of the outer planets gives rise to long-wavelength radiowaves whose intensity varies with a periodicity equal to the planet's rotation period. A third type of interaction is that of synchrotron radiation, caused by the motions of electrons accelerated by the magnetic field and sometimes visible in short-wavelength radio (microwave) emissions from the planet.

Rotation of the outer planets will cause their shapes to depart from being precisely spherical. The resulting oblateness (polar flattening and equatorial bulging) is generally sufficient to be observed from Earth-based telescopes. The degree of distortion is dependent not only on the rotation rate, but also on the internal structure of the planet. If the rotation rate can also be measured (either by observing the periodicity in long-wavelength radiowave intensity or by observing the average rate of cloud motions across the disk of the planet), the planetary oblateness can serve as another constraint on interior models of the planet.

One additional constraint is generally applied to models of the giant planet interiors – namely, that of consistency with accepted models for the origin and evolution of the giant planets. This constraint is less confining than others in that it constitutes a two-way street: realistic models of the outer planet interiors can also be used to test the validity of cosmogonic (origin of the 'cosmos') models.

4.1.1   Pre-Voyager models

Prior to the volumes of data returned by Voyager 2 from its encounter with Neptune in August of 1989, there was not enough evidence to determine which of several proposed planetary interior models best described Neptune. The mean density of the planet was known to within a small fraction of a per cent, but the precise dimensions and shape were poorly understood. Furthermore, accurate determination of the planet's rotation period proved to be an almost hopeless task (Table 4.1), and estimates of the rotation period ranged from less than 8 h to more than 24 h. In 1984 Hubbard [3] had described the state of Uranus's interior modeling prior to the Voyager encounter, and the same questions still hold true for Neptune's internal models. Some of the main questions that interior models of Neptune must attempt to answer are:

- What is the bulk chemical composition of Neptune? What respective fractions of Neptune's total mass are hydrogen, helium, methane, ammonia, water, and heavier rocky materials? How does each of these compare with the fractions of each presumed to exist in the primordial solar nebula?

Table 4.1. Early observations to determine Neptune's rotation rate

Date	Observer(s)	Technique	Period (h)
1872	Flammarion	Direct observation	10.97
1884	Hall	Direct observation	7.914
1924	Öpik and Liviander	Direct observation	7.7
1928	Moore and Menzel	Classical spectroscopy	15.8 ± 1.0
1930	Moore and Menzel	Classical spectroscopy	15.8 ± 1.0
1977	Hayes and Belton	Classical spectroscopy	22 ± 4
1978	Cruikshank	Photometric light curves	18.17 or 19.58
1978	Slavsky and Smith	Photometric light curves	18.44 ± 0.01
1980	Münch and Hippelein	Classical spectroscopy	11.8 ± 1.2
1980	Belton et al.	Classical spectroscopy	15.4 ± 3
1980	Brown et al.	Time-resolved spectroscopy	17.95
1983	Terrile and Smith	CCD obs. cloud motion	17.83 ± 5
1984	Belton and Terrile	CCD obs. cloud motion	18.2 ± 0.4

- What is the chemical composition of the observable layers in the atmosphere of Neptune? What does this tell us about the internal composition of the planet?
- To what degree are the different components within the planet separated, and how does this compare with the other giant planets of the Solar System? What processes could lead to this chemical differentiation within Neptune?
- How much heat is coming from the interior of Neptune? What processes contribute to this heat flux? How are they related to the planet's formation processes? How are they related to the planet's atmospheric and interior dynamics? Do circulation patterns in the interior of Neptune give rise to a planetary magnetic field?

It is fortunate that at least partial answers to many of these questions were known for Jupiter and Saturn. The initial studies of these two planets by spacecraft were carried out via the Pioneer 10 and 11 spacecraft in the mid to late 1970s. These were supplemented by the Voyager 1 and 2 encounters with Jupiter in 1979 and with Saturn in 1980 and 1981. Voyager 2 added to this planetary dataset with observations of Uranus in 1986. Extrapolation from these three planets, accounting for the greater distance and slightly larger mass of Neptune over Uranus, provided some leverage for attacking an otherwise extremely complicated problem.

The relative abundance of elements in the early solar nebula has been deduced [4] on the basis of direct observation of the outer atmosphere of the Sun and from chemical analyses of meteorites collected from Earth's surface. In order of relative abundance (in mass fraction), the ten most abundant elements in the early solar nebula are hydrogen (74.4%), helium (23.7%), oxygen (0.87%), carbon (0.39%), neon (0.19%), iron (0.14%), nitrogen (0.09%), silicon (0.08%), magnesium 0.07%, and sulfur (0.04%). The remaining 0.03% includes all the rest of the elements, none of which constitutes more than 0.01% of the total mass. Relative abundances are often stated as ratios of numbers of atoms of the given element to the number of hydrogen atoms. Stated in these terms, abundances are even more heavily weighted toward the lighter elements. For every 1,000,000 hydrogen atoms, there are 80,000 helium atoms, 739 oxygen atoms, 441 carbon atoms, 128 neon atoms, 91 nitrogen atoms, 39 magnesium atoms, 37 silicon atoms, 33 iron atoms, and 19 sulfur atoms.

Most Neptune model calculations make a simplifying assumption by grouping the elements into three categories: gas (G), ice (I), and rock (R). These groupings are according to the state of the matter in the solar nebula at the distance of Neptune, but just before Neptune was formed. The G group contains those components that were gaseous then and remain gaseous now, primarily hydrogen and helium. The I group contains those components whose condensation temperatures are close to the estimated existing temperatures at the onset of Neptune formation. These include water (H_2O, containing most of the oxygen), methane (CH_4, containing most of the carbon), and ammonia (NH_3, containing most of the nitrogen). Much of the sulfur might also be tied up in hydrogen sulfide (H_2S) 'ice'. The R group comprises the remaining heavier elements and compounds, primarily those of magnesium, silicon, and iron. A given model attempts to specify relative amounts of G, I and R within Neptune. Although the relative amounts of each of these groups are not necessarily constrained to conform to solar abundances, most modelers assume that relative solar abundances are maintained within the I and R groups. The bulk density of Neptune is such that only a minor fraction of the total planetary mass can be composed of G component. Models must specify the relative amounts and radial distribution of I and R components.

If the I/R mass ratio is assumed to be consistent with solar abundance, it would have a value of about 3.5. This is the largest plausible I/R for any Neptune model that starts with a solar-abundance nebula. Any variation from this value would involve loss of the more volatile ices and a reduction of the I/R ratio. For example, if temperatures were such that most of the ammonia and methane were in the G component at the time of Neptune's formation, and the I component was primarily water, the bulk density of Neptune leads to an I/R ratio of $\sim$2.0. A suggestion by Lewis and Prinn [5] was that carbon and nitrogen in the primordial nebula may have been tied up in carbon monoxide (CO) and nitrogen gas (N_2), and that reactions which combine the carbon and nitrogen with hydrogen may have been inhibited. In this process, more than half the oxygen would have been tied up in gaseous carbon monoxide, greatly reducing the oxygen available to form water. The resulting I/R would be $\sim$0.6, a practical lower limit to the ratio. Cosmochemical considerations and the mean density of Neptune thus serve to constrain the I/R ratio for Neptune to values between 0.6 and 3.5.

Another clue to internal composition is the ratio of deuterium (D) to normal hydrogen. Deuterium is a form of hydrogen which contains an extra neutron in the atomic core. It reacts to form water, methane, and ammonia more readily than it bonds to another hydrogen atom. Thus, if chemical equilibrium was attained during Neptune formation, a larger-than-normal amount of deuterium should have been trapped in the I component of Neptune in the form of deuterated water (HDO), deuterated methane (CH_3D) and deuterated ammonia (NH_2D). As Neptune cooled, there should have been free exchange between the deuterium in the I component and hydrogen in the thin G component, resulting in a marked atmospheric enhancement of deuterium relative to solar abundances. Such deuterium enhancement was not observed in the thick hydrogen atmospheres of Jupiter and Saturn. Rough measurements of the deuterium/hydrogen ratio from Earth seem to show no significant differences between Neptune and the larger giant planets.

If indeed the deuterium abundance was not enhanced over solar values, there were some possible explanations. It is possible that deuteration of water, methane, and ammonia did not proceed to chemical equilibrium at the low temperatures associated with condensation of the I component. Or perhaps poor circulation within Neptune since its formation has not permitted deuterium to chemically equilibrate between the G and I components. A third possibility is that chemical equilibrium did occur with a much more extensive G component that the planet has since lost. The helium/hydrogen ratio could similarly have helped to resolve some of the unanswered questions about the origins and internal structure of Neptune. Present theories generally explain the thick hydrogen atmospheres of Jupiter, Saturn, and Uranus as captured atmospheres after the formation of the R-component and I-component cores. This would account for the similarity between the helium abundances on these planets and those of the pre-planet solar nebula. In the case of Neptune, it is possible that the hydrogen atmosphere came from the partial decomposition of methane rather than from a capture mechanism. Such an atmosphere would have a somewhat reduced abundance of helium as compared to Jupiter and Saturn. The extreme difficulty of making Earth-based observations of Neptune's helium abundance implied that planetary astronomers would have to wait for the scrutiny of Voyager 2.

Models of Jupiter and Saturn include well-differentiated interiors. An inner core of molten R-component material was overlain by an outer core of molten I-component material. Because they constitute a relatively small percentage of the total mass of Jupiter and Saturn, little external difference in the gravitational harmonics would be seen between these two-layered core models and well-mixed cores. The extensive G-component layer of these planets constitutes the majority of their mass.

A differentiated three-layered model of the type successfully employed for Jupiter and Saturn does not work well for Uranus or Neptune. Too large a fraction of the mass is near Uranus's center, and the resultant gravity harmonics do not match those deduced from observations of its rings. Zharkov and Trubitsyn [6] resolved this dilemma by assuming that the R and I components of Uranus and Neptune were uniformly mixed. When adjusted for higher compressibility of water at high pressures, this two-layer model adequately represented the externally observed data except for the fourth-order gravity harmonic coefficient, J_4. This model is depicted schematically in Figure 4.1. Calculation of the temperatures and pressures at depth in the dense atmosphere model led to the conclusion that the I component was likely to be liquid rather than gaseous. Hence it was supposed that Uranus and Neptune had deep superheated oceans.

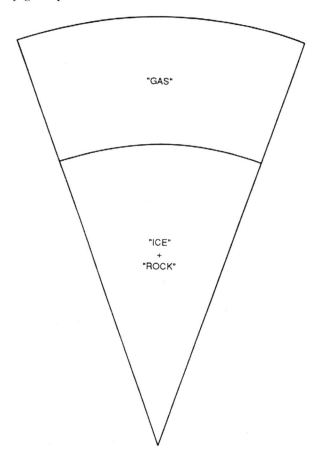

Figure 4.1. Neptune interior model of Zharkov and Trubitsyn. R   and I components are assumed to be uniformly mixed. This model adequately represented the externally observed data except for the fourth-order gravity harmonic coefficient J_4.

The models presented are not unique in their ability to match the observational data and to fit well with cosmogonic models of Solar System formation. Perhaps the interior of Neptune is not layered, but consists of a mixture of materials whose relative abundances change smoothly with depth. More complex models with many layers may also be possible. Even the Voyager 2 data from the Uranus encounter did not provide enough constraints to uniquely define the interior structure of that distant planetary neighbor.

4.1.2   Rotation period

Because of Neptune's great distance from the Sun and the difficulty of seeing cloud features, the only way to ascertain the planet's rotation rate was to look for variations that would be caused by rotational effects. One of the first approaches utilizing this technique was proposed by Deslandres [7] in 1902. He pointed out that sunlight reflected from the approaching limb of a rotating planet would show spectral features, which are shifted to shorter (bluer) wavelengths than for the center of the disk. Similarly, spectral features seen in sunlight reflected from the receding limb will be shifted to longer (redder) wavelengths. These shifts are due to the Doppler effect [8]. Though attempted at Uranus, this technique was exceedingly difficult to use for Neptune due to the small angle subtended by the planet. The small angle results in a very weak signal from which to observe the shifted spectral features.

In 1928 Moore and Menzel, using classical spectroscopy, determined Neptune's rotation rate to be 15.8 ± 1 hours [9]. Eight years later, Cambell published an alternative method for determining a planet's rotation rate [10]. His approach, first used on Uranus, assumed that periodic changes in light levels could be used to determine an upper atmospheric rotation rate. Unfortunately, re-analysis of his data in the late 1970s showed that its quality was very poor.

The great difficulty of determining a rotation rate for Neptune was a deterrent for astronomers from even attempting to make measurements for many decades. In 1977 Hayes and Belton, using classical spectroscopic techniques, obtained a rotation rate of $22 \pm 4 \, \text{h}$ [11]. By the early 1980s progress was beginning to be made in another method called time-resolved photometry. This technique also used reflected sunlight, and 1980 observations of Neptune showed the anticipated variations in light intensities. These variations were interpreted as being due to an upper atmospheric rotation rate of 17.95 h for Neptune [12, 13]. One shortcoming inherent in this method is aliasing, due to the transient nature of atmospheric clouds of methane. The appearance and disappearance of methane clouds alters the measured light levels and leads to an erroneous period determination. Time-resolved photometry at Neptune also revealed rapid changes in the planet's brightness at infrared wavelengths, often taking months to return to steady-state conditions [14, 15].

Telescopic observations of upper atmospheric motion in Neptune's atmosphere provided very little useful information until the advent of charge-coupled device (CCD) detectors (see Figure 4.2). These detectors have the sensitivity and dynamic range to measure faint details, even near the $0.89 \, \mu\text{m}$ band of methane, and were first used in April 1976 to obtain infrared images of Uranus. With these detectors astronomers were able to make direct measurements of cloud features, verifying Neptune's direct rotation and providing evidence for rapid zonal atmospheric outflows [16, 17].

It should be noted that the angular rate of motion of atmospheric features can and does vary with latitude on the other giant planets. Hence, measurements of the speed of rotation of atmospheric clouds may not accurately represent the rotation rate of the planet's interior. A more accurate approach is to measure the periodicity of long-wavelength radiowaves emitted by charged particle interactions within the planet's rotating magnetosphere. Such measurements require that the spacecraft be close enough to the Neptune magnetosphere for a long enough period to detect such emissions. Thus the measurement of these long-wavelength radiowaves is one of the

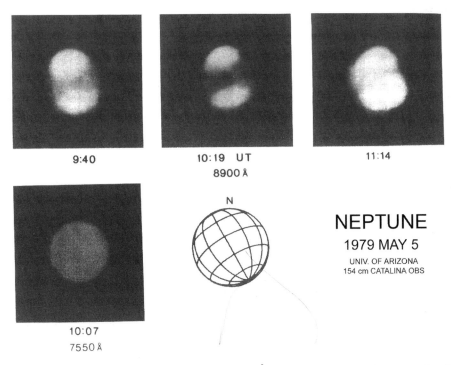

Figure 4.2. CCD image of Neptune taken at 8900 Å from the 154-cm reflector at Catalina Observatory (University of Arizona), May 5, 1979, by B. A. Smith, H. Reitsema and S. Larson.

numerous important observations Voyager 2 attempted to acquire during its August 1989 encounter.

4.1.3 Polar flattening

Optical oblateness (ε), or polar flattening, is defined as $\varepsilon = 1 - (R_{\mathrm{p}}/R_{\mathrm{e}})$, where R_{p} and R_{e} are the polar and equatorial radii of Neptune, respectively. Here, too, Neptune's extreme distance from Earth made ground-based direct measurements of its oblateness virtually impossible with the techniques available to astronomers prior to Voyager 2's encounter. Fortunately, stellar occultations offered an alternate approach for understanding, to first order, Neptune's polar flattening. Determining the planet's oblateness would provide additional constraints on the planet's rotation rate and its internal mass distribution.

   An early estimate of Neptune's oblateness was obtained when Neptune occulted a star, BD174388, in 1968. By this technique, Kovalevsky and Link determined an oblateness of 0.021 ± 0.004 [18]. The difficulties inherent in such analyses are apparent when one notes that Freeman and Lyngå, using the same data set, obtained a different oblateness, 0.0259 ± 0.0051 [19]. The difference in the results is due in part to differences in the data analysis techniques and differences in the assumptions of the two teams. The data was recorded on a strip chart, and the determination of starting and ending times was somewhat subjective. Furthermore, it was assumed that Neptune did not have a ring system to provide a reference system from which to determine with any accuracy the star's position relative to the center of the planet [20]. Stellar occultation measurements in 1983 produced results that were consistent with those from the 1968 occultation, but with a significantly smaller uncertainty.

4.1.4 Internal heat

Uranus has a mean distance from the Sun of 19.21 AU; Neptune's mean distance is 30.11 AU. Because of its greater distance, the intensity of the sunlight falling on Neptune is only 41% of that falling on Uranus. Yet the temperatures of the reflected back into space, namely $(1 - A^*)$. It is this fraction which is responsible for the solar heating of the planet. Any additional heat required to give the planet its observed temperature must originate from the interior of the planet.

The bolometric Bond albedo, A^*, of Neptune cannot easily be determined from Earth as our atmosphere blocks many important wavelengths, especially in the ultraviolet and infrared regions of the solar spectrum. Light at wavelengths more readily transmitted by the atmosphere is also distorted (seeing effects) and selectively absorbed (transparency effects), each by time-varying amounts. Some of the problem can be overcome by using high-altitude observatories (e.g., the Mauna Kea Observatory on the island of Hawaii), airplane-mounted telescopes (e.g., Kuiper Airborne Observatory) or balloon-borne telescopes (e.g., Stratoscope 11). Still greater altitudes can be achieved with sounding rockets (e.g., the Aerobee or Scout rockets) or Earth-orbiting observatories (e.g., the InfraRed Astronomy Satellite and the Hubble Space Telescope). Nearly all of these techniques have been used to collect information on the brightness of both the Sun and Neptune at a large variety of wavelengths.

Determination of the light-scattering properties of the atmosphere of Neptune has even greater inherent difficulties. Because it is so distant from the Sun, Earthbound (including Earth-orbiting) observers and instruments always view a fully illuminated disk. The maximum angle between Sun and Earth as viewed from Neptune (i.e., the maximum phase angle) never exceeds $1.9°$. From the vicinity of Earth it is impossible to view Neptune under a variety of lighting and viewing geometries and thereby deduce the Bond albedo. It is possible to watch for longitudinal brightness variations as the planet rotates, and seasonal or latitudinal changes as it revolves around the Sun, but only for a planetary disk that is fully illuminated.

The ratio of the Bond albedo, A (monochromatic) or A^* (bolometric), to the corresponding albedos at $0°$ phase angle (geometric albedos) is known as the phase integral, q (monochromatic) or q^* (bolometric). Before the advent of space probes to the outer planets, astronomers were forced to rely on theoretical models of lightscattering to estimate the value of q and q^*. Harris [24] tackled the problem of estimating the phase integral for the giant planets, deriving a value of 1.65 (for visual wavelengths). Combined with his estimate for the visual geometric albedo of Neptune, 0.509, (Uranus = 0.565), this yielded a visual Bond albedo, A_V of 0.84 (Uranus = 0.93). Harris's estimate was that Neptune absorbed 16% of the visual sunlight incident on it. He pointed out that Uranus's albedo was higher than for any other planet and that it implied that it was only absorbing 7% of the visual sunlight falling on it. If these ratios were accurate and had applied at other wavelengths as well, they might have explained the curious similarity in the temperatures of Uranus and Neptune.

The additional piece of data needed to answer the question of how much heat is coming from the interior of Neptune is an accurate measure of the amount of energy being emitted by the planet. Most of the solar energy used to heat the planet is in the form of ultraviolet, visible and near-infrared light. Because of its low temperature, most of the energy reradiated by Neptune is in the far-infrared part of the spectrum. Earth-based observers again are limited by the transparency of Earth's atmosphere and by the small range of phase angles (less than 1.9°) observable from Earth. The limitation in phase angle is not a severe problem in temperature determinations because the outer planets with their deep atmospheres change temperatures at a rate much slower than a Neptune day or even a Neptune year. In other words, contrasted with terrestrial planets like Earth, Neptune's temperature at midnight is identical to its temperature at noon, and its temperature in midsummer is identical to its temperature in midwinter.

If Neptune did not have an internal heat source, its temperature would be approximately 44 K. But observations made by many astronomers indicated that Neptune could have one of the largest internal energy sources (relative to incoming solar energy) among the planets [25]. The abundance of horizontal inhomogeneities (i.e., weather) apparent in images supported this conjecture. Additional support was found in the systems of zonal winds observed by Belton et al. [26]. Neptune appeared to be emitting approximately 2.4 times as much energy as it received from the Sun [27–30]. One early mechanism thought to explain this abundance of heat was the frictional dissipation of Neptune's tides produced by the gravitational pull from Triton, but in 1974 Trafton proved that this could not be the case [31]. Another suggestion was that, within the giant planets, a slow planet-wide contraction, combined with relatively heavy material in the planet's mantle sinking towards the core, might produce the observed heat. If this were the case, the model would have to explain the lack of an internal heat source in Uranus. The only way to be certain was to acquire more information about the planet, which only a planetary encounter could provide.

4.2   THE ATMOSPHERE OF NEPTUNE

There were several early indications that the atmosphere of Neptune was similar to that of Uranus. Firstly, there was the similar color noted by many early observers.

Uranus was blue-green while Neptune appeared a somewhat deeper blue. Secondly, there was the similarity in size, with Neptune's radius differing from Uranus by only 3% and a mass difference of 15% (in Neptune's favour). Thirdly, its overall density was comparable but slightly larger than that of either Uranus or Jupiter. This, combined with the fact that Neptune was much smaller than Jupiter, led to the conclusion that Neptune had much lower pressures in its interior than Jupiter. The low-density elements that constituted a major fraction of Jupiter must contribute a substantially smaller fraction of Neptune's mass.

4.2.1 Chemical composition

Hydrogen (H) and helium (He) had earlier been suggested as the major components of the Neptunian atmosphere, but spectroscopic evidence for their existence was difficult to obtain. Since the solar spectrum was dominated by absorptions from these two elements, the reflected solar spectrum from Neptune would have to show substantial changes in those absorption lines to prove the existence of H and He. Some had suggested that a line of atomic hydrogen at 0.4861 μm was enhanced, but based on Slipher's methane results, Russell [32] showed that the atmosphere of Neptune was too cold to permit atomic hydrogen to exist in large enough amounts to be detectable from Earth. He also showed that temperatures were too low to permit water vapor to exist in the visible atmosphere in measurable quantities.

In 1948–9, Kuiper utilized improved infrared photography techniques to find a number of new absorption bands in the Uranus spectrum. He, together with Phillips [33], proceeded to show that these hydrogen absorption bands were not due to methane or any of its chemical byproducts, nor to ammonia, nor to other suggested molecular components of the atmosphere of Uranus. Herzberg used a gas cell with a path length of 80 m filled with hydrogen at 100 atm pressure to obtain a spectrum, which, he then showed, reproduced the newly discovered Uranian spectral lines of Kuiper. The same spectral signature was found by Herzberg in the atmosphere of Neptune [34]. Although atomic hydrogen (H) had been ruled out for the visible atmosphere, molecular hydrogen (H_2) was obviously present in large quantities. Astronomers soon came to realize that molecular hydrogen was the dominant constituent of the atmospheres of all the giant planets. It was a natural step in logic to assume that the second most abundant gas was helium, although spectroscopic evidence for the amount would have to await the Voyager encounter.

Other than hydrogen and helium, the next most abundant gas in Neptune, methane, was first detected by Rupert Wildt in 1937 [35]. The relative amount of methane is several times greater than for the Sun; and methane and ethane were found in the emission spectra of Neptune, but not in Uranus [36]. To confirm Wildt's detection, two things were necessary: firstly, to obtain spectra of Neptune that extended to a wavelength of 0.886 μm, and, secondly, to obtain definitive laboratory spectra of methane at visible wavelengths and compare them with the Neptune spectrum. Adel and Slipher of the Lowell Observatory in Arizona accomplished both [37, 38]. They used high-pressure tubing 5 cm in diameter and 12 m in length

Table 4.2. Pre-Voyager estimates of Neptune gas abundances

Gas	Neptune	Solar	Neptune/Solar
Molecular hydrogen (H_2)	300–600 km atm	H/H = 1.000	–
Helium (He)	<1	He/H = 0.080	0 to >1
Methane (CH_4)	$1.5 \times 10^{-4} < C/H < 5 \times 10^{-3}$	$C/H = 4.4 \times 10^{-4}$	0.3 to 11
Deuterated hydrogen (HD)	Unknown	$D/H = 1.5 \times 10^{-4}$	–
Ammonia (NH_3)	Less than solar	$N/H = 1.3 \times 10^{-4}$	<1
Hydrogen sulfide (H_2S)	Unknown	$S/H = 1.7 \times 10^{-5}$	–
Ethane (C_2H_6)	Detection	–	–

to hold the methane gas, which they then pressurized to 40 atm (40 times Earth's sea-level atmospheric pressure). This provided an effective path length of 480 'meter-atmospheres' of methane. Glass windows at each end of the rigid tube enabled them to illuminate the gas at one end and place the spectrograph at the other, and by this technique they obtained a superb methane spectrum that covered wavelengths from the violet to the infrared.

Earlier, Adel and Slipher had used the relatively high-altitude (to eliminate most of the absorption from Earth's atmosphere) telescope site at Lowell Observatory to obtain high-quality violet-through-infrared spectra of both Uranus and Neptune. They found that their laboratory spectrum of methane was almost a perfect match with each of the two planetary spectra, and had succeeded in proving that methane was a major component of the atmospheres of both Uranus and Neptune.

It should be emphasized that the atmospheric composition details given above refer to a level in the Neptune atmosphere that is above a reflecting cloud layer, probably composed of methane ice. The three main components of the atmosphere had been shown (or estimated, in the case of helium) to be molecular hydrogen, helium and methane. Later observers were to detect trace amounts of ethane (C_2H_6). By analogy with Jupiter, Saturn, and Uranus, ammonia (NH_3) was also presumed to be present. Pre-Voyager abundance estimates for the atmospheric gases above the cloud deck are given in Table 4.2, which has been adapted from data given by Hunten [39], Atreya [40], and Miner [41].

4.2.2   Temperatures

Neptune's atmospheric temperatures vary substantially with altitude. Two general types of Earth-based observations have been used to infer temperatures at different levels in the atmosphere. The first of these is stellar occultation measurements, which provide information on temperatures relatively high in the atmosphere. The second is based on the observed intensity of Neptune radiation at various wavelengths in the infrared-through-microwave parts of the spectrum.

During a stellar occultation, changes in the apparent brightness of the star can be monitored as the atmosphere (of Neptune) passes in front of it. From such measurements and a general knowledge of the chemical composition (expressed as a mean molecular weight [42]) of the atmosphere, it is possible to deduce the variation of gas density with altitude. At high temperatures, gas density decreases slowly with altitude; at lower temperatures, the density decreases more rapidly with altitude. With some simplifying assumptions, one may then calculate the temperature and pressure of the gas as a function of altitude in the atmosphere.

The atmospheric pressure levels sensed by stellar occultations are almost 1 microbar (1 μbar = 0.000001 of Earth's sea-level atmospheric pressure). These correspond to an altitude of about 400 to 500 km above the 1-bar reference level (which is near the Neptunian cloud tops). Temperatures (near the 1-μbar level) derived from stellar occultations by Neptune generally fall between 120 and 160 K [25], and very little variation with latitude has been observed [43].

Determination of altitude–temperature profiles from infrared and microwave data is not as straightforward. Only data at wavelengths of 5 μm or longer (generally referred to as the thermal spectrum) are useful in this determination, since shorter wavelengths are dominated by reflected sunlight. An absorption or emission line or band in the thermal spectrum of Neptune is not formed at a single altitude within the atmosphere, but is the result of absorption over a range of altitudes. In order to reconstruct an altitude profile, one must construct a mathematical model of the atmosphere with a postulated temperature-versus-pressure profile and a chemical composition. The model must also allow for the presence of haze layers or clouds of ice particles. With such a model, it is then possible in principle (though difficult in practice) to predict the amount of light that will escape Neptune at each point in the thermal spectrum.

The problem is simplified somewhat by the realization that, at wavelengths longer than about 100 μm, the observed temperatures increase uniformly with wavelength and are representative of greater and greater depths within the atmosphere. For each of these longer wavelengths, one may determine an effective depth (and temperature) that is more nearly independent of small variations in the assumed chemical composition.

Deviations of the observed thermal spectrum from model predictions lead to adjustments of the model. When the fit is considered close enough, the model represents a best estimate by its author of the temperature–pressure profile and composition of the atmosphere. A number of pre-Voyager models are presented by Orton and Appleby [25], which generally show a temperature minimum of 53 to 55 K near a pressure level of 0.2 bar (about 50 to 100 km above the 1-bar reference level), as shown in Figure 4.3. Temperatures increase with depth below the 0.1-bar level, reaching 70 to 80 K at 1 bar. The rate of increase with depth below the 1-bar pressure level cannot be determined from Earth-based observations, but the 'lapse rate' (change in temperature per kilometer of depth) is generally assumed to be adiabatic [44].

Temperature–pressure relationships within the Neptune atmosphere (Figure 4.4) deviate from those expected for a purely gaseous atmosphere if haze or cloud layers are present. The pre-Voyager estimates of these effects are discussed below.

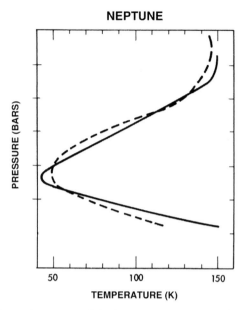

Figure 4.3. A typical pre-Voyager model of Neptune's atmospheric temperature at different pressure levels.

4.2.3   Aerosols

Tiny liquid or solid particles suspended in a gas are termed 'aerosols' and can alter the temperature of the surrounding gaseous atmosphere. Smog, which is familiar to most urbanites, is an example of an aerosol in Earth's atmosphere. The nature and magnitude of the effects of such aerosol hazes in Neptune's atmosphere depend on how much sunlight they reflect, absorb, and transmit. Reflection of sunlight will reduce the amount of energy available for heating the lower atmosphere. Absorption will also reduce the transmitted sunlight, but additionally cause local heating as the particles transfer their absorbed heat to the surrounding gases. In general, it was expected that the aerosol particle concentrations in the upper atmosphere of Neptune would be small enough that such haze layers would have only a limited effect on the temperature profile.

It was presumed that the aerosols of importance in the atmosphere of Neptune are all chemical byproducts of methane (CH_4), primarily from interaction with ultraviolet light from the Sun ('photolysis'). Ethane (C_2H_6) and acetylene (C_2H_2) both condense into ice particles at temperatures found just above the level of the temperature minimum (the 'tropopause') in Neptune's atmosphere. Methane gas is present in large enough quantities in the stratosphere (the first few hundred kilometers above the tropopause) that ethane and acetylene are produced by photolysis. Some of the ethane and acetylene then migrates slowly downward until the temperatures are low enough and the concentrations are high enough to form detectable ice hazes. Although the Earth-based estimates of the temperature–pressure profile are

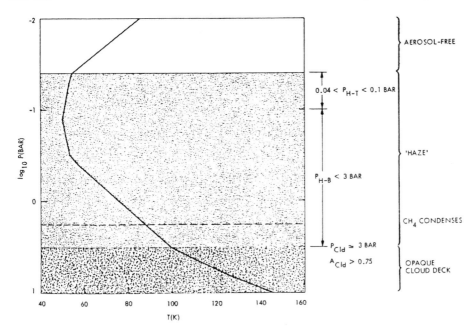

Figure 4.4. Schematic of Neptune model atmosphere, illustrating pressure–temperature structure (Appleby, private communication) and aerosol parameters to be constrained.

not well constrained at these levels in the atmosphere, such condensation was expected to take place in the pressure range from 1 to 100 mbar (50 to 200 km above the 1-bar reference level).

The lack of detailed models does not prevent planetary scientist from making some educated guesses about the clouds and hazes on Neptune. In general, because Neptune's spin axis is slightly tilted to its orbital plane, it should exhibit clouds and hazes opposite to that of Uranus. That is, Neptune's poles are expected to be 'more' cloudy and hazy than those of Uranus [40].

4.2.4   Clouds and cloud motions

Clouds in the atmosphere of Neptune are formed by a process different from that which produces aerosols. Clouds occur at deeper levels within the atmosphere and are the result of condensation of the 'ice' constituents of the interior of the planet, namely methane (CH_4), ammonia (NH_3), water (H_2O), and possibly hydrogen sulfide (H_2S). Buoyant bubbles of warmer gas circulate upward from the interior of the planet, eventually reaching levels where temperatures are cold enough that condensation of the icy component of the gas occurs. The heavier condensates reverse the upward flow and sink back to levels where the process begins again with the evaporation of condensates. An equilibrium cloud base is formed at the level in the atmosphere where the temperature is near the freezing point for each icy constituent.

In order of increasing depth below the Neptune troposphere, cloud layers of methane, ammonia and water were expected to exist. The base of the methane cloud was predicted to occur near a temperature of 80 K, somewhat below the 1-bar pressure level in the atmosphere. Ammonia would form a cloud base near a temperature of 130 K, at a pressure level near 4 bar. Water ice clouds would occur much deeper in the atmosphere, at pressure levels in excess of 30 bars [45] (see Figure 4.5).

Observations made from April 1975 to March 1976 revealed that high-altitude cloud layers formed over the planet and then partially dissipated [14, 46]. This was the first high-quality observational data to suggest that Neptune had weather. Later observations by Terrile and Smith in 1983 proved that Neptune cloud features were so prominent that they could be seen in CCD images [47]. These cloud features were then used to estimate upper atmosphere rotation rates at various latitudes. Earth-based observations, which found that the planet changed in brightness at various wavelengths, implied that Neptune's atmosphere changed on at least four different timescales [48].

- Diurnal variations.
- Day-to-day variations.
- Neptune/solar cycle variation.
- Bright outbursts.

Diurnal variations were used initially to determine the length of a Neptune day. Astronomers would make observations between the strongest absorption and

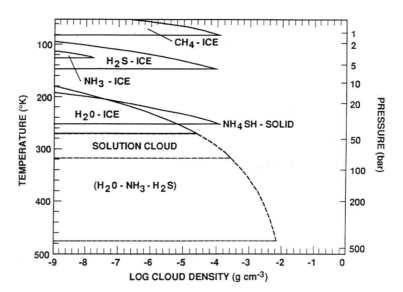

Figure 4.5. Predicted locations and densities of various tropospheric condensate cloud layers within Neptune.

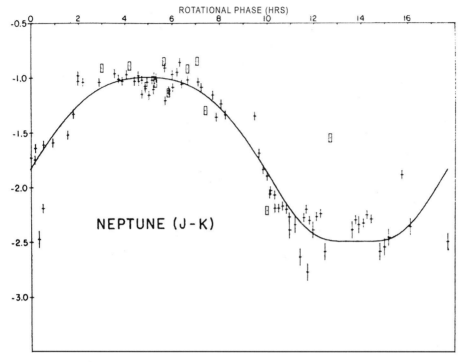

Figure 4.6. Near-infrared observations in the J–K color (1.25–2.2 μm) by Brown, Cruikshank, and Tokunaga revealed a rotation rate of 17.73 ± 0.1 h.

non-absorption spectral regions. This contrast occurs in the near-infrared between 1 and 3 μm and showed brightness variations that changed by more than 1 magnitude (a brightness ratio of 2.5) in the J–K color (1.25–2.2 μm) [49] (see Figure 4.6). As previously stated, rotation periods were found to be about 18 hours (see Section 4.1.2). Other observations revealed several distinct atmospheric rotation periods that implied large-scale atmospheric currents with relative zonal velocities of at least 109 m/s [17].

Day-to-day variations indicated a non-uniform distribution of high-altitude hazes and suggested that the atmosphere of Neptune was very active. Even with this level of movement, haze levels could remain stable and persist for a number of days [48]. This implied a very complicated meteorology that required more detailed observations in order to begin to understand its dynamics.

Additional observations indicated that Neptune's brightness varied on an approximately 10-year cycle, possibly anti-correlated with solar activity. That is, between 1972 and 1978, Neptune's brightness increased approximately 4% in evident anti-correlation with a sunspot minimum activity period on the Sun [50]. The anti-correlation with solar activity is not well established, due to the relatively short time that has passed since the paper was published. With or without a solar cycle correlation, Neptune's variability on an approximately 10-year cycle seemed to be well established.

Finally, Neptune may have periodic outbursts, the first of which was observed by Richard Joyce in 1976 [14]. In that year, Joyce measured a Neptune brightness at near-infrared wavelengths that was much greater than that measured the previous year. Subsequent observations revealed a decrease in brightness over the next few months. Analysis of the brightening led observers to conclude that it could be explained by the formation of a cloud with particles of at least 1 μm in diameter at high altitudes [46]. The upwardly convective movement of methane and other hydrocarbons could result in stratospheric clouds as these gases move through Neptune's cold trap. Convection in some parts of Neptune's atmosphere could be so strong that methane gas 'overshoots' the regular radiative-convective boundary, injecting methane high into the stratosphere [25]. Once produced, these methane ice clouds would dissipate by a gradual settling of particles deeper into Neptune's atmosphere.

Indications of other outbursts in 1980 and 1983 were reported, but no supporting evidence was found by other investigators [48].

4.3   THE ATMOSPHERE AND SURFACE OF TRITON

Triton is a massive and bitterly cold satellite. Just how massive and how cold were a matter of conjecture. The temperature of the surface would be highly dependent on the reflectivity of the surface. A highly reflective surface would absorb less sunlight and would be colder; a darker surface would absorb more sunlight and have a higher temperature. The predicted dimensions of the satellite were related to both the mass estimates and the temperature estimates. The pre-Voyager mass estimate, which was based on very difficult measurements of the wobble in Neptune's motion, was erroneously high, and it seemed unreasonable to assume a density lower than about $2 \, g/cm^3$ or higher than about $4 \, g/cm^3$. These two extremes then set bounds on the range of possible diameters for Triton. The integrated brightness of Triton from Earth-based measurements could then be used to estimate the reflectivity of the surface. Because the mass estimate was too high, the size estimates were similarly too high and the corresponding estimates of surface reflectivity too low. In fact, there was no reasonable set of mass, size and surface reflectivity numbers for Triton prior to the Voyager 2 encounter. The actual density of Triton was found to be near the lower limit mentioned above, but the mass was much lower and the surface reflectivity much higher than pre-Voyager estimates.

If Triton had an atmosphere, it would be unlike any atmosphere previously studied. Only Titan, the large satellite of Saturn, was known to have a substantial atmosphere composed primarily of nitrogen. The surface pressure of the Titan atmosphere is 1.6 bar, and the corresponding temperature is 92 K. One of the first astronomers to search for an atmosphere about Triton was Spinrad, who made his observations in 1969, but his search unfortunately found no evidence of an atmosphere [51]. In 1979, Cruikshank and Silvaggio [52] described Triton as follows:

We picture a surface that is largely covered with rocky material with a few patches of frozen CH_4, probably away from the sub-solar point, where the temperature is lower. The dark side of Triton may act as a cold trap; because of the inclination of the satellite's orbital plane to that of Neptune, portions of the body are in perpetual darkness during part of Neptune's orbit around the Sun. The variability of the geometry over Neptune's 165-year orbital period, together with the satellite's apparent 5.877-day rotation period, may result in a methane meteorology, the details of which merit further study.

Astronomers made their spectroscopic observations of Triton (Figure 4.7) at near-infrared wavelengths, due to the presence of strong methane, nitrogen, and water ice absorption bands at these wavelengths. Such near-infrared spectrophotometry succeeded in revealing the presence of methane. By 1988, planetary scientists believed that Triton had a thin atmosphere, probably composed of methane and nitrogen. Observational astronomy was providing compositional information about Triton's atmosphere, while computers were being used to model it. If Triton had a surface temperature consistent with its distance from the Sun, and if its atmosphere was produced by the sublimation of methane ice, then the partial pressure of methane near its surface would be about $(1.0 \pm 0.5) \times 10^{-4}$ bar.

The observations seemed to indicate that there was more to Triton's atmosphere than just a rarified methane and nitrogen gas surrounding a frigid surface. The combination of Triton's atmosphere, its orbital period, and its icy conditions could potentially produce periodic variations (i.e., seasons). This led Trafton [53] to state:

> Condensed phases of gases making up the bulk of Triton's atmosphere are likely to exist on Triton's surface in the form of solid or liquid 'polar caps' which extend as far as 55 degrees from the poles. The mass of Triton's atmosphere is governed by the energy balance between the sunlight these caps absorb and the heat they radiate to space. The polar cap temperatures should be approximately equal and uniform over their surfaces. Because of the rapid precession of Triton's orbit around Neptune's pole, the insulation and, therefore, the temperature of the polar caps must vary in a complex fashion. This will cause the mass of Triton's atmosphere to undergo a sinusoidal seasonal variation with an amplitude which ranges sinusoidally from mild to extreme in extent. The variations in the temperature of the polar caps will also cause seasonal variations in the mixing ratio of the volatile atmospheric gases owing to the different behaviors of their saturation vapor pressures with temperature.

Some investigators predicted that because of Triton's great distance from the Sun and the observed chemistry, Triton could also have lakes or pools of liquid nitrogen co-existing with surrounding surfaces of methane ice and water ice. In this case the observed reddish color of the satellite might be due to the organic compounds dissolved in nitrogen lakes.

In summary, the surface of Triton was thought to be characterized by solid methane, either as a continuous surface or as icebergs floating in a sea of

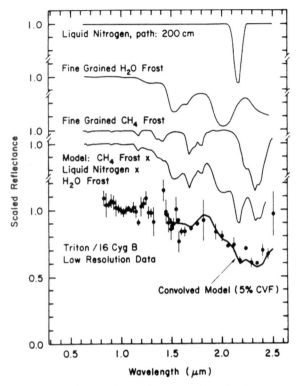

Figure 4.7. Earth-based observation of Triton.

unknown depth of liquid nitrogen. Reddish photochemical products might give the surface a slight coloration, and water ice might occur as crystals suspended in the liquid nitrogen or as a solid mixed with the methane frosts on expanses of solid surface (spectral modeling favors a suspension of fine crystals in the liquid). In this scenario, the satellite would have an atmosphere of nitrogen with other possible minor constituents; the surface pressure may be regulated by the vapor pressure at the local temperature, but strong diurnal and seasonal effects were thought to be probable [54].

The state of Triton's surface was still in question, right up to the time of the Voyager 2 Neptune encounter. Scientists did not know if Voyager would find a surface that was solid, liquid, or some combination of both (see Figure 4.8).

4.4   THE MAGNETOSPHERE OF NEPTUNE

In 1966, Kellerman and Pauliny-Toth detected the first radio emissions from Neptune [55]. Even as late as 1981, some researchers suggested that Neptune would not have a magnetic field because of the high density and pressure at the core of the planet, in spite of the presence of a strong internal heat source [56].

Figure 4.8. This artist drawing is taken from the perspective of being on Triton looking up towards Neptune. Notice that the surface was drawn with puddles of some kind of liquid, which, if they did exist, could be liquid nitrogen. (See color plate 8.)

Others reasoned that it might be possible to generate a magnetic field from a layer outside the core, if a model could explain such a conducting convection layer surrounding the central core. If this were the case, Neptune might be the first known planet with an internal magnetic field that was not generated from the core, but from a layer surrounding it.

4.4.1   The solar wind at Neptune's distance from the Sun

The Sun's magnetic field (the heliosphere) extends well beyond the orbit of Pluto. The magnetic fields of the planets exist as magnetic bubbles (magnetospheres) within the larger heliosphere, and the shapes and sizes of these planetary magnetospheres are greatly altered by the solar wind – the stream of charged particles (mainly electrons and protons) which flows outward from the Sun at hypersonic speeds. The solar wind is neither constant with time nor with distance from the Sun, but ebbs and flows depending on conditions near the outer boundaries of the Sun. The Sun itself undergoes an 11-year activity cycle, during which the observed daily number of sunspots grows to a maximum and then declines to a minimum. Near the time of minimum sunspot activity, the Sun's magnetic field actually reverses polarity; the north and south magnetic poles have the opposite orientation during any given 11-year activity cycle than in the adjacent 11-year periods. All these characteristics must be taken into account if we are to understand more fully the magnetic field and charged particle environment near a planet.

The characteristics of the solar wind can be studied effectively only from interplanetary spacecraft. The Earth itself and most Earth-orbiting spacecraft are entirely within Earth's magnetosphere, which excludes all but the highest-energy charged particles in the solar wind. Only Pioneer 10 and 11 spacecraft preceded the two Voyager spacecraft into the outer Solar System beyond Mars. On the basis of Pioneer data, Barnes and Grazis [57] predicted the characteristics of the solar wind at Neptune at the time of the Voyager 2 encounter. The solar wind was predicted to be about 400 km/s, with short periods where departures of up to

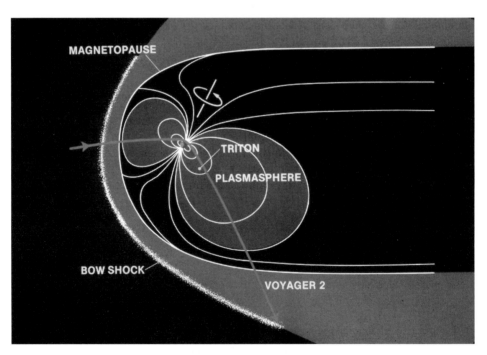

Figure 4.9. A schematic representation of the pre-Voyager view of the Neptunian magnetic field. Interaction with the solar wind reduces the sunward extent of the Neptunian field and also results in an anti-sunward magnetotail.

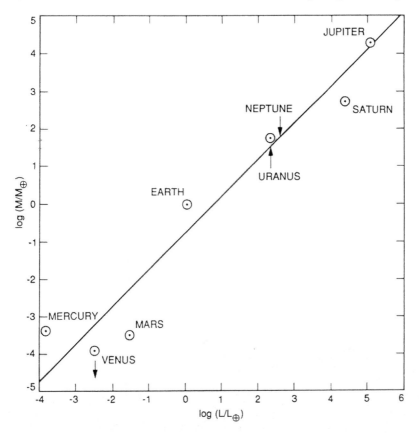

Figure 4.10. The magnetic Bode's law relates the magnetic field strength of a planet to its rotational angular momentum. Saturn's magnetic field was found by Pioneer 11 to be close to the value, and the magnetic Bode's law was used to predict approximate field strengths that Voyager 2 might encounter at Uranus and Neptune.

50 km/s might occur. They also made predictions of typical values for the number of protons per cubic centimeter (0.013), the proton energy (expressed as a temperature of 10,000 K), and the dynamic pressure exerted by the solar wind on a Neptunian magnetosphere (4×10^{-11} dyne/cm^2).

Interaction of charged particles in the solar wind with the magnetosphere also results in an anti-sunward 'magnetotail' that behaves in certain respects like an airport wind sock. The planetary field is stretched downstream until it is eventually indistinguishable from the solar wind (see Figure 4.9). The planetary field also rotates with the planet.

4.4.2   The magnetic Bode's law applied to Neptune

On the basis of magnetic field data obtained prior to 1975, Hill and Michel [58] proposed a 'magnetic Bode's law'. The purely empirical magnetic Bode's law relates the magnetic field strength of a planet to its rotational angular momentum (determined from the mass and spin rate of the planet). In contrast to Johann Bode's law of planetary distances, there exists a weak theoretical basis for such a relationship, even when one considers how little is known about how and why planetary magnetic fields are generated. The strengths of the magnetic fields of Saturn and Uranus were found to be close to the values predicted (Figure 4.10), and it seemed appropriate to use the magnetic Bode's law to predict approximate field strengths for a Neptune magnetic field. The field was predicted by Hill [59] to be that of a magnetic dipole approximately aligned with the rotational axis of Neptune and intermediate in strength between the magnetic fields of Saturn and Earth. Although the Uranus magnetic field strength was found to be close to that predicted by the magnetic Bode's law, the magnetic field orientation was aligned nowhere near the rotation axis of Uranus. Was Uranus a maverick among Solar System planetary magnetic fields, or might Neptune give scientists another look at a new class of planetary magnetic fields?

NOTES AND REFERENCES

1. Apsidal precession refers to a slow turning of the orientation of an elliptical orbit. The line of apsides is a line connecting the periapsis (closest) point of the orbit to the apoapsis (most distant) point. For the giant planets, the sense of this precession is prograde, or in the same general direction as the orbital motion of the ring particles or satellites.

2. Nodal regression refers to an apparent retrograde motion of the points where an inclined orbit crosses the planet's equatorial plane. The 'line of nodes' is a line through the center of the planet which connects the two nodes, or crossing points. The sense of motion of the line of nodes is retrograde, or in a general direction opposite that of the orbital motion of ring particles or satellites. Its rate is measured in angular distance per unit time (e.g., degrees per day) in the planet's equatorial plane.

3. Hubbard, W. B. (1984) Interior structure of Uranus. In Bergstralh, J. T. (ed.) *Uranus and Neptune*. NASA Conference Publication 2330, National Aeronautics and Space Administration, Scientific and Technical Information Branch, pp. 291–325.

4. Anders, E. and Ebihara, M. (1982) Solar-system abundances of the elements. *Geochimica et Cosmochimica Acta*, **46**, 2363–80.

5. Lewis, J. S. and Prinn, R. G. (1980) Kinetic inhibition of CO and N_2 reduction in the solar nebula. *Astrophysical Journal*, **238**, 357–64.

6. Zharkov, V. N. and Trubitsyn, V. P. (1978) *Physics of Planetary Interiors*, translated and edited by W. B. Hubbard. Pachart Publishing House, Tucson.

7. Deslandres, H. (1902) *Comptes Rendus hebdomaires des Seances de l'Academie des Sciences*, **135**, 472.

8. The Doppler effect was named after Christian J. Doppler, who showed that motion of a light, radio, or soundwave source toward (or away from) an observer will cause the observed waves to be shifted to higher (or lower) frequencies. He also showed that the magnitude of the shift was directly proportional to the speed of the source relative to the observer. The Doppler effect is responsible for the apparent reduction in pitch in the whistle of a rapidly moving train as it passes by a stationary listener standing at the railroad crossing.

9. Moore, J. H. and Menzel, D. H. (1928) Preliminary results of spectroscopic observations for rotation of Neptune. *Publications of the Astronomical Society of the Pacific*, **40**, 234–38.

10. Cambell, L. (1936) The rotation of Uranus. *Harvard College Observatory Bulletin*, **904**, 32–5.

11. Hayes, S. H. and Belton, M. J. S. (1977) The rotational periods of Uranus and Neptune. *Icarus*, **32**, 383–401.

12. Belton, M. J. S., Wallace, L., Hayes, S. H., Price, M. J. (1980) Neptune's rotation period: a correction and a speculation on the difference between photometric and spectroscopic results. *Icarus*, **42**, 71–8.

13. Brown, R. H., Cruikshank, D. P. and Tokunaga, A. T. (1981) The rotation period of Neptune's upper atmosphere. *Icarus*, **47**, 159–65.

14. Joyce, R. R., Pilcher, C. B., Cruikshank, D. P. and Morrison, D. (1977) Evidence for weather on Neptune. I. *Astrophysical Journal*, **214**, 657–62.

15. Cruikshank, D. P. (1985) Variability of Neptune. *Icarus*, **64**, 107-111.

16. Smith, B. A., Reitsema, H. J. and Larson, S. M. (1979) Discrete cloud features on Neptune. *Bulletin of the American Astronomical Society*, **12**, 704 (abstract).

17. Belton, M. J. S. and Terrile, R. J. (1984) Rotational properties of Uranus and Neptune. In Bergstralh, J. T. (ed.) *Uranus and Neptune*. NASA Conference Publication 2330, National Aeronautics and Space Administration, Scientific and Technical Information Branch, Washington, DC, pp. 327–47.

18. Kovalevsky, J. and Link, F. (1969) Dramètre, aplatissement et propriétés optiques de la haute atmosphère de Neptune d'après l'occultation de l'étoile BD-17°4388. *Astronomy and Astrophysics*, **2**, 398–412.

19. Freeman, K. C. and Lyngå, G. (1970) Data for Neptune frm occultation observations. *Astrophysical Journal*, **160**, 767–80.

20. French, R. G. (1984) Oblateness of Uranus and Neptune. In Bergstralh, J. T. (ed.) *Uranus and Neptune*. NASA Conference Publication 2330, National Aeronautics and Space Administration, Scientific and Technical Information Branch, Washington, DC, pp. 349–55.

21. Stier, M. T. and Traub, W. A. (1978) Far-infrared observations of Uranus, Neptune, and Ceres. *Astrophysical Journal*, **226**, 347–9.

22. Moseley, H., Conrath, B. and Silverberg, R. F. (1985) Atmospheric temperature profiles of Uranus and Neptune *Astrophysical Journal Letters*, **292**, L83–L86.

23. Orton, G. S., Baines, K. H., Bergstralh, J. T., Brown, R. H., Caldwell, J. and Tokunaga, A. T. (1987) Infrared radiometry of Uranus and Neptune at 21 and 33 μm. *Icarus*, **69**, 230–8.

24. Harris, D. L. (1961) Photometry and colorimetry of planets and satellites. In Kuiper, G. P. and Middlehurst, B. M. (eds) *Planets and Satellites* (Volume III of a series on the Solar System). The University of Chicago Press, Chicago, pp. 272–342.

25. Orton, G. S. and Appleby, J. F. (1984) Temperature structures and infrared-derived properties of the atmospheres of Uranus and Neptune. In Bergstralh, J. T. (ed.) *Uranus and Neptune*. NASA Conference Publication 2330, National Aeronautics and Space Administration, Scientific and Technical Information Branch, Washington, DC, pp. 89–155.

26. Belton, M. J. S., Wallace, L. and Howard, S. (1981) The periods of Neptune: evidence for atmospheric motions. *Icarus*, **46**, 263–74.

27. Murphy, R. E. and Trafton, L. M. (1974) Evidence for an internal heat source in Neptune. *Astrophysical Journal*, **193**, 253–5.

28. Stier, M. T., Traub, W. A., Fazio, G. G. and Low, F. J. (1977) *Bulletin of the American Astronomical Society*, **9**, 511 (abstract).

29. Neff, J. S. (1985) Bolometric Albedos of Titan, Uranus and Neptune. *Icarus*, **62**, 425.

30. Hubbard, W. B. (1978) Comparative thermal evolution of Uranus and Neptune. *Icarus*, **35**, 177–81.

31. Trafton, L. (1974) The source of Neptune's internal heat and the value of Neptune's tidal dissipation factor. *Astrophysical Journal*, **193**, 477–80.

32. Russell, H. N. (1935) Atmospheres of the planets. *Science*, **81**, 1–9.

33. Kuiper, G. P. and Phillips, J. (1949) New absorptions in the Uranus atmosphere. *Astrophysical Journal*, **109**, 540–41.

34. Herzberg, G. (1952) Spectroscopic evidence of molecular hydrogen in the atmospheres of Uranus and Neptune. *Astrophysical Journal*, **115**, 337–40.

35. Wildt, R. (1937) Decomposition of methane in the atmosphere of Uranus and Neptune. *Astrophysical Journal*, **86**, 321.

36. Marcy, W. and Stinton, W. (1977) Ethane and methane emission by Neptune. *Bulletin of the American Astronomical Society*, **9**, 537 (abstract).

37. Adel, A. and Slipher, V. M. (1934) The identification of the methane bands in the solar spectra of the major planets. *Physical Review*, **46**, 240–41.

38. Adel, A. and Slipher, V. M. (1934) The constitution of the atmospheres of the giant planets. *Physical Review*, **46**, 902–6.

39. Hunten, D. M. (1984) Atmospheres of Uranus and Neptune. In Bergstralh, J. T. (ed.) *Uranus and Neptune*. NASA Conference Publication 2330, National Aeronautics and Space Administration, Scientific and Technical Information Branch, Washington, DC, pp. 27–54.

40. Atreya, S. K. (1984) Aeronomy. In Bergstralh, J. T. (ed.) *Uranus and Neptune*. NASA Conference Publication 2330, National Aeronautics and Space Administration, Scientific and Technical Information Branch, Washington, DC, pp. 55–88.

41. Miner, E. D. (1998) *Uranus: the Planet, Rings and Satellites* (2nd edn), Wiley–Praxis, Chichester, UK, p. 77.

42. The molecular weight of a gas is roughly the sum of the numbers of protons and neutrons in the nuclei of its constituent atoms. A molecule of hydrogen gas (H_2), with one proton in the nucleus of each of its two hydrogen atoms, has a molecular weight of 2. Helium, with two protons and two neutrons in each atom, has a molecular weight of 4. A gas mixture of 90% hydrogen and 10% helium would then have a mean (average) molecular weight of $(2 \times 90\%) + (4 \times 10\%) = 2.2$. For a gas with 80% hydrogen and 20% helium, the molecular weight would be 2.4.

43. Wallace, L. (1984) The seasonal variation of the thermal structure of the atmosphere of Neptune. *Icarus*, **59**, 367–75.

44. If a volume of gas confined in an insulated container (one which allows no heat flow through its walls) is compressed, the absolute temperature of the gas will rise by an amount that is determined by the changes in the volume and the pressure of the gas. Furthermore, for a gas composed of 90% H_2 and 10% He, the increase in pressure is related to the decrease in volume by the relationship, $pV^{1.44} = $ constant. For an ideal gas $p_i V_i T_f = p_f V_f T_i$, where 'i' and 'f', respectively, stand for the initial and final values of the pressure (p), volume (V) and absolute temperature (T). An adiabatic lapse rate is one for which the temperature of the gas increases with pressure by precisely the same amount as it would if it were confined in such an insulated container. Mathematical combination of the two above relationships yields $T_f = T_i (p_f/p_i)^{0.30}$. Thus for an adiabatic lapse rate in a

gas which is 90% H_2 and 10% He, the temperature approximately doubles for each decade increase in pressure.

45. Baines, K. H., Hammel, H. B., Rages, K. A., Romani, P. N. and Samuelson, R. E. (1995) Clouds and hazes in the atmosphere of Neptune. In Cruikshank, D. P. (ed.) *Neptune and Triton*. University of Arizona Press, pp. 489–546.

46. Joyce, R. R., Pilcher, C. B., Cruikshank, D. P. and Morrison, D. (1977) Evidence for weather on Neptune. II. *Astrophysical Journal*, **214**, 663–6.

47. Terrile, R. J. and Smith, B. A. (1983) The rotation rate of Neptune from ground-based CCD imaging. *Bulletin of the American Astronomical Society*. **15**, 858 (abstract).

48. Cruikshank, D. P. (1984) Variability of Neptune. In Bergstralh, J. T. (ed.) *Uranus and Neptune*. NASA Conference Publication 2330, National Aeronautics and Space Administration, Scientific and Technical Information Branch, Washington, DC, pp. 279–87.

49. Cruikshank, D. P. (1978) On the rotation period of Neptune. *Astrophysical Journal Letters*, **220**, L57–L59.

50. Lockwood, G. W. and Thompson, D. T. (1979) A relationship between solar activity and planetary albedos. *Nature*, **280**, 43–5.

51. Spinrad, H. (1969) Lack of a noticeable methane atmosphere on Triton. *Publications of the Astronomical Society of the Pacific*, **81**, 895.

52. Cruikshank, D. P. and Silvaggio, P. (1979) Triton: a satellite with atmosphere. *Astrophysical Journal*, **233**, 1016–20.

53. Trafton, L. M. (1984) Seasonal variations in Triton's atmospheric mass and composition. In Bergstralh, J. T. (ed.) *Uranus and Neptune*. NASA Conference Publication 2330, National Aeronautics and Space Administration, Scientific and Technical Information Branch, Washington, DC, pp. 481–93.

54. Cruikshank, D. P. (1984) Physical properties of the satellites of Neptune. In Bergstralh, J. T. (ed.) *Uranus and Neptune*. NASA Conference Publication 2330, National Aeronautics and Space Administration, Scientific and Technical Information Branch, Washington, DC, pp. 425–36.

55. Kellermann, K. and Pauliny-Toth, I. I. K. (1966) The detection of thermal radio emission from Uranus, Neptune and other planets at 1.9 cm. *Astronomical Journal*, **71**, 390 (abstract).

56. Smoluchowski, R. and Torbett, M. (1981) Can magnetic fields be generated in the icy mantles of Uranus and Neptune? *Icarus*, **48**, 146–8.

57. Barnes, A. and Grazis, P. R. (1984) The solar wind at 20–30 AU. In Bergstralh, J. T. (ed.) *Uranus and Neptune*. NASA Conference Publication 2330, National Aeronautics and Space Administration, Scientific and Technical Information Branch, Washington, DC, pp. 527–40.

58. Hill, T. W. and Michel, F. C. (1975) Planetary magnetosphere. *Reviews of Geophysics and Space Physics*, **13**, 967–74.

59. Hill, T. W. (1984) Magnetospheric structures: Uranus and Neptune. In Bergstralh, J. T. (ed.) *Uranus and Neptune*. NASA Conference Publication 2330, National Aeronautics and Space Administration, Scientific and Technical Information Branch, Washington, DC, pp. 497–525.

BIBLIOGRAPHY

Bergstralh, J. T. (ed.) (1984) Uranus and Neptune, NASA Conference Publication 2330, National Aeronautics and Space Administration, Scientific and Technical Information Branch, Washington, DC.

Cruikshank, D. P. (ed.) (1995) *Neptune and Triton*. University of Arizona Press, Tucson.

Miner, E. D. (1998) *Uranus: the Planet, Rings and Satellites* (2nd edn). Wiley–Praxis, Chichester, UK.

Moore, P. (1988) *The Planet Neptune*. Wiley–Praxis, Chichester, UK.

5

The saga of Voyager 2

5.1 CALIFORNIA INSTITUTE OF TECHNOLOGY'S JET PROPULSION LABORATORY

The California Institute of Technology (Caltech) has a long and proud history of pushing the boundaries of knowledge forward, but it did not start as an institute for science and engineering. In 1891, the Honorable Amos G. Troop founded a school to provide an education for students from elementary through college and a teacher-training program. Troop University, as it was called, began in a small area of land now centrally located in Pasadena, California. Under the guidance of George E. Hale (1868–1938), the university decided in 1907 to discontinue its undergraduate and teacher-training programs, but continue its college of science and technology. Hale wanted the university to be a world-class institute capable of attracting world-class talent. Arthur A. Noyes (1866–1936) and Robert A. Millikan (1868–1953) joined Hale in his vision and together guided the institute to the frontiers of science. Hale led the way in astronomy, Noyes in physical chemistry, and Millikan in the physical sciences. By 1920 the university had 359 undergraduates, 9 graduate students, and 60 faculty, and in that same year was renamed the California Institute of Technology.

One would think that Caltech would be the ideal place for the birth of astronautics (the engineering of rockets and space vehicles). Unfortunately that honor does not go to California but to Kaluga in Russia. Here, prior to the end of the nineteenth century, a self-taught schoolteacher named Konstantin E. Tsiolkovsky (1857–1935) was beginning to establish the fundamentals of space travel. His contemporaries believed that space travel was a foolish endeavor, so his work remained unnoticed and his ideas of liquid-fuel rockets, space suits, staging of rockets, robotic satellites, and colonization of the Solar System received very little attention. However, he had confidence in his ideas and funded and published his own work. On his tombstone is written, 'Mankind will not remain tied to Earth forever' [1]. Meanwhile, work in astronautics continued in many countries by many individuals.

Figure 5.1. Von Kármán and his students posing at the first rocket site, which is now the present site for the Jet Propulsion Laboratory. (P-9007A)

Of particular note are the works of Hermann Oberth (1894–1989) in Germany and Robert H. Goddard (1882–1945) in the USA. Tsiolkovsky, Oberth and Goddard are considered to be the fathers of astronautics.

Of these three, it was Goddard who took the field from a theoretical exercise to an engineering endeavor by building hardware. On March 16, 1926, at 2:30 pm, Goddard launched the world's first liquid-fueled rocket. It rose 41 feet and traveled 184 feet downrange in 2.5 seconds [2]. This engineering accomplishment was the precursor of every other rocket system ever built, from Germany's V-2 rocket to the European Space Agency's Ariane and the USA's Space Shuttle. In 1904, in a prophetic speech at South High School in Worcester, Massachusetts, Goddard stated, '... the dreams of yesterday are the hope of today and the reality of tomorrow' [3]. By the 1930s research in astronautics was moving forward in countries across the world, and Caltech in the USA was no exception. There, Professor Theodore von Kármán, head of Guggenheim Aeronautical Laboratory (GALCIT), was interested in space travel, and began working on the institute's own liquid-fueled rockets. However, as this work was deemed too dangerous to be performed at Caltech, von Kármán and his students were asked to move to a less populous location north of Pasadena (Figure 5.1).

Von Kármán's team of graduate student scored a success on October 31, 1936, with the launch of their first liquid-fueled rocket. This success led von Kármán to seek funding from the US Army Air Corps for future work. The program was to design, build, and test rockets that could be safely placed under the wings of airplanes. Their task was to assess the feasibility of using rockets to allow airplanes to take off in a shorter time and on a shorter runway.

Upon the completion of the 'jet-assisted take-off' (JATO) program, von Kármán's group was asked to provide technical analysis of Germany's V-2 rockets, which had recently been discovered by military intelligence. His team countered with a more ambitious proposal to build a US version of the V-2 and enhance its capabilities. In this proposal, von Kármán referred to his organization for the first time as the Jet Propulsion Laboratory (JPL), a division of Caltech. By May 1947, JPL launched the Corporal rocket, which attained altitudes in excess of 100 km.

To reach the goals required for the Corporal rocket, JPL had to develop technologies in missile guidance and control, radio telemetry, ground radars, supersonic aerodynamic design and testing, and a myriad of other technologies. These technologies were to be crucial in America's first forage into space in 1958. The Soviet Union had successfully launched Sputnik 1 on October 4, 1957 and this event, coupled with the failure of the Vanguard rocket project, led the US government to enlist the aid of Werner von Braun (1912–77). Braun was asked to increase the capability of his Redstone rocket to enable it to inject a spacecraft into Earth orbit.

On January 31, 1958, a Redstone rocket lifted off its pad, boosting JPL's Explorer 1 satellite skyward (Figure 5.2). Explorer 1 carried an instrument

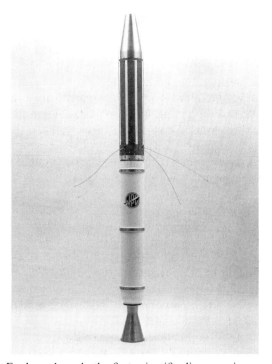

Figure 5.2. Explorer 1 made the first scientific discovery in space. (P-977A)

designed by James A. Van Allen, chairman of the University of Iowa's Physics Department. This instrument made the first scientific discovery in space: it detected 'belts' of charged particles in orbit around the Earth, which are now known as the Van Allen Radiation Belts.

In October 1958, the US government established the civilian space agency known as the National Aeronautics and Space Administration (NASA). JPL was transferred administratively from the Army to NASA but was still managed and operated by Caltech. The laboratory's primary responsibility was NASA's lead center for the robotic exploration of space. Virtually all deep space missions for the USA have been developed, tested, and/or operated in some way by JPL since that time.

5.2   A 'GRAND TOUR' OF THE OUTER PLANETS PROPOSED

The art of orbital mechanics follows exacting laws of motion. The general concept for planetary rendezvous is to change a spacecraft speed by injecting it onto an elliptical orbit such that its trajectory is aligned with the Earth's velocity vector. The spacecraft is launched in the same direction as Earth's motion to reach a planet further from the Sun, and in the opposite direction to reach a planet closer to the Sun. In either case, the primary source of velocity for the spacecraft is the orbital velocity of the Earth itself.

To accomplish a planetary rendezvous with the least amount of fuel, the spacecraft is placed on an elliptical trajectory known as a Hohmann transfer ellipse, named after its developer, Walter Hohmann (1880–1945). Hohmann realized that fuel was by far the most massive part of a rocket, so he determined that a trajectory that left Earth tangential to its orbit and arrived at the target planet tangential to its orbit would require the least amount of fuel (and thus the smallest cost) for the rocket. For an inner planet rendezvous, the Hohmann transfer ellipse has its aphelion at Earth's orbit with its perihelion at the orbit of the target planet. This situation is reversed for an outer planet rendezvous (Figure 5.3).

Throughout the 1950s and early 1960s astrodynamicists used this technique to design trajectories. They realized that chemical rockets could not attain the energies needed to reach planets that were more distant than Jupiter and, to make matters worse, the travel time on these minimum energy orbits continued to grow. As an example, the travel time on a Hohmann transfer orbit from Earth to Neptune would take about 30 years. Consequently, mission planners designed missions only for the terrestrial planets.

In the summer of 1961, Michael A. Minovich (1936–), a mathematics graduate student, was hired to work in the mission design section at JPL. He noticed that the gravitational field of a planet could change the velocity of a spacecraft relative to the Sun. The speed of the spacecraft would be decreased if it passed in front of the planet in its orbit, and increased if it passed behind. In either case, angular momentum of the combined planet and spacecraft is conserved. However, since the spacecraft has such a small mass compared to that of the planet, the velocity change imparted to the spacecraft could be quite large. For example, in 1979 the Voyager 1 spacecraft

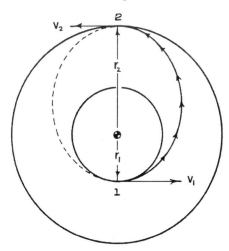

Figure 5.3. A spacecraft on a Hohmann transfer trajectory to an outer planet follows an ellipse with perihelion at Earth at launch and aphelion at the target planet at spacecraft arrival.

gained approximately 48,000 km/h as it passed behind Jupiter. However, to conserve angular momentum, Jupiter lost about 30 centimeters every trillion years!

By 1962 Minovich was using the principle of gravity assist to construct outer planet tours based on close encounters of Jupiter. Since its mass was highest among the planets and its orbital period was relatively short, its gravity could be used to achieve rendezvous with each of the outer planets in relatively short periods of time. He calculated that by launching either between 1962 and 1966 or between 1976 and 1980, a Jupiter gravity-assist trajectory could send a spacecraft to multiple outer planets. He showed that for a 1976 launch this technique could achieve encounters with Jupiter, Saturn, and Neptune with a single spacecraft. By this means, Neptune could be reached in only 12 years instead of the 30 years required without the use of gravity assists.

In 1966, Gary A. Flandro modified Minovich's work to develop a Grand Tour trajectory that could reach Jupiter, Saturn, Uranus, and Neptune [4]. He also noted that this special planetary alignment occurs only once every 176 years. This launch opportunity would open in 1976 and close in 1978.

The prospect of launching a single spacecraft that would take advantage of this opportunity was too great to ignore. Scientific enthusiasm for such a mission was immediately apparent, and two separate studies were funded by the National Academy of Sciences to refine and evaluate it. One study, chaired by the same James Van Allen who designed Explorer 1's scientific instrumentation, set out the specific scientific goals for such a mission. The report also noted that:

> professional resources for full utilization of the outer-solar-system mission opportunities in the 1970s and 1980s are amply available within the scientific community, and there is a widespread eagerness to participate in such missions.

The second study, headed by F.S. Johnson, was equally enthusiastic:

> An extensive study of the outer solar system is recognized by us to be one of the major objectives of space science in this decade [1970s]. This endeavor is made particularly exciting by the rare opportunity to explore several planets and satellites in one mission using long-lived spacecraft and existing propulsion systems. We recommend that ... spacecraft be developed and used in Grand Tour missions for exploration of the outer planets in a series of four launches in the late 1970s.

JPL proposed such a mission to NASA in 1970. The mission was to include two spacecraft to be launched in 1976 and 1977 to Jupiter, Saturn, and Pluto, followed by two spacecraft launches to Jupiter, Uranus, and Neptune in 1979. NASA's 'Grand Tour' plans were received enthusiastically by the US Congress and funding was recommended, but fiscal considerations led to less favorable reception for several other NASA-proposed missions. Faced with a difficult decision, NASA asked JPL to propose a version of the Grand Tour that might meet most of the same objectives at a lower total cost.

5.3   SCALING DOWN TO A VOYAGER MISSION TO FIT NASA'S BUDGET

In response to requests from NASA Headquarters for a less expensive version of the Grand Tour, JPL proposed a two-spacecraft Mariner Jupiter–Saturn (MJS) mission. The spacecraft were to be designed specifically for scientific investigation of the two planets only, with no additional instrumentation or requirements for Uranus, Neptune, or Pluto observations. The spacecraft components were to be reliable enough to provide a high probability of successful spacecraft operation for four years – the period from launch through the Saturn encounters. The cost for such a mission was estimated to be about $250 million, only one-third of the estimated cost of the earlier, more ambitious Grand Tour proposal. NASA and the US Congress approved funding for MJS, and the project was officially started July 1, 1972. Harris M. Schurmeier of JPL was appointed Project Manager for the new project; and Caltech physicist Edward C. Stone was appointed Project Scientist. From 31 scientific investigation proposals submitted to NASA, nine were selected with their associated Principal Investigators (PIs) and Co-Investigators (Co-Is). Each of these nine teams of scientists would provide the scientific instrumentation needed for their investigations, which would then be integrated into the overall spacecraft design. An additional 19 scientists were selected from 46 individual proposers to participate in imaging or radio science investigations for which the instrumentation was to be provided by NASA. A Team Leader (TL) was chosen for each of these two investigations. Stone and 11 PIs and TLs constituted the Science Steering Group (SSG). The first meeting of the SSG was in December 1972 at JPL.

The initial selection of instruments included particulate material detectors. On December 3, 1973, just 17 months after the MJS project start date, Pioneer 10

PIONEER SATURN VOYAGE

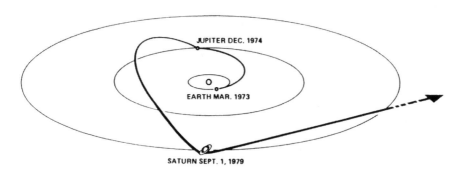

Figure 5.4. Trajectory of the Pioneer 11 spacecraft from Earth to Saturn.

encountered Jupiter. It had detected 67 hits on its micrometeoroid detector during the flight, with ten of these hits near Jupiter [5], but the rate of such impacts showed no increase during the passage through the asteroid belt between Mars and Jupiter. By mid-1974, Pioneer 11 had also passed through the asteroid belt without detecting any increase in the number of particles. Pioneer trajectory specialists had discovered that Pioneer 11 could utilize gravity assist at Jupiter to direct the spacecraft to the opposite side of the Sun where it could make a pre-Voyager encounter of Saturn (see Figure 5.4). Since instrument selection for MJS had been made at least partially contingent upon the scientific findings of Pioneers 10 and 11, a decision was made to drop the particulate material detector investigation.

Pioneers 10 and 11 also served to emphasis the severity of the Jupiter radiation environment. By early 1975, a decision had been made to 'harden' the spacecraft and science instruments to make them less vulnerable to severe radiation effects. NASA also agreed to authorize inclusion of a plasma wave investigation that would be diagnostic of the radiation environment.

The MJS spacecraft design was developing into something unlike any prior Mariner spacecraft because of the longer cruise times, the greater distances from the Sun, and the harsher environments it would have to survive to complete its mission. It was decided to give recognition of that fact by calling the spacecraft Voyager. The name became official, and the older MJS designation was dropped by mid-1977. It is interesting to note that an early version of the Viking Mission to Mars was also known as Voyager, but that designation was not used after 1974.

Voyager management, with the assistance of its Mission Analysis and Engineer-ing team, provided a rich array of potential trajectories for consideration by the

SSG. Close encounters of Jupiter's large moons (Io, Europa, Ganymede, and Callisto) and of Saturn's largest moon (Titan) were considered extremely important. Another vital consideration at Saturn was the provision for a trajectory that would permit the spacecraft as viewed from the Earth to traverse the full radial extent of the rings, as such a trajectory would allow the Radio Science team to obtain a valuable occultation scan of the rings. Initially, it seemed that the Titan and ring requirements were mutually exclusive. It was then discovered that a single spacecraft, with a properly timed arrival, could perform both experiments, leaving open the geometry of the encounter of the second spacecraft. The logical choice seemed to be a Voyager 1 Saturn encounter trajectory which would allow Voyager 2 to continue on to Uranus, once the successful completion of the Saturn objectives of Voyager 1 were assured.

All but one of the scientific instruments could be used without revision for observations of Uranus. The sole exception was the Infrared Interferometer Spectrometer and Radiometer (IRIS). IRIS could adequately measure the thermal radiation of Uranus, but would have poor sensitivity at near-infrared wavelengths owing to the much lower intensities of reflected sunlight at the distance of Uranus. In July 1976, NASA gave approval for building a more sensitive modified IRIS (MIRIS), with the condition that completion of its construction and testing should not delay the scheduled launch dates. In the same letter, approval for consideration of a continuation of the second Voyager spacecraft on to Uranus was granted. The actual choice of the Voyager 2 Saturn encounter trajectory would remain dependent on successful Voyager 1 Saturn and Titan encounters.

The science objectives of the basic mission, outlined by the SSG in December 1972, were:

- to conduct comparative studies of the Jupiter and Saturn systems: (a) the environment, atmosphere, surface, and body characteristics of the planets; (b) one or more of their satellites; (c) the nature of the rings of Saturn;
- to perform investigations in the interplanetary and interstellar media.

With the conditional approval for a Uranus mission, two scientific objectives for an extended mission were added:

- to extend interplanetary and interstellar media investigations well beyond the orbit of Saturn;
- to extend these investigations to the Uranus systems, if conditions should permit implementation of the Uranus targeting option.

The following types of information would be investigated at each of the bodies: (a) physical properties, dynamics, and compositions of the atmospheres; (b) surface features; (c) thermal regimes and energy balances; (d) charged particles and electromagnetic environments; (e) periods of rotation, radii, figures, and other body properties; and (f) gravitational fields.

The Voyager Project Plan, dated April 1, 1977, further states [6]:

> The spacecraft will continue to escape from the solar system toward the solar apex, and communications could be maintained as long as the spacecraft continues to function. If the spacecraft continues to function past Saturn encounter, an extended mission could be conducted in anticipation of penetrating the boundary between the solar wind and the interstellar medium, allowing measurements to be made of interstellar fields and particles unmodulated by the solar plasma. Also, the second spacecraft to arrive at Saturn could be targeted for a gravity-assist aiming point which would permit an encounter with Uranus in January of 1986. Objectives at Uranus would be similar to those at Jupiter and Saturn, including its physical properties, atmosphere and methane bands, satellites and surrounding environment.

5.4 THE LAUNCH OF VOYAGERS 1 AND 2

The two spacecraft were scheduled to be launched from the same pad at Cape Canaveral, Florida. The launch period opened in late August and continued through mid-September 1977. Voyager 2 would follow a slower trajectory so that passage outside the visible rings of Saturn would still permit sufficient gravitational bending of the trajectory to direct the spacecraft on to Uranus. Voyager management also wanted to avoid passing Voyager 2 as deeply through the intense Jupiter radiation environment as Voyager 1, and the slower trajectory accommodated that desire.

Three spacecraft (including a spare) took the long and arduous trip via specially equipped moving vans from JPL in Pasadena, California, to Cape Canaveral, Florida, arriving in June 1977. At the Cape, the spacecraft underwent extensive testing, including final integration with the Titan IIIE/Centaur launch vehicle (Figure 5.5).

During the last few weeks before launch a late addition was made to each spacecraft. Imaging team member Carl Sagan (1934–96) had been requested in December 1976 by Project Manager John R. Casani to organize an effort to place a message on each of the two spacecraft. The message was intended for intelligent beings from other worlds outside our Solar System who might some day find one of the Voyager spacecraft. Sagan accepted the challenge, selected five team members, and with help from a large number of other individuals, his team prepared a two-sided copper record. The record contains 118 digitized images of typical scenes on Earth, greetings in 54 languages, a variety of music, and recordings of some of the sounds of Earth. The history and contents of the record are described in detail in the book, *Murmurs of Earth* [7]. Two identical records were made and encased in a polished aluminum housing, gold in color; instructions on the use of the records were etched on the housing. One encased record with accompanying stylus, cartridge, and mounting bracket was affixed to the exterior of each of the two Voyager spacecraft (Figure 5.6). It is anticipated that these sights and sounds of

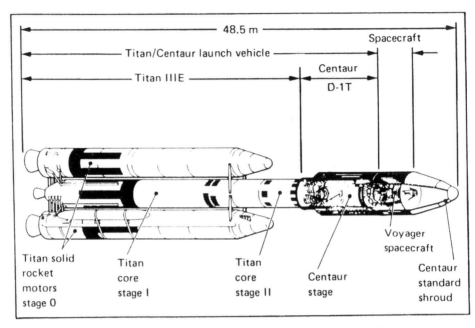

Figure 5.5. Schematic of the Titan IIIE/Centaur launch vehicle.

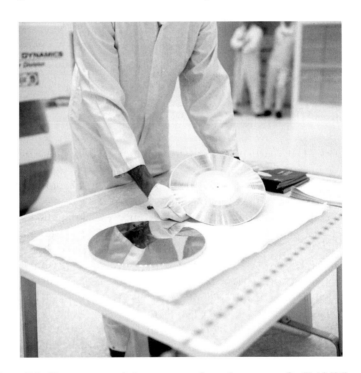

Figure 5.6. Voyager record, later mounted on the spacecraft. (P-19404)

Figure 5.7. Voyager 2 launch from Cape Canaveral on August 20, 1977. (373-7511A)

Earth will remain decodable for millions, perhaps even billions of years into the future.

While the spacecraft were being prepared for launch, the infrared investigations team and the Texas Instruments contractors were struggling to complete fabrication and testing of the two MIRIS instruments in time to include them in place of the IRIS instruments already on the spacecraft. The launch period for Voyager 2 opened on August 20, 1977. The MIRIS instruments were completed, but testing disclosed some problems that had to be corrected before they could be shipped to the Cape. Since Voyager 2 was the spacecraft destined to have the opportunity to go on to Uranus, it was especially important that, if at all possible, a MIRIS should be included. However, nothing occurred to delay the launch, and Voyager 2 began its journey to the stars on the first day of the launch period without MIRIS. The launch was perfect (Figure 5.7), though later analysis of the engineering data showed that the Titan rocket gave Voyager 2 a much rougher ride than anticipated. There was still hope that MIRIS might be completed in time for the launch of Voyager 1. The launch pad was refurbished, and Voyager 1 was mounted in the nose cone of the Centaur booster. On September 5, 1977, just 16 days after the launch of its twin, Voyager 1 lifted off atop the giant Titan/Centaur rocket. The total thrust of the Titan was less than expected, resulting in a lower altitude release for the Centaur upper stage and its precious cargo. The first burn of the upper stage had to have a longer duration to make up for the lower-than-expected performance of the Titan. Once in a parking orbit about the Earth, mission planners tried to determine if the Centaur stage had enough fuel for its second and final burn. Although there was nothing they could do about it, the mission would be lost if the Centaur did not have enough fuel to inject Voyager 1 on its trajectory to Jupiter. Irrespective of the answer, the operations team had to wait to see if the burn would be successful. Almost an hour later the operations team was rewarded with a successful injection. Voyager 1 was on its way to Jupiter, but unfortunately, like Voyager 2, without MIRIS.

5.5   PROBLEMS ALONG THE WAY

The launch environment is generally the most hazardous part of a spacecraft's journey. More spacecraft are lost in the first five minutes of their journey than in any individual one-month period thereafter. For a mission with the complexity and longevity of Voyager, however, there was a high probability that some electronic or mechanical components would fail at some time during the post-launch period. It is a tribute to Voyager designers and engineers that none of the problems encountered was fatal to the spacecraft operation or to the achievement of major mission science objectives.

Because of the enormous distances from which the Voyager spacecraft would be required to communicate their problems to Earth, radio signals traveling at the speed of light (299,292 km/s) would eventually take hours to reach Earth (see Figure 5.8 for a diagram of trajectories of Voyagers 1 and 2 through the Solar System). Assuming that immediate analysis and generation of corrective commands by operations personnel were possible, it would still take a similar amount of time for those commands to reach the spacecraft. At Neptune distances, a command from Earth would take over four hours to reach the spacecraft, and some problems could develop beyond the point of recovery in the required 'round trip light time' needed to correct it. For this reason, automatic responses to certain classes of anomalous spacecraft behavior were built into the spacecraft computers, so that by the time news of the problems reached Earth, corrective action would have been taken.

Although the idea was sound, pre-programmed responses by a non-reasoning robot often led to difficulties. Spacecraft acceleration when Voyager 2 was separating from its final propulsive rocket was larger than anticipated by the spacecraft's computers, which initiated a number of corrective actions. As a result, the antenna was slightly mispointed, and several commands that were sent by ground

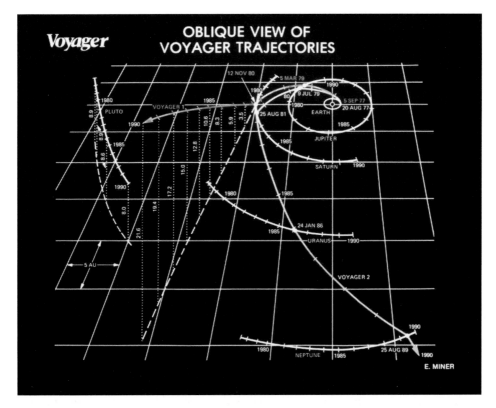

Figure 5.8. Trajectories of Voyager 1 and 2 through the Solar System. (JPL-10687A)

controllers were ignored by the computers, giving engineers cause for concern. Once they understood what had happened, normal operation was restored.

During the launch vehicle ascent, several long appendages were either folded alongside the spacecraft or wound into small containers awaiting commands that would cause them to be released and unfurled. The two largest appendages were the opposing booms which support the electrical power supplies (Radioisotope Thermoelectric Generators, or RTGs) on one side and instrumentation for 7 of the 11 science investigations on the other side of the spacecraft body (Figure 5.9). The RTG boom deployed successfully, but data from the spacecraft indicated that the science boom was not fully deployed and latched. Careful analysis later led to the conclusion that the science boom was fully extended and probably latched, but that the electronic sensor which signaled latching had given an erroneous signal.

Several of Voyager's critical components exist in duplicate pairs on each space-craft. There was, for example, a backup radio receiver to intercept commands from Earth in case the primary receiver failed. Since a failed receiver could not react to commands from Earth, Voyager's computers were programmed to restart a timer each time the radio sensed a command receipt. This timer could be set to any

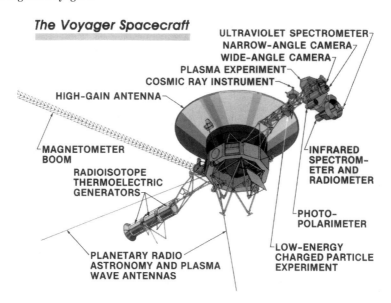

The Voyager Spacecraft

ULTRAVIOLET SPECTROMETER
NARROW-ANGLE CAMERA
WIDE-ANGLE CAMERA
PLASMA EXPERIMENT
COSMIC RAY INSTRUMENT
HIGH-GAIN ANTENNA
MAGNETOMETER BOOM
RADIOISOTOPE THERMOELECTRIC GENERATORS
INFRARED SPECTROMETER AND RADIOMETER
PHOTO-POLARIMETER
LOW-ENERGY CHARGED PARTICLE EXPERIMENT
PLANETARY RADIO ASTRONOMY AND PLASMA WAVE ANTENNAS

Figure 5.9. The two largest appendages on the Voyager spacecraft are the RTG boom, which supports the radioisotope thermoelectric generators, and the science boom, which supports instrumentation for 7 of the 11 scientific investigations. (P-29473A)

duration but was initially set for seven days. If the time elapsed without a received command, the computers were to assume that the primary receiver had failed and switch to the backup receiver. A problem on Voyager 1 in early April 1978 diverted everyone's attention away from Voyager 2. Operations personnel were trying to determine the severity of the fault when Voyager 2's seven-day timer completed its timeout without detecting a command. Responding correctly to its programmed instructions, the computer switched to the backup receiver. Voyager personnel, somewhat chagrined, generated the commands to switch back to the primary receiver and transmitted them back to the spacecraft. When Voyager 2 at first failed to respond to commands from Earth, an additional problem was discovered. Each receiver was equipped with electronic circuitry designed to detect the frequency of the incoming radio signal automatically and tune the receiver to that frequency. A critical part of that circuitry was a Tracking Loop Capacitor (TLC). Voyager 2's backup TLC had apparently failed, so that the backup receiver would respond only to commands that arrived at precisely the right frequency. The most probable failure mechanism was electrical shorting caused by the migration of tiny conducting particles into the space between the capacitor plates. Continual operation of the primary receiver had prevented the buildup of such particles in its TLC.

The calculation of the appropriate radio frequency to be transmitted was not a simple task. To ensure that the appropriate frequency arrived at Voyager 2's receiver, one had to account for (1) the motion of the transmitting tracking station on Earth due to the Earth's rotation, (2) the motion of the Earth around the center of mass of the Earth–Moon system, (3) the orbital motion of the Earth around the Sun, and (4) the velocity of the spacecraft relative to the Sun. As a further complication, the radio frequency of the receiver changed with its temperature; a temperature difference of $\frac{1}{4}$°C ($\frac{1}{2}$°F) shifted the receiver frequency by 96 Hz (cycles per second), half of its total bandwidth. Each time the electrical load being used by the spacecraft changed by as little as 2 W, the receiver's 'Best Lock Frequency' (BLF) had to be monitored for at least 24 hours before reliable commanding of the spacecraft could continue.

The command to switch back to Voyager's primary receiver was transmitted at a range of frequencies, and this time the spacecraft responded. The response was not very satisfying, however, for during the switching process, a short developed in the primary receiver which blew both of the redundant fuses. Another seven days had to pass before Voyager 2's computers again switched to the backup receiver.

Several responses to the backup TLC failure were developed. Procedures were followed to regularly determine the BLF of the receiver. Elaborate computer programs assisted Deep Space Network (DSN) personnel to determine the precise frequency to be transmitted so that commands would arrive at the spacecraft at a frequency very close to the BLF. When power loading on the spacecraft had to be changed, a 'command moratorium' of 24 to 72 hours was observed, the duration depending on the nature and magnitude of the power change. Since no usable receiver for backup purposes existed, the computers were reprogrammed after each planetary encounter to execute a minimum set of observational sequences during the period surrounding close approach to the next planet. Fortunately, the receiver continued to operate, and none of the Backup Mission Loads (BMLs) was used. The lack of automatic frequency tracking by the receiver caused no significant science losses during the Jupiter, Saturn, Uranus, and Neptune encounters of Voyager 2.

Electronic sensors on the spacecraft and its science instruments regularly monitored temperatures, voltages, and mechanical positions of movable parts. The readouts of these sensors were transmitted in a low-rate engineering data stream, in the science data stream, or in both. The low-rate engineering stream was multiplexed in such a way that some of the engineering readouts were sampled every 12 seconds, while others are sampled at 2-, 3-, 6-, 12-minute intervals, or not at all. An electronic failure of one of Voyager 2's 'tree switches' late in 1977 rendered 72 of several hundred engineering channels inaccessible. Fortunately, none of the lost engineering data was critical to the success of the Voyager missions.

The science instruments themselves were not without problems. The infrared instrument successfully jettisoned its telescope cover shortly after launch, but by early December 1977 its sensitivity had degraded markedly. The problem was traced to crystallization of some of the bonding material which held the mirrors internal to the spectrometer. The crystallization caused misalignment or warping of the mirrors which led in turn to the loss of sensitivity. The instrument was furnished with a 'Flash-Off Heater' (FOH) designed to heat the external primary telescope mirror and promote rapid evaporation of any impurities deposited during launch. After trying a number of alternative, but ineffective, solutions the Principle Investigator (PI) elected to turn on the FOH and the heating 'miraculously' restored almost all of the pre-launch sensitivity. Apparently the heat reversed the crystallization process, allowing the bonding material to relax to its former shape. The instrument had to be cold to operate properly, but prior to the 1989 Neptune encounter it was the practice to keep the instrument warm between planetary encounters in order to minimize the rate of sensitivity degradation.

Inside the photopolarimeter (a very sensitive light meter) are three commandable wheels. The first wheel contains a series of colored filters; the second contains drilled holes of various sizes to control the angular view of the instrument; the third has polarizing filters with a variety of orientations. The photopolarimeter was designed to be able to use any combination of positions of the three wheels. Excessive use of the wheels prior to the Jupiter encounter, combined with radiation damage during that encounter, caused a complete failure of the Voyager 1 instrument and degradation of the Voyager 2 instrument. Of the eight wheel positions available for each wheel at launch, only three filter positions, four aperture positions and four polarization analysis positions were usable after the Jupiter rendezvous.

A host of other problems occurred on each of the spacecraft. Most of the remaining problems occurred only on Voyager 1 or subsequent to the start of the Voyager 2 Jupiter encounter period, and others were less significant than those chronicled in this section. Radiation damage will be discussed in connection with the Jupiter encounter; scan platform seizure will be discussed in the Saturn section; and computer memory loss will occupy a portion of the next section.

5.6   THE VOYAGER TEAM

The accomplishments of Voyagers 1 and 2 were due primarily to the ingenuity and dedication of a large number of people. The Science Steering Group (SSG) has already been mentioned. It consisted of a chairman, Team Leaders (TLs) for the Imaging Science Subsystem (ISS) and Radio Science Subsystem (RSS) investigations, and Principal Investigators (PIs) for the nine other science investigations (see Figure 5.10). Edward C. Stone, who continues as of this writing as the Project Scientist, has ably filled the role of SSG Chairman since the inception of the Voyager Project in 1972. Bradford A. Smith of the University of Arizona's Lunar and Planetary Laboratory was ISS TL, with Laurence Soderblom of the United States Geological Survey in Flagstaff, Arizona, as his deputy. Von R. Eshleman led the RSS team until 1978, when G. Leonard Tyler assumed the role of RSS TL. Both are associated with Stanford University's Center for Radar Astronomy in California.

The Cosmic-Ray (CRS) investigation was led by Rochus E. Vogt of Caltech until 1984, when Stone was given the responsibility of CRS PI in addition to his roles as SSG Chairman and Project Scientist. Rudolph A. Hanel guided the Infrared Interferometer Spectrometer (IRIS) investigation until after the Uranus encounter. Barney J. Conrath became IRIS PI in 1987. Both worked at NASA's Goddard Space Flight Center in Maryland. Low Energy Charged Particle (LECP) PI was (and is) S.M. 'Tom' Krimigis of Johns Hopkins University's Applied Physics Laboratory in

Figure 5.10. Members of the Voyager Science Steering Group and other Voyager Project Management personnel at JPL during the Neptune encounter preparations. Pictured, from left to right, are 'Tom' Krimigis (LECP PI), John Belcher (PLS PI), Leonard Tyler (RSS TL), Pieter de Vries (FSO Mgr), Don Gurnett (PWS PI), Norman Ness (MAG PI), Lyle Broadfoot (UVS PI), Norm Haynes (Proj. Mgr), James Warwick (PRA PI), Edward Stone (Proj. Sci. and CRS PI), Bradford Smith (ISS TL), Rudolph Hanel (former IRIS PI), Barney Conrath (IRIS PI), and Herbert Bridge (former PLS PI). Absent were Lonne Lane (PPS PI) and Ellis Miner (APS). (See color plate 6.)

Maryland. Norman F. Ness, PI for the Magnetometer (MAG) investigation, was initially associated with NASA's Goddard Space Flight Center, but later headed the Bartol Research Institute at the University of Delaware. The Plasma Subsystem (PLS) investigation was headed by Herbert S. Bridge until after the Uranus encounter, when John W. Belcher was named PLS PI. Both were associated with Massachusetts Institute of Technology. Excessive problems with the Photopolari-metery Subsystem (PPS) investigation led to the dismissal of Charles F. Lillie of the University of Colorado's Laboratory for Atmospheric and Space Physics (LASP) as PI in 1978. Charles W. Hord (also of LASP) was named interim PI until the appointment of JPL's Arthur L. Lane in 1979. James W. Warwick of Radiophysics, Incorporated, in Colorado was the sole PI for the Planetary Radio Astronomy (PRA) investigation. The Plasma Wave Subsystem (PWS) investigation was led by Frederick L. Scarf of TRW Defense and Space Systems in California until his untimely death in 1988; Donald A. Gurnett, of the University of Iowa was appointed PWS PI in the same year. A. Lyle Broadfoot, initially associated with the University of Southern California, later joined the University of Arizona's Lunar and Planetary Laboratory. He was the PI for the Ultraviolet Spectrometry (UVS) investigation.

Nine different men have served as Voyager Project Manager, beginning with Harris M. Schurmeier, followed (in order) by John R. Casani, Robert J. Parks, Raymond L. Heacock, Esker K. Davis, Richard P. Laeser, Norman R. Haynes, George P. Textor, and Willis Meeks. The Project Manager oversees the full-time effort at JPL to plan, implement, and monitor the activities of the two Voyager spacecraft. The Project Manager also serves as liaison with NASA Headquarters personnel in matters of policy and funding.

The Mission Director works under the Project Manager to manage the detailed activities of the Voyager Flight Team. Patrick Rygh, Richard P. Laeser, George P. Textor, and Richard Rudd served successively as Mission Director. Long-range overview planning and establishment of guidelines and constraints for operation of the spacecraft are under the direction of the Mission Planning Office (MPO). Ralph F. Miles, Jr, held that responsibility very early in the pre-launch time period. Charles E. Kohlhase then served as the MPO Manager from two years before the Voyager launches until after the Neptune encounter. A Ground Data System Engineering Office (GDSEO), managed by Allan L. Sacks during Saturn–Uranus cruise and Uranus encounter time periods, was in operation when there were problems associated with the multitude of computer programs needed to handle the spacecraft. These programs were of two general types: those associated with generating commands to be sent to the spacecraft ('uplink'), and those associated with data being returned by the spacecraft ('downlink').

The remainder of the Voyager Project organization at JPL was contained in three organizations: the Flight Science Office (FSO), the Flight Engineering Office (FEO), and the Flight Operations Office (FOO), as depicted in Figure 5.11.

The FSO Manager was J. Pieter de Vries, who had been preceded by James E. Long and Charles H. Stembridge. There were four teams in the FSO (Figure 5.12) and the Project Scientist (PS) and Assistant Project Scientist (APS) also formed an integral part of the FSO staff.

The Science Investigation Support Team (SIS) handled liaison with each of the 11 investigation teams and ensured that encounter science, encounter preparations, and cruise science requirements were appropriately planned and implemented. The individuals who took the lead for each investigation were the respective Experiment Representatives (ERs) and Assistant ERs. The Assistant Project Scientist (APS) had the responsibility of coordinating those efforts for the encounter and encounter preparation observations. The APS also arbitrated when there were conflicting requirements and ensured that the overall science observation plans would return the maximum available science within the capabilities of the spacecraft and ground support systems. The SIS Team chief had the responsibility to see that the work of SIS was efficiently performed. He and others of his staff made certain that both intrateam and interteam efforts were properly coordinated. The Radio Science Support Team (RSST) coordinated efforts to provide appropriate instrumentation and procedures for the collection of radio science data at the DSN tracking stations. The Imaging Science Data Team (ISDT) coordinated the massive amounts of pro-

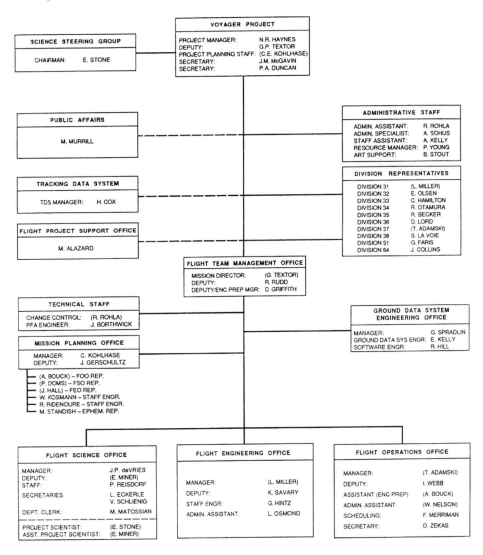

Figure 5.11. Voyager Project organization chart during the Neptune encounter period.

cessing and cataloging of imaging data. Preliminary data processing and sorting for the other nine investigations was done by the General Science Data Team (GSDT).

The FEO was managed by William I. McLaughlin for the Uranus encounter time period. Francis Sturms, Jr, Edward L. McKinley, and Saterios (Sam) Dallas preceded McLaughlin in that responsibility, and Lanny J. Miller occupied that post during the Neptune preparations and encounter. FEO consisted of three teams: the

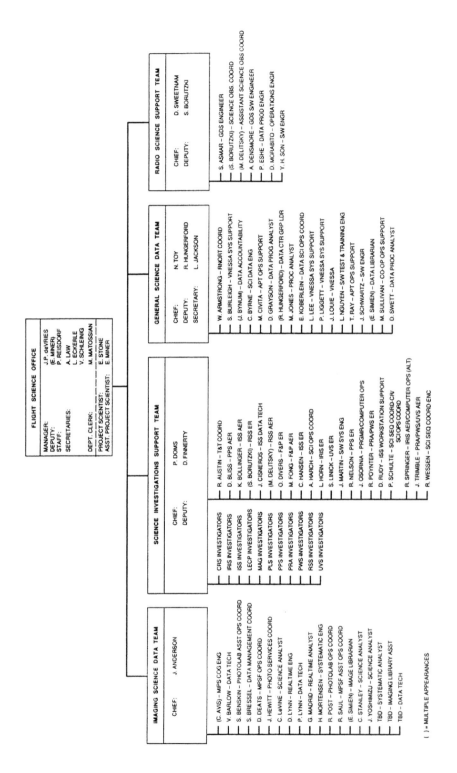

Figure 5.12. Voyager Flight Science Office organization chart during the Neptune encounter period.

Sequence Team (SEQ), the Navigation Team (NAV), and the Spacecraft Team (SCT), plus the Advanced Software Development Group (ASD), as show in Figure 5.13.

SEQ assisted SIS in the early development of the spacecraft science and engineering sequence of events. Once that process reached a certain level of development, SEQ took the lead and was assisted by the other elements of the Voyager Project until the commands were fully prepared and ready to transmit to the spacecraft for execution. SEQ also validated these sequences to ensure that they were consistent with operational, hardware, and software constraints.

NAV was charged with the responsibility of determining the precise trajectory being followed by the spacecraft, both relative to Earth and relative to the target planet and its satellites, and with making required adjustments to the trajectory by designing corrective propulsive maneuvers. NAV also provided information to the rest of the Voyager Project about the range of possible trajectories, their characteristics, and the associated delivery and knowledge uncertainties at each epoch where decisions had to be made.

The tools NAV used for these tasks included Earth-based observations of the planets and their satellites, radio tracking of the spacecraft to determine both range and velocity, and two other more esoteric processes. Delta Differential One-way Ranging (DDOR) utilized simultaneous comparison of the spacecraft frequency and the frequency of a celestial radio source from two widely separated tracking stations to provide an independent determination of the spacecraft position in the plane of the sky. Optical Navigation (OPNAV) used high-resolution ISS camera to image the planet and/or its satellites against a background of stars whose directions in the sky had been accurately determined from Earth. This provided an accurate direction and an improved range from the spacecraft to the planet as well as an improved orbit for each of the satellites imaged. For the distant outer planets, the position of Voyager relative to Earth and Sun was generally much better known than the position of the planet Voyager was approaching, so OPNAV provided a powerful and proven technique for determination of the spacecraft position relative to the planet.

SCT was the largest of the Voyager teams. These were the individuals who were charged with the responsibility of knowing the electronic and mechanical structure, health, capabilities, and limitations of each of the science and engineering subsystems. It was this knowledge base and the expertise of SCT members that led to enormous improvements in spacecraft capability utilization during the course of the mission. The SCT was also largely responsible for finding innovative and insightful ways to overcome the limitations imposed by spacecraft hardware failures and the continually increasing distances of the spacecraft from Sun and Earth.

There are three pairs of computers aboard each Voyager spacecraft. The two Attitude and Articulation Control Subsystem (AACS) computers control the attitude of the spacecraft, keeping the antenna pointed at Earth by use of Sun and star sensors and an array of attitude control jets. These computers were reprogrammed several times in flight by SCT personnel to improve the stability of the spacecraft and the pointing accuracy of the science instruments. The AACS also

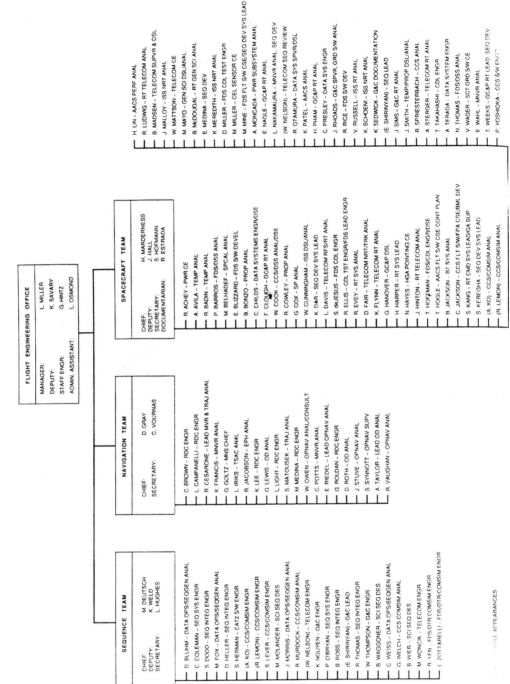

Figure 5.13. Voyager Flight Engineering Office organization chart during the Neptune encounter period.

controlled the steerable scan platform which pointed the ISS, PPS, UVS, and IRIS instruments. The two Flight Data Subsystem (FDS) computers provided science instrument timing and control functions; they also formatted the science and engineering data for transmission to Earth. FDS programming was the responsibility of the ASD, who worked closely with SCT in this task. Portions of the FDS memory on each spacecraft are no longer usable, but clever use of the remaining memory locations minimized the impact of the memory loss. In the event one of the two FDS computers fails completely, a 'Single Processor Program' was developed for Neptune (similar to one developed earlier for Uranus) by the ASD to minimize science data loss from such a failure. The two Computer Command Subsystem (CCS) computers contained the second-by-second timed command sequences to be executed by the spacecraft in performing its observations and calibrations. They also controlled access to the FDS and AACS computers and provided the logic for the automatic spacecraft fault responses mentioned earlier.

The FOO Manager during the Neptune encounter periods was Terrence P. Adamski. He was preceded by Douglas G. Griffith, Michael W. Devirian, and Raymond J. Amorose. FOO was directly responsible for communications with the spacecraft and consisted of four teams (Figure 5.14): the Mission Control Team (MCT), the Mission Control and Computing Center (MCCC), the Deep Space Network Operations Control Team (NOCT), and the Data Management Team (DMT).

MCT directed and controlled real-time Voyager operations, including transmission of commands to the spacecraft, monitoring the overall status of the spacecraft and the DSN tracking stations, and initial collection and dissemination of data from the spacecraft. MCCC provided the working interface between the Voyager Project and the various computer facilities used by, but not directly a part of, the Voyager Project for both uplink and downlink products. NOCT interfaced with DSN operational elements to ensure that appropriate DSN tracking support was provided and to monitor the status of radiometric (tracking) and telemetry (science and engineering data) streams coming from the spacecraft. The DMT prepared the experiment data records, which were magnetic tape records of spacecraft data for each of the individual investigations. The DMT also provided routine graphical displays of minimally processed science and engineering data for 'quick-look' analysis.

These, then, were the people of Voyager, exclusive of the science investigation team members at various institutions in the USA, the UK, France, and Germany. Since those science team members are all listed as co-authors in the preliminary science reports for the Neptune encounter [8], their names are not listed here. Their individual contributions to the success of the Voyager 2 Neptune encounter are gratefully acknowledged.

5.7 THE EYES AND EARS OF VOYAGER

Many parts of the human body contribute to our sensory perception of the world around us, the eyes and ears playing major roles in that process. Touch, taste, and

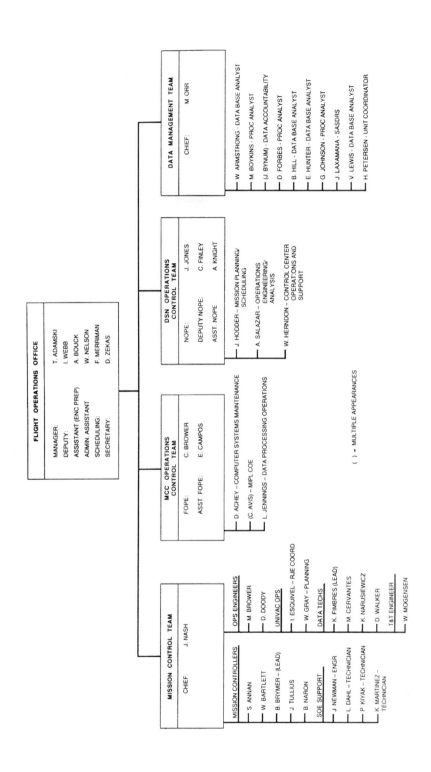

Figure 5.14. Voyager Flight Operations Office organization during the Neptune encounter period.

smell are also important senses, but are limited to the perception of things within our immediate vicinity. Because of their ability to detect waves generated by processes occurring at remote distances from our bodies, the eyes and ears play unique roles in our sensory perception, and so it is with the Voyager spacecraft. The science instruments served as our 'eyes and ears' for conditions extant at the remote distances of Neptune, enabling us to 'see' things never before seen by humans. It is likely that future generations may have the opportunity to travel to vicinities of the major planets to see such marvels personally, but until then robot exploration will do much to temporarily sate our intellectual appetites for knowledge about these distant worlds.

The location of instrumentation for the 11 scientific investigations on the Voyager spacecraft is shown in Figure 5.15. Voyagers 1 and 2 are basically identical, except where they have been altered by hardware failures during their mission or by differences in the contents of their computers. Where differences exist, the description here will relate to Voyager 2, since that is the spacecraft that was sent to Uranus and Neptune. All 11 science investigations on Voyager 2 were operative throughout the period of the mid-1989 Neptune encounter.

The prominent feature of the spacecraft is its 3.66-m diameter parabolic antenna, which serves both to receive commands from Earth and transmits science and engineering data back to Earth. The RSS investigation [9] utilizes the X-band and S-band carriers (the radiowave signals onto which the modulated telemetry data are superimposed). When using the internal ultrastable oscillator as a frequency standard, X-band and S-band signals are transmitted simultaneously. The X-band frequency is 8420.4 MHz (wavelength 3.560 cm); the S-band frequency is 2296.5 MHz (wavelength 13.054 cm). Voyager 2's S-band receiving frequency is 2113.3 MHz. When Voyager's radio system is 'in lock', the transmitted frequencies are determined by the received frequency, and are closer to 8415.0 MHz and 2295.0 MHz, respectively. The transmitted frequencies are in the ratio 11/3 (X-band/S-band). Much of the necessary instrumentation for the RSS investigation was co-located with the DSN tracking stations. The primary RSS science objectives at Neptune were:

- to study by means of radio occultations the atmosphere and ionosphere of Neptune, including atmospheric turbulence;
- to determine by means of occultations and scattering experiments the existence and nature of Neptunian rings, including radial structure, particle size distribution, and radio optical depth;
- to study the gravitational fields of Neptune, Triton, and other satellites, including mass and density determinations and higher-order gravity harmonics.

The ISS [10] is a modified version of the slow-scan vidicon camera designs that were used in earlier Mariner missions (Figure 5.16). The system consists of two cameras, a high-resolution narrow-angle (NA) camera and a more sensitive low-resolution wide-angle (WA) camera. The NA camera has a focal length of 1499.125 mm and a field-of-view of 7.40 mrad (0.424°) on a side. The WA camera's focal length is 201.568 mm; and its field-of-view is 55.31 mrad (3.169°)

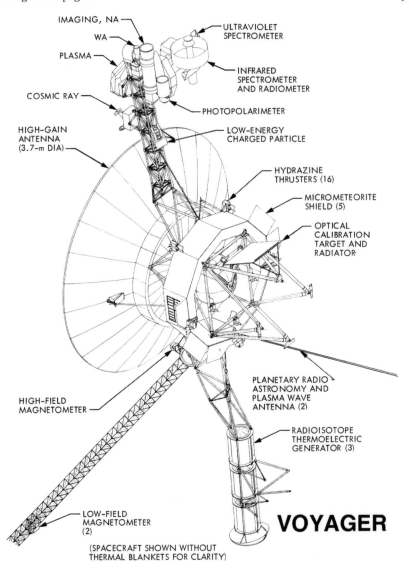

IMAGING, NA
WA
PLASMA
COSMIC RAY
HIGH-GAIN
ANTENNA
(3.7-m DIA)
HIGH-FIELD
MAGNETOMETER
LOW-FIELD
MAGNETOMETER
(2)
(SPACECRAFT SHOWN WITHOUT
THERMAL BLANKETS FOR CLARITY)

ULTRAVIOLET
SPECTROMETER
INFRARED
SPECTROMETER
AND RADIOMETER
PHOTOPOLARIMETER
LOW-ENERGY
CHARGED PARTICLE
HYDRAZINE
THRUSTERS (16)
MICROMETEORITE
SHIELD (5)
OPTICAL
CALIBRATION
TARGET AND
RADIATOR·
PLANETARY RADIO
ASTRONOMY AND
PLASMA WAVE
ANTENNA (2)
RADIOISOTOPE
THERMOELECTRIC
GENERATOR (3)

VOYAGER

Figure 5.15. Diagram of the Voyager spacecraft, showing the location of the science instrumentation for the 11 scientific investigations. (P-19409)

square. The output of each camera is an array of 800 lines of video by 800 picture elements per line, or 640,000 'pixels' per image. Pixel brightness level is represented by an eight-digit binary number (i.e., the brightness can be any decimal number between 0 and 255). Exposure times at Neptune could vary from 0.005 to 61.44 s. The available filters are listed in Table 5.1. The frame readout time was as short as 48 s (for recorded images) or as long as 480 s (for non-recorded, i.e., real-time, images). Most of the non-recorded images were passed through a data-compression

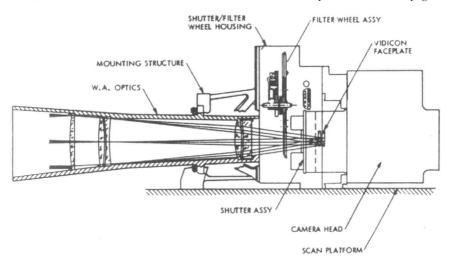

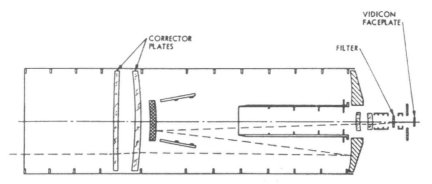

Figure 5.16. Diagram of the Voyager Imaging Science Subsystem slow-scan vidicon camera optics. The WA camera is shown above and the NA camera below.

routine in the FDS to reduce the total number of transmitted bits by, approximately, a factor of 3 without loss of information. ISS objectives at Neptune were:

- to observe and characterize the circulation of the Neptune atmosphere, provide limits on atmospheric composition, and determine wind velocities in the regions observed;
- to map the radial and azimuthal distribution of material in the ring plane and search for new rings and ring arcs;
- to obtain global multi-spectral coverage of all satellites, establish their rotation rates and spin axis orientations, study their surface morphologies, and search for undiscovered satellites;
- to provide support images to assist IRIS, PPS, and UVS in their data reduction.

The IRIS [11] combines a Michelson interferometer for measurements in the mid-infrared region and a radiometer for measurements in the visible and

Table 5.1. Imaging spectral filters

Filter number and name	Characteristics and brief description	Effective wavelength (μm)	Filter factor
NA camera			
(#0) Clear	Broadband	0.497	1.0
(#1) Violet	Wideband, centered at 0.400 μm	0.416	7.4
(#2) Blue	Wideband, centered at 0.480 μm	0.479	3.5
(#3) Orange	Cut-on at 0.570 μm	0.591	7.0
(#4) Clear	Broadband	0.497	1.0
(#5) Green	Cut-on at 0.530 μm	0.566	3.3
(#6) Green	Cut-on at 0.530 μm	0.566	3.3
(#7) Ultraviolet	Wideband, centered at 0.323 μm	0.346	46
WA camera			
(#0) Methane, J/S/T	Narrowband, centered at 0.619 μm	0.6184	60
(#1) Blue	Wideband, centered at 0.480 μm	0.476	3.1
(#2) Clear	Broadband	0.470	1.0
(#3) Violet	Wideband, centered at 0.400 μm	0.426	7.2
(#4) Sodium-D	Narrowband, centered at 0.589 μm	0.589	2.5
(#5) Green	Cut-on at 0.530 μm	0.560	3.5
(#6) Methane, U	Narrowband, centered at 0.541 μm	0.541	40
(#7) Orange	Cut-on at 0.590 μm	0.605	15

near-infrared range (see Figure 5.17). The two share a 50-cm diameter reflecting telescope and have a 0.25° (4.4 mrad) diameter circular field-of-view, boresighted with the imaging cameras. IRIS requires 48 s (including mirror flyback time) to obtain a spectrum from 180 to 2500 cm^{-1}, corresponding to wavelengths from 55 to 4 μm. The radiometer is sampled every 6 s and covers from 2 to 0.33 μm. The primary IRIS objectives at Neptune were:

- to determine the atmospheric vertical thermal structure, which aids modeling of atmospheric dynamics;
- to measure the abundances of hydrogen and helium as a check on theories regarding their ratio in the primary solar nebula;
- to determine the balance of energy radiated by Neptune to that absorbed from the Sun to help to investigate planetary origin, evolution, and internal processes.

The UVS [12] covers the wavelength range of 0.05 to 0.17 μm with 0.001 μm resolution (see Figure 5.18). There are no moving parts: spectral coverage is obtained by use of a reflective diffraction grating which spreads the light and focuses it onto an array of 128 adjacent detectors. Brightness levels for each channel range from 0 to 1023. A spectral scan is normally completed in 3.84 s; in a special occultation data mode, each scan is completed in 0.32 s. No lenses are used to define the pointing direction: 13 identical aperture plates are placed at pre-calculated distances from each other such that off-axis light is effectively rejected. The field-of-view thus

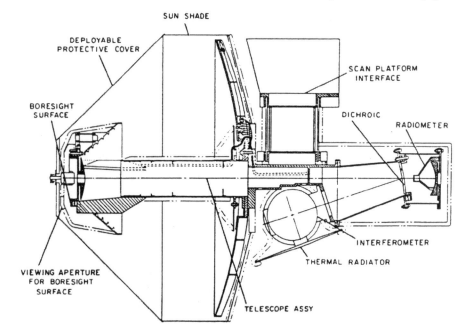

Figure 5.17. The Infrared Interferometer Spectrometer and Radiometer (IRIS) consists of two Michelson interferometers and a wideband visible and infrared radiometer.

defined is 0.1° by 0.86° (1.7 by 15 mrad) and is boresighted with the ISS. A smaller aperture is offset by 20° by use of a small mirror. The field-of-view for this solar occultation port is 0.25° by 0.86° (4.4 by 15 mrad). At UVS wavelengths between 0.05 and 0.12 μm, Earth's atmosphere is opaque; prior to the launch of the Galileo spacecraft, these wavelengths were covered by no other spacecraft-borne ultraviolet spectrometers. The UVS was therefore a useful tool for studying stars and distant galaxies. The primary science objectives for UVS at Neptune were:

- to determine the scattering properties of the lower atmosphere of Neptune;
- to determine the distribution of atmospheric constituents with altitude;
- to determine the extent and distribution of hydrogen coronae of Neptune and its satellites;
- to investigate night airglow and auroral activity;
- to determine the ultraviolet scattering properties and optical depths of rings and ring arcs; and
- to search for emissions from the rings and from any ring 'atmosphere.'

The PPS [13] instrument was partly described near the end of Section 5.5. It consists of a sensitive photoelectric photometer mounted behind a 15.2-cm diameter f1/4 Cassegrain telescope (see Figure 5.19). It has a four-position aperture wheel which allows for fields-of-vew of 0.12° (2.1 mrad), 0.33° (5.8 mrad), 1.0° (17 mrad), and 3.5° (61 mrad). It can access the three filter wheel positions and four polarizing

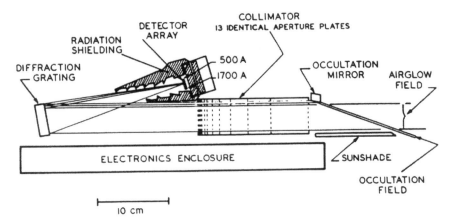

Figure 5.18. Diagram of the Voyager Ultraviolet Spectrometer.

analyzer positions shown in Table 5.2. It also has two sensitivity levels separated by a factor of about 50, J-mode on for high light levels and J-mode off for low light levels. Each sample is represented as a brightness level from 0 to 1023. The PPS sampling rate was dependent on the FDS mode, normally returning a sample every 0.6 s, but sampling 100 times per second in the special occultation mode. The PPS scientific objectives at Neptune were:

- to determine the vertical distribution of cloud particles (atmospheric aerosols) down to an optical depth of unity;
- to determine the scattering and polarizing characteristics of the cloud particles and ring particles to obtain information on size, shape, and probable composition;
- to determine the Neptune atmospheric optical depth as a function of altitude, and to search for satellite atmospheres;
- to determine the Bond albedos of Neptune, it satellites, and its rings; and
- to use stellar occultation measurements to determine ring optical depths, radial distribution, and particle sizes rings.

The PRA [14] utilizes two vertically mounted 10-m electric antenna elements to monitor radio waves in two frequency bands (see Figure 5.20). These antennas were extended by Earth-based commands after separation from the launch vehicle. The low-frequency PRA band covers from 1.2 kHz (250-km wavelength) to 1228 kHz (0.24-km wavelength) at intervals of 19.2 kHz and a bandwidth of 1 kHz. The high-frequency band covers a range of 1.228 MHz (240-m wavelength) to 40.5504 MHz (7.4-m wavelength) at intervals of 0.3072 MHz and a bandwidth of 0.2 MHz. In its basic mode, the PRA scans a total of 198 frequencies in 6 s, representing each frequency level as a logarithmic intensity of 0 to 255. On occasion it may also sample one or more frequencies at much higher rates. At its highest rate (6,400 bits every 0.06 s), PRA data output is identical to the highest ISS data output and must be recorded for more leisurely playback at a later time. The PRA cannot

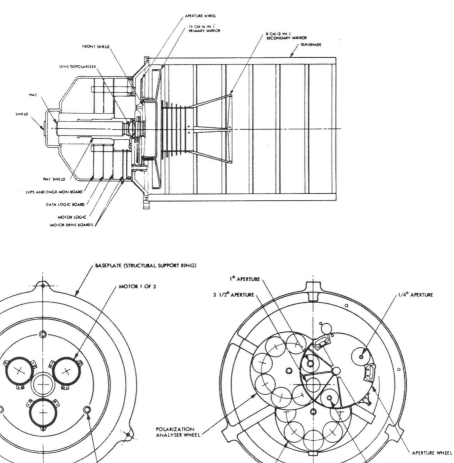

Figure 5.19. The Photopolarimeter is a shielded photoelectric photometer with three motor driven wheels to control the size of the field-of-view, the filter passband, and the polarizer.

operate simultaneously in this high data rate and its lower data rate mode. The PRA science objectives for the Neptune encounter were:

● to locate and explain radio emissions from Neptune at wavelengths ranging from tens of meters (decametric) to kilometers (kilometric);

● to detect radio emissions indicative of a Neptunian magnetic field, and to determine the rotation rate of such a field;

● to describe planetary radio emissions and their relationship to Neptune's satellites;

Table 5.2. Photopolarimeter apertures, filters, and analyzers

	Configuration	Purpose	Number and comments
Aperture	0.12°, 0.33°, 1.0°, 3.5°	Determines field-of-view of instrument	(#0) 0.33°, (#1) 1.0°, (#2) 3.5°, (#3) 0.12°
Filter	(#0) 0.590 μm (#4) 0.265 μm (#6) 0.750 μm	Photometry	Only these three filter positions were available at Neptune
Analyser	(#2) 60° (#3) 120° (#6) 135° (#7) 45°	Polarization	Only these four analyser positions were available at Neptune

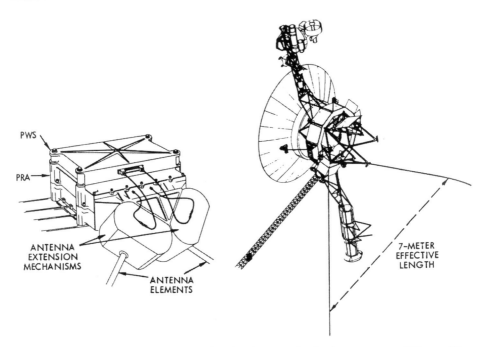

Figure 5.20. Instrumentation packages for the Planetary Radio Astronomy and Plasma Wave Subsystem investigations are mounted together and utilize the same pair of 10-m antennas.

- to measure plasma resonances near Neptune;
- to detect lightning in Neptune's atmosphere.

The PWS [15] uses the same two 10-m antennas used by PRA, but connected in such a way that they act as a single 7-m dipole antenna (see Figure 5.20). The PWS consists of two instruments: a spectrum analyzer, which operates continuously, and a wideband waveform receiver, which operates for brief periods on command. The

SPACECRAFT CONFIGURATION

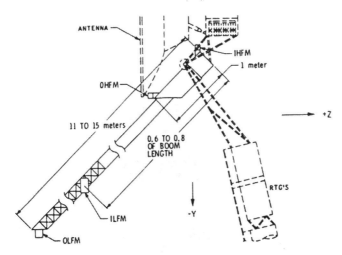

Figure 5.21. Four magnetometers are mounted at the base of and along the 13-m magnetometer boom.

spectrum analyzer samples frequencies of 0.0100, 0.0178, 0.0311, 0.0562, 0.100, 0.178, 0.311, 0.562, 1.00, 1.78, 3.11, 5.62, 10.0, 17.8, 31.1, and 56.2 kHz every 4 s with logarithmic intensity levels of 0 to 255. The bandwidth of each frequency channel is 15% of the sampled frequency. The wideband waveform receiver is used to provide occasional views of the full frequency range from 0.050 to 10 kHz with higher time resolution. It is sampled 1,600 times each 0.06 s with logarithmic intensity levels of 0 to 15. PWS science objectives at Neptune were:

- to determine the role of interactions between charged particles and wave motions in various parts of the magnetic field of Neptune;
- to determine the mechanisms that control the interactions of the rotating magnetic field of Neptune with its satellites;
- to determine the strengths and frequencies of variation of electric fields generated by flows of interacting charged particles (i.e., by 'plasmas') within Neptune's magnetic field;
- to map variations in plasma density along the spacecraft trajectory;
- to monitor low-frequency radio waves generated by atmospheric lightning discharges.

The MAG [16] sensors consist of two high-field magnetometers (HFMs) mounted on the spacecraft body and two low-field magnetometers (LFMs) located, respectively, at distances of 7.4 m and 13.0 m along a special magnetometer boom that was unfurled after separation of Voyager from the launch vehicle (see Figure 5.21).

The separation of the LFMs from the spacecraft body was necessary because of the small magnetic field generated by electrical currents flowing within the science and engineering components of the spacecraft. Each magnetometer measures the ambient magnetic field strengths along three mutually perpendicular axes. The HFMs have two ranges, ± 0.5 Oe (Oersted) and ± 20 Oe. For comparison, Earth's surface magnetic field strength varies from about 0.29 Oe near the equator to about 0.68 Oe near the South magnetic pole. Gain changes occur automatically when signals reach levels within 256 data numbers of the low end or the high end of the 0 to $\pm 2{,}047$ data number range. There are eight separate LFM ranges: $\pm 8.8 \times 10^{-5}$ Oe, $\pm 26 \times 10^{-5}$ Oe, $\pm 79 \times 10^{-5}$ Oe, $\pm 240 \times 10^{-5}$ Oe, $\pm 700 \times 10^{-5}$ Oe, $\pm 2{,}100 \times 10^{-5}$ Oe, $\pm 6{,}400 \times 10^{-5}$ Oe, $\pm 50{,}000 \times 10^{-5}$ Oe. Note that the highest range for the LFM is almost the same as the lowest range for the HFMs. The total data rate of the MAG during the Neptune encounter period was 750 bits/s. The scientific goals of the MAG at Neptune were:

- to measure and analytically represent the Neptunian magnetic field;
- to determine the magnetospheric structure of Neptune;
- to investigate the basic physical mechanisms and processes involved both in interactions between the solar wind and the magnetosphere and in internal magnetosphere dynamics;
- to investigate the interactions of the satellites of Neptune with its magnetosphere;
- to study the solar wind in the vicinity of Neptune.

The PLS [17], LECP [18], and CRS [19] study electrons and other charged particles (ions of elements from hydrogen to iron), and utilize detectors that can only sense particles that strike them. Each investigation is designed to discriminate between charged particles with different energies and each instrument has some ability to determine the incoming direction of such particles. The scientific goals of the three instruments at Neptune were:

- to study energetic particles in the solar wind near Neptune;
- to study the composition, sources, and other characteristics of charged particles within the Neptunian magnetosphere;
- to study the interactions of these charged particles with rings and satellites;
- to search for evidence of galactic cosmic-ray particles.

The PLS consists of two Faraday cup plasma detectors, one pointed in the general direction of Earth and the other at right angles to a line from the spacecraft to Earth (Figure 5.22). The Earth-facing sensor has three separate apertures, which in combination permit determination of the plasma velocity, density, and pressure. This cluster has a combined field-of-view that covers the entire Earth-facing hemisphere; the overlap between the apertures is a cone of half angle $45°$. Each field-of-view (i.e., four, including a side-looking sensor) is approximately conical with half angle $60°$. Electrons can be detected by the side-looking sensor. The PLS senses singly charged particle energies of 10 to 5950 eV, where eV is the kinetic energy acquired by an electron when it is being accelerated across an electrical potential of 1 V (an energy of about 1.602×10^{-12} erg). The instrument has four operating modes: M ($=$ medium and high-resolution ion mode), L ($=$ low-resolution ion mode), E1 ($=$ high-resolu-

Figure 5.22. The Plasma Wave Subsystem is mounted on the spacecraft science boom. (260-786B)

tion electron mode), and E2 (= low-resolution electron mode). The PLS cycled through the four modes every 12 minutes during the Neptune encounter, providing a data output rate of only four samples (32 bits) per second.

The Low Energy Charged Particle (LECP) instrument consists of two arrays of detectors on a rotating platform (see Figure 5.23). The first array, with eight solid-state detectors, is the Low Energy Magnetospheric Particle Analyzer (LEMPA). The detectors in the LEMPA are designed for detection of particles with energies as low as 10 to 15 keV, discriminating between ions and electrons, high sensitivity, and a large range of energies. The second array, with seven solid-state detectors and an anti-coincidence shield of eight active detectors, is the Low Energy Particle Telescope (LEPT), but the smallest of the Voyager 2 detectors (called 'D1c') failed shortly before the Saturn encounter. LEPT detectors are designed to measure the distributions of charge and energy of ions in the range 0.1 MeV/nucleon to 500 MeV/nucleon. The data rate output of the LECP was 600 bits/s during the Neptune encounter. The LECP telescope normally makes one 45° step each 48 s during an encounter. Within 24 h of closest approach, the LECP stepping occurred in a 12-min cyclic fashion, such that two eight-position cycles were completed in 1.6 min, followed by a 10.4 min period during which no stepping occurred. This minimized electronic interference with the other instruments caused by the LECP's stepping motor.

The CRS consists of three particle telescope systems. The High Energy Telescope (HET) system consists of two double-ended telescopes, each with 15 solid-state particle detectors and seven anti-coincidence detectors. The HETs measure the energy spectra of electrons and all elements from hydrogen to iron

Figure 5.23. The Low Energy Charged Particle instrument consists of two particle telescope systems mounted on a rotating platform. (260-781B)

over a broad range of energies. They can also discriminate between isotopes [20] of hydrogen through oxygen. The four Low Energy Telescopes (LETs) each have four surface barrier detectors designed to determine the three-dimensional flow patterns of cosmic-ray particles and to extend element discrimination to lower energies than detectable by the HETs. The electron telescope (TET) consists of eight solid-state detectors and six tungsten absorbers, all surrounded by a grid of six solid-state anti-coincidence detectors. TET measures the energy spectrum of electrons between 5 and 110 MeV. The combination of HET and LET measures ions in the energy range 1 to 500 MeV/nucleon. The data rate of the CRS is 260 bits/s. Both CRS and LECP completely sample their respective arrays of detectors every 96 s.

NOTES AND REFERENCES

1. Asimov, I. (1982) *Asimov's Biological Encyclopedia of Science and Technology*. Doubleday & Company, New York, Second Revised Edition, p. 568.
2. Goddard, R. H. (1970) *The Papers of Robert H. Goddard*, Volume II: 1925–1937. McGraw-Hill Book Company, p. 580.
3. Goddard, R. H. (1970) *The Papers of Robert H. Goddard*, Volume I: 1898–1924. McGraw-Hill Book Company, preface.
4. Flandro, G. A. (1966) Fast reconnaissance missions to the outer Solar System utilizing energy derived from the gravitational field of Jupiter. *Astronautica Acta*, **12**, 329–37.
5. Humes, D. H., Alvarez, J. M., O'Neal, R. L. and Kinard, W. H. (1974) The interplanetary and near-Jupiter meteoroid environments. *Journal of Geophysical Research*, **79**, 3677–84.

6. Voyager Project Plan, Voyager Document 618-5, Revision B, dated 1 April 1977. Jet Propulsion Laboratory, California Institute of Technology, Pasadena, p. 2–2.

7. Sagan, C., Drake, F. D., Druyan, A., Ferris, T., Lomberg, J. and Sagan, L. S. (1978) *Murmurs of Earth.* Random House, New York.

8. A series of 12 papers on the Neptune encounter, including an overview by Stone and Miner, and preliminary reports from each of the twelve science investigation teams. Appears in *Science*, **246**, 1417–1501.

9. Eshleman, V.R., Tyler, G. L., Anderson, J. D., Fjeldbo, G., Levy, G. S., Wood, G. E. and Croft, T. A. (1977) Radio Science investigations with Voyager. *Space Science Reviews*, **21**, 207–32.

10. Smith, B. A., Griggs, G. A., Danielson, G. E., Cook, A. F. II, Davies, M. E., Hunt, G. E., Masursky, H., Soderblom, L. A., Owen, T. C., Sagan, C. and Suomi, V. E. (1977) Voyager Imaging experiment. *Space Science Reviews*, **21**, 103–27.

11. Hanel, R., Conrath, B., Gautier, D., Gierasch, P., Kumar, S., Kunde, V., Lowman, P., Maguire, W., Pearl, J., Pirraglia, J., Ponnamperuma, C. and Samuelson, R. (1977) The Voyager Infrared Spectroscopy and Radiometry investigation. *Space Science Reviews*, **21**, 129–57.

12. Broadfoot, A. L., Sandel, B. R., Shemansky, D. E., Atreya, S. K., Donahue, T. M., Moos, H. W., Bertaux, J. L., Blamont, J. E., Ajello, J. M., Strobel, D. F., McConnell, J. C., Dalgarno, A., Goody, R., McElroy, M. B. and Yung, Y. L. (1977) Ultraviolet Spectrometer experiment for the Voyager mission. *Space Science Reviews*, **21**, 18–205.

13. Lillie, C. F., Hord, C. W., Pang, K., Coffeen, D. L. and Hansen, J. L. (1977) The Voyager mission Photopolarimeter experiment. *Space Science Reviews*, **21**, 159–81.

14. Warwick, J. W., Pearce, J. B., Peltzer, R. G. and Riddle, A. C. (1977) Planetary Radio Astronomy experiment for Voyager missions. *Space Science Reviews*, **21**, 309–27.

15. Scarf, F. L. and Gurnett, D. A. (1977) A Plasma Wave investigation for the Voyager mission. *Space Science Reviews*, **21**, 289–308.

16. Behannon, K. W., Acuna, M. H., Burlaga, L. F., Lepping, R. P., Ness, N. F. and Neubauer, F. M. (1977) Magnetic field experiment for Voyagers 1 and 2. *Space Science Reviews*, **21**, 235–57.

17. Bridge, H. S., Belcher, J. W., Butler, R. J., Lazarus, A. J., Mavretic, A. M., Sullivan, J. D., Siscoe, G. L. and Vasyluinas, V. M. (1977) The Plasma experiment on the 1977 Voyager mission. *Space Science Reviews*, **21**, 259–87.

18. Krimigis, S. M., Armstrong, T. P., Axford, W. I., Bostrom, C. O., Fan, C. Y., Gloeckler, G. and Lanzerotti, L. J. (1977) The Low Energy Charged Particle (LECP) experiment on the Voyager spacecraft. *Space Science Reviews*, **21**, 329–54.

19. Stone, E. C., Vogt, R. E., McDonald, F. B., Teegarden, B. J., Trainor, J. H., Jokipii, J. R. and Webber, W. R. (1977) Cosmic Ray investigation for the Voyager missions; energetic particle studies in the outer heliosphere and beyond. *Space Science Reviews*, **21**, 355–76.

20. An isotope is an atom with the same number of protons in its nucleus but a different number of neutrons. Deuterium (also called heavy hydrogen) is an isotope of normal hydrogen. Atoms with the same number of protons in their nucleus (i.e., the same nuclear charge) also have the same 'atomic number', Z. Different isotopes of the same element will have the same atomic number, but different 'atomic mass'. Atomic mass units (emu) are scaled such that a carbon-12 atom, consisting of six protons, six neutrons, and six electrons, weighs precisely 12.0000 emu.

BIBLIOGRAPHY

Kohlhase, C. (ed.) (1989) *The Voyager Neptune Travel Guide*. NASA, JPL Publication 89-24.

Miner, E. D. (1990) *Uranus: the Planet, Rings and Satellites*. Wiley–Praxis, Chichester, UK.

Stone, E.C., Kohlhase, C. E. and the authors listed in references 9 through 19 above (1977) *Space Science Reviews*, **21**, 75–376. (Contains a series of 13 papers, including a science overview, a mission description, and individual papers on each of the 11 scientific investigations.)

Caltech historical Sketch, http://www.caltech.edu/catalog/geninfo/history.html

6

The pre-Neptune scientific results of Voyager

6.1 'JUPITER THE GIANT'

Voyager 1 reached its closest approach with Jupiter on March 5, 1979; Voyager 2 reached that milestone on July 9, 1979. The paths of the spacecraft through the Jupiter system are shown in Figures 6.1 and 6.2. Each spacecraft received a substantial gravity assist from Jupiter, which increased their respective velocities sufficiently to enable each to completely escape the Sun's gravity and eventually enter interstellar space (see Figure 6.3). Some effects, due to the intense radiation field surrounding Jupiter, were noted on both spacecraft. The Voyager 1 PPS, partially crippled before the encounter, failed completely and has been unusable since that time. The UVS on each spacecraft returned few useful data near its closest approach to Jupiter, but each recovered shortly after the spacecraft exited the most intense radiation period. The clock internal to the FDS computers on Voyager 1 were 'reset' 40 times within a period of several hours, causing them to lose synchronization with clocks on the CCS. Some of the ISS, PRA, and PWS data was degraded as a result, but no permanent damage was done to the computers or these three instruments. The radio transmitter frequencies on both spacecraft were shifted by radiation effects, but these shifts had no effect on subsequent operation. The optics in the Canopus Star Trackers (CSTs) on both spacecraft were darkened, resulting in an effective 13% reduction in sensitivity. The CST on Voyager 1 additionally lost some pointing capability, resulting in a reduction in the number of usable post-Jupiter reference stars.

Despite these difficulties, the Voyager encounters with Jupiter were unqualified successes. Dedicated issues or large groupings of scientific papers resulting from the Voyager encounters appeared in *Science, Nature, Geophysical Research Letters*, and in the *Journal of Geophysical Research* [1]. An excellent narrative containing the highlights of the scientific findings was given by Morrison and Samz [2] and a very brief summary of the major scientific findings is given below.

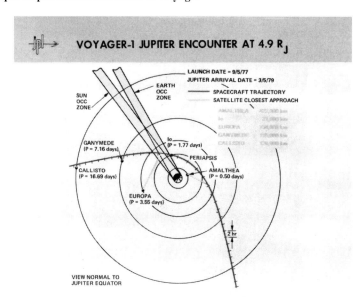

Figure 6.1. A trajectory-plane view of the path of Voyager 1 through the Jupiter system. Voyager 1's closest approach to Jupiter occurred on March 5, 1979. (260-284B)

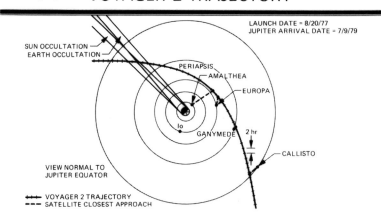

Figure 6.2. A trajectory-plane view of the path of Voyager 2 through the Jupiter system. Voyager 2's closest approach to Jupiter occurred on July 9, 1979. (260-533A)

6.1.1   Jupiter's atmosphere

1. The atmosphere of Jupiter above the cloud tops is composed almost exclusively of hydrogen and helium. There are $26 \pm 5\,\mathrm{g}$ of helium for every $100\,\mathrm{g}$ of hydrogen. (Since improved to $23.4 \pm 0.5\,\mathrm{g}$ by Galileo.)

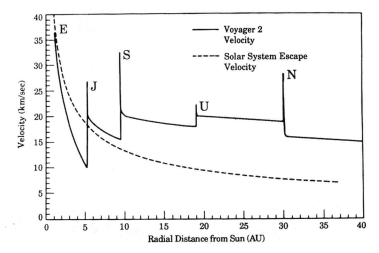

Figure 6.3. The Sun-centered velocity vector of Voyager 2 (solid line) is plotted against radial distance from the Sun in astronomical units, where 1 AU = 149,597,900 km. The dashed line represents the velocity needed to escape the Sun's gravity. The large effects of 'gravity assists' at Jupiter and Saturn are apparent. Jupiter increased Voyager's velocity beyond that necessary to escape the Sun's gravity. (JPL-12723)

2. The Great Red Spot (Figure 6.4), the white oval (Figure 6.5), and other major cloud disturbances rotate clockwise in the northern hemisphere and counterclockwise in the southern hemisphere, indicating that they are regions of higher pressure within the atmosphere.

3. There exists a stable zonal pattern of winds (Figure 6.6) which is an indication that the wind patterns are a more fundamental characteristic of the atmosphere than the bright zones and dark belts.

4. The zonal wind pattern extends into the polar regions which were previously thought to form a region where vertical motions would dominate over horizontal flows.

5. The interactions of the Great Red Spot and the zonal wind flows are extremely complex, and major changes in those patterns occurred over the four-month period between the encounters of Voyager 1 and Voyager 2.

6. Lightning superbolts (Figure 6.7) are commonly seen near the cloud tops and give rise to radio wave emissions called 'whistlers', so called because the generated radio frequencies change in much the same way as an audible whistle decreasing in pitch.

7. The presence of a stratospheric temperature inversion was verified. Temperatures rise from a minimum of about 110 K near the 100-mbar level to about 160 K near 10 mbar.

8. A concentration of ultraviolet-absorbing haze was observed in the polar regions of the planet.

Figure 6.4. The Great Red Spot, an anti-cyclonic storm which has been observed in Jupiter's southern hemisphere for nearly 400 years. (P-21742) (See color plate 1.)

Figure 6.5. Although the white ovals are smaller and less spectacular than the Great Red Spot, they too are enormously large anti-cyclonic storms in Jupiter's southern hemisphere. (P-21754)

9. Acetylene (C_2H_2) and ethane (C_2H_6) are present in ratios that vary both with time and with latitude on the planet.
10. Both visible and ultraviolet emissions (Figure 6.8) are seen in the nightside polar atmosphere. They are probably due to charged particles generated near Io's orbit and streaming down Jovian magnetic field lines.

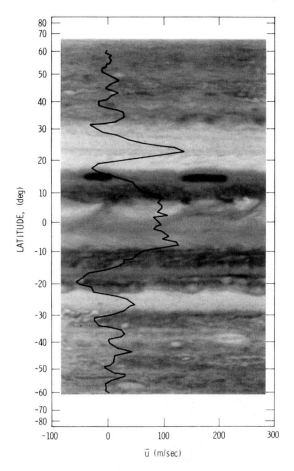

Figure 6.6. The zonal pattern of winds in Jupiter's atmosphere is relatively stable and of much longer duration than most cloud features. (260-1126A)

11. There is a strong ultraviolet emission from the entire sunlit disk of the planet. The large-scale height of this emission is indicative of extreme upper atmospheric temperatures in excess of 1000 K.

6.1.2   Jupiter's ring system

1. An equatorial ring of finely divided material (Figure 6.9) surrounds Jupiter. The main component of this ring extends from about 51,000 km to 57,000 km above the cloud tops.

2. A gossamer extension of the main ring may extend to Amalthea's orbit, about 109,000 km above the cloud tops. A halo of diffuse ring material whose source

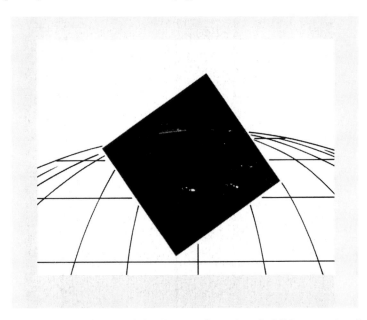

Figure 6.7. Lightning superbolts (bright clusters of spots) and visible auroral emissions (near the edge of the disk) are seen in this Voyager image of Jupiter's dark side. (260-929A)

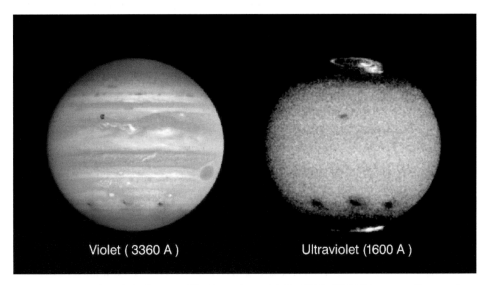

Violet (3360 A)   Ultraviolet (1600 A)

Figure 6.8. Far-ultraviolet image of Jupiter taken with the Wide-Field Planetary Camera-2 on the Hubble Space Telescope on 17 July 1994 showing the emissions from the auroral ovals around the north and south magnetic poles. (The dark spots in Jupiter's southern hemisphere were caused by the impacts of fragments of Comet Shoemaker–Levy 9 with the planet.)

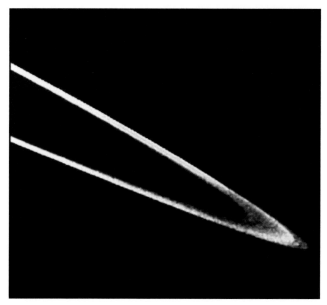

Figure 6.9. This equatorial ring of finely divided material around Jupiter was discovered by Voyager. (260-674)

may be the main ring, extends downward to within about 20,000 km of the cloud tops.

3. Typical ring particles are only a few micrometers across, and scatter light predominantly in a forward direction. The ring would be extremely difficult to detect from Earth-based telescopes.

6.1.3   Jupiter's satellites

1. Two previously undetected satellites (Metis and Adrastea) orbit near the outer edge of the ring. Metis (Figure 6.10) has a diameter of about 40 km; Adrastea (Figure 6.11) is smaller and non-spherical, ranging from 15 to 25 km in diameter.
2. Amalthea (Figure 6.12) was shown to be an irregular, elongated, dark, and very red object, ranging in size from 150 to 270 km. Its surface is probably contaminated by sulfur from Io.
3. Thebe (Figure 6.13), with a diameter of almost 100 km, was found between the orbits of Amalthea and Io.
4. Nine nearly continuous volcanic eruptions (Figure 6.14) were observed on Io, apparently generated by a combination of orbit perturbations by Europa and tidal distortions from Jupiter. At least seven of the nine were still erupting when Voyager 2 passed four months later. (One had stopped, and one could not be observed by Voyager 2.)

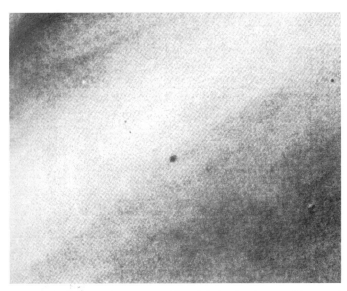

Figure 6.10. Jupiter's innermost known satellite, Metis, is shown silhouetted against the planet's clouds.

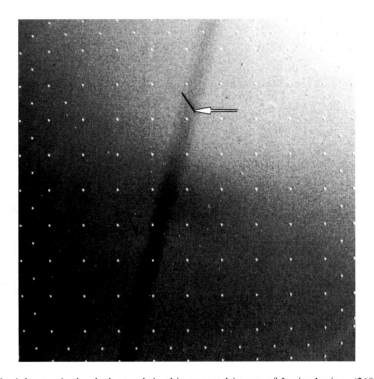

Figure 6.11. Adrastea is the dark streak in this smeared image of Jupiter's ring. (260-806)

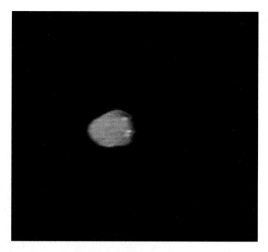

Figure 6.12. The innermost of the satellites discovered from Earth-based telescopes is Amalthea. It is seen here to be a dark, irregular body. (P-21223)

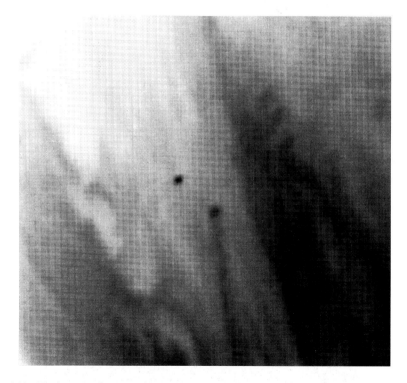

Figure 6.13. Thebe was discovered by Voyager. It orbits between Amalthea and Io. With its diameter of about 100 km, it is the largest of the three Jupiter satellites discovered by Voyager. (P-22580)

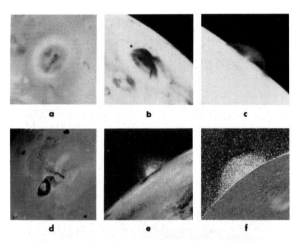

Figure 6.14. Active volcanic eruptions were observed on Jupiter's Io, evidence of the strong tidal heating caused by Jupiter's gravity. (260-451)

Figure 6.15. Io is the innermost of the four Galilean satellites of Jupiter. (P-34590) (See color plate 2.)

5. The surface of Io (Figure 6.15) is void of impact craters, extremely young, colored by sulfur and sulfur compounds, and able to undergo large-scale surface changes in a few months.

6. At least two hot spots on Io's surface were detected, further confirming the presence of active volcanism.

7. Io possesses a thin and probably transient atmosphere of sulfur dioxide (SO_2).

8. The surface of Europa (Figure 6.16) is very smooth and nearly void of impact craters. A deep-water ocean may exist beneath a 10-km thick crust of water ice.

9. A network of light and dark lines criss-cross Europa. These suggest major stressing of the surface in the geological past.

Figure 6.16. Streaks across Europa's surface may be fossil remnants of ancient fractures. (P-21751)

10. The surface of Ganymede (Figure 6.17) is highly variegated. Some are very dark, while others are relatively bright. Heavily cratered and more lightly cratered areas both exist.
11. Extensive parallel ridge and valley systems (Figure 6.18) were seen on Ganymede, perhaps indicating past periods of subsurface activity or motion of crustal plates.
12. The surface of Callisto (Figure 6.19) is covered by impact craters, and no indications of significant recent geological activity were seen.
13. The sizes, masses, and densities of the four largest satellites have been determined (Table 6.1). Ganymede is the largest satellite in the Solar System. There is a downward progression of densities with distance from Jupiter, implying that heat from Jupiter may have helped to remove water from the inner satellites.

6.1.4   Jupiter's magnetosphere

1. An electric current of more than a million amperes flows along the magnetic flux tube linking Jupiter and Io.
2. A doughnut-shaped torus (Figure 6.20) containing sulfur and oxygen ions surrounds Jupiter at the orbit of Io. This torus emits ultraviolet light, has temperatures of up to 100,000 K, and is populated by more than 1,000 electrons/cm^3.

Figure 6.17. Ganymede is the largest of the Galilean satellites, and the largest satellite in the Solar System. (P-21271)

3. A 'cold' (i.e., forced to rotate with the magnetic field) plasma exists between Io's orbit and the planet (see Figure 6.21). It has larger than expected amounts of sulfur, sulfur dioxide, and oxygen, all probably derived from Io's volcanic eruptions.
4. The Sun-facing magnetopause (outer edge of the magnetosphere) responds rapidly to changing solar wind pressure, varying from less than 50 Jupiter radii to more than 100 Jupiter radii in distance from the planet's center.
5. A region of 'hot' (i.e., not forced to rotate with the magnetosphere) plasma exists in the outer magnetosphere. It consists primarily of hydrogen, oxygen, and sulfur ions.
6. Jupiter emits low-frequency radio waves (wavelengths of one to several kilometers). The amount of radiation is strongly latitude-dependent.
7. There exists a complex interaction between the magnetosphere and Ganymede. This results in deviations from a smooth magnetic field and charged particle distributions which extend up to 200,000 km from the satellite.
8. About 25 Jupiter radii behind the planet, the character of the magnetosphere changes from the 'closed' magnetic field lines to an extended magnetotail without line closure. This occurs as a result of downstream interaction with the solar wind.

Figure 6.18. These parallel ridge and valley systems on Ganymede may be evidence for crustal movements on this large satellite. (P-21752)

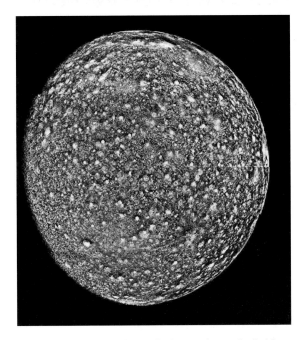

Figure 6.19. Callisto appears to have changed little over its geologic history. Its surface bears the unobscured markings of an early bombardment by meteor and other cosmic debris. (P-21746)

Table 6.1. Earth, Jupiter, Saturn, Uranus major satellite data

Satellite	Diameter (km)	Mass (10^{19} kg)	Density (g/cm^3)
Moon (E)	$3,476 \pm 1$	$7,349 \pm 7$	3.342 ± 0.001
Io (J)	$3,630 \pm 10$	$8,920 \pm 40$	3.55 ± 0.03
Europa (J)	$3,138 \pm 20$	$4,870 \pm 50$	3.01 ± 0.07
Ganymede (J)	$5,262 \pm 20$	$14,900 \pm 60$	1.95 ± 0.02
Callisto (J)	$4,800 \pm 20$	$10,750 \pm 40$	1.86 ± 0.02
Mimas (S)	392 ± 6	4.55 ± 0.54	1.44 ± 0.18
Enceledus (S)	500 ± 20	7.4 ± 3.6	1.13 ± 0.57
Tethys (S)	$1,060 \pm 20$	75.5 ± 9.0	1.21 ± 0.16
Dione (S)	$1,120 \pm 10$	105.2 ± 3.3	1.43 ± 0.06
Rhea (S)	$1,530 \pm 10$	249 ± 15	1.33 ± 0.08
Titan (S)	$5,150 \pm 4$	$13,457 \pm 3$	1.882 ± 0.004
Hyperion (S)	300 ± 30	Not determined	Not determined
Iapetus (S)	$1,460 \pm 20$	188 ± 12	1.15 ± 0.09
Phoebe (S)	220 ± 20	Not determined	Not determined
Miranda (U)	271.6 ± 2.4	6.3 ± 0.7	1.15 ± 0.15
Ariel (U)	$1,157.8 \pm 2.2$	127 ± 1.4	1.56 ± 0.09
Umbriel (U)	$1,169.4 \pm 8.0$	127 ± 1.4	1.52 ± 0.11
Titania (U)	$1,577.8 \pm 5.6$	69.8 ± 1.4	1.70 ± 0.05
Oberon (U)	$1,522.8 \pm 3.0$	60.6 ± 1.6	1.64 ± 0.06

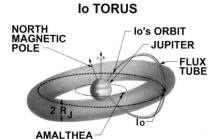

Figure 6.20. Encircling Jupiter is a doughnut-shaped torus of charged particles near the orbit of Io. The source of these particles may be Io's volcanic eruptions. (P-21218)

9.   Jupiter's magnetotail (Figure 6.22) probably extends to, and may extend beyond, the orbit of Saturn – more than 700 million km 'downwind' of Jupiter.

6.2 'SATURN THE GEM'

The second leg of the journeys of Voyager 1 and 2 culminated with the Saturn encounters of November 12, 1980, and August 26, 1981, respectively. The trajectories of the two spacecraft through the Saturn system are shown in Figures 6.23 and

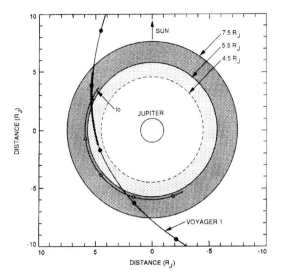

Figure 6.21. Between Io's orbit and the planet is a region of 'cold' plasma.

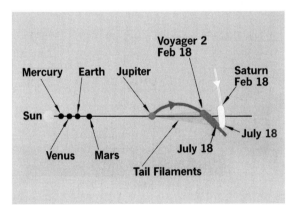

Figure 6.22. Jupiter's magnetotail may extend more than 700 million km 'downward' of Jupiter to the orbit of Saturn. (260-1264A).

6.24. Two of the primary objectives of the Voyager 1 encounter with Saturn required a close approach to Titan (Figure 6.25) and a spacecraft passage behind the full radial extent of the rings (Figure 6.26). Because of the geometry of those two requirements, Voyager 1 was pulled by Saturn's gravity into a path northward of the orbital planes of the planets. Voyager 1's post-Saturn mission will be discussed later.

Following the successful acquisition of critical Titan and ring data by Voyager 1, project personnel were given permission by NASA Headquarters to target Voyager 2 for a Uranus encounter. The appropriate gravity-assist path required that Voyager 2

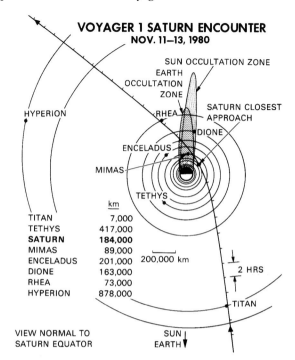

Figure 6.23. A trajectory-plane view of the path of Voyager 1 through the Saturn system. Voyager 1's closest approach to Saturn occurred on November 12, 1980. (260-845A)

pass very close to Saturn's G ring. (Saturn's rings, from the innermost to the outermost, are called D, C, B, A, F, G, and E.) There were some concerns that the spatial density of ring particles might be sufficient to damage the spacecraft, but the potential gains were judged to outweigh the risks. Ring-particle and radiation environments during the Saturn encounters proved to be relatively benign. No damage was done to the spacecraft or to any of its instrumentation.

One major problem did occur during the Saturn encounter. As the experience of the flight team increased, more efficient use of the spacecraft resources resulted. The 48-hour period surrounding the closest approach to Saturn was the most active observing period attempted. In some respects, the period was almost frenetic, as the scan platform was pointed in rapid succession at a dozen different satellites, at the rings, and at Saturn itself. To accomplish this, the platform was moved extensively at its most rapid rate, 1° per second. About 110 minutes after the closest approach to Saturn and an hour after passing the G ring, the platform slowed, then stopped. Voyager 2 was behind Saturn at the time, and news of the platform problem did not reach JPL until after the spacecraft had exited the planet's shadow and the additional 90-min one-way light time had elapsed. By that time, the spacecraft had automatically disabled further 'slewing' of the platform. Commands were sent to the spacecraft to try to free the platform by moving it at its low rate (5° per

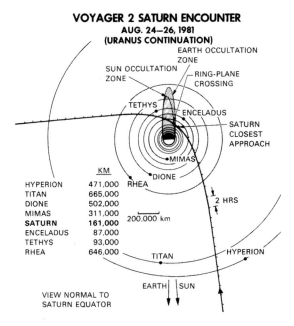

Figure 6.24. A trajectory-plane view of the path of Voyager 2 through the Saturn system. Voyager 2's closest approach to Saturn occurred on August 25, 1981. (P-23349)

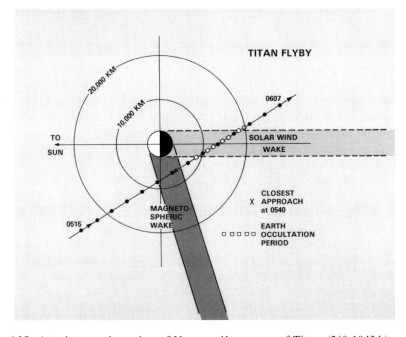

Figure 6.25. A trajectory-plane view of Voyager 1's passage of Titan. (260-1043A)

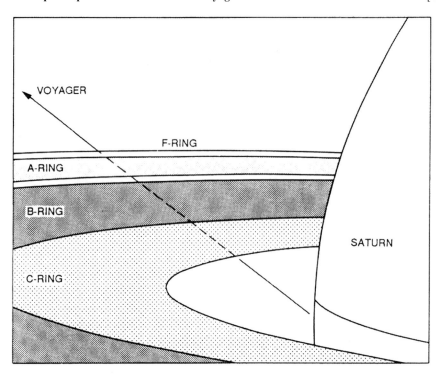

Figure 6.26. As viewed from Earth, Voyager 1 passed behind the full radial extent of Saturn's rings.

minute), but the spacecraft refused to respond. Platform slewing commands were stripped from the computer sequences scheduled for transmission to Voyager 2. Special commands to alternately heat and cool the faulty scan platform azimuth actuator were generated and transmitted to the spacecraft. In the meantime, Voyager 2 was receding rapidly from Saturn, and the disabled platform was pointed uselessly at dark sky. At the urging of the Voyager scientists, commands were generated to try to turn the platform back in the direction of Saturn. Project management agreed to the transmission of the commands and a unanimous sigh of relief occurred as Saturn reappeared in the NA camera field-of-view about 2.8 days after closest approach (Figure 6.27).

One other important objective was dependent on the ability to point the ISS cameras. About 10 days after Saturn closest approach, Voyager 2 would pass within 2 million km of Phoebe, Saturn's outermost satellite. A special sequence of commands was generated to obtain images of Phoebe during the close passage, and that sequence also executed successfully. Following the Phoebe imaging, the only platform slewing allowed for more than a year was that required to analyze the problem and to deduce the failure mechanism. The prognosis was guarded, but encouraging. Extensive high-rate slewing had driven lubricant out of the critical parts of the platform gears, and galling of the gear shafts had caused them to

Figure 6.27. This dim image of Saturn about 2.8 days after closest approach served as verification that Voyager 2's scan platform was again responding to commands from Earth. (P-23969)

seize. Temperature cycling had squeezed the shafts sufficiently to create some clearance and free the gears; the migration of lubricant back into the interface between the gear and the shaft also helped. Further use of high-rate slewing was prohibited, and the platform operated flawlessly from 1983 until it was shut down in 1998.

When asked what percentage of its anticipated science Voyager 2 returned prior to its platform difficulties, Project Scientist Stone quickly replied, 'About 200%!' It is true that some pictures of Tethys and some darkside Saturn data were not obtained, but what was obtained far exceeded the expectations of all associated with the Voyager Project. Analysis of the data is likely to continue for decades. Formal Voyager papers on the Saturn results are contained in the following dedicated issues or large groupings of scientific papers in *Science, Nature, Icarus,* and the *Journal of Geophysical Research* [3]. The University of Arizona published, *Saturn* [4] and *Planetary Rings* [5], which draw heavily on the Voyager results at Saturn. Morrison [6] was once again instrumental in documenting the highlights of the scientific findings, summarized below.

6.2.1   Saturn's atmosphere

1. As was the case with Jupiter, Saturn's atmosphere above the cloud tops is composed almost exclusively of hydrogen and helium. There are $22\pm4\,g$ of

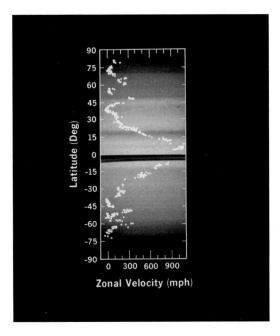

Figure 6.28. Zonal winds on Saturn showed a greater degree of north–south symmetry than Jupiter. (260-1529B)

helium for every 100 g of hydrogen, much closer to solar abundance than the original estimate of 6±5 g.

2. Excess thermal energy is radiated by Saturn at a rate of 1.82±0.09 of that received from the Sun, probably due to the gravitational separation of helium and hydrogen in the interior.

3. Oval cloud features with clockwise rotation in the northern hemisphere and counterclockwise rotation in the southern hemisphere are limited to latitudes poleward of ±45°.

4. Maximum and minimum zonal wind speeds are located in the middle, rather than at the edges, of the dark belts and bright zones.

5. A definite north–south symmetry of the zonal winds was observed (Figure 6.28). This symmetry may imply deep circulation within the atmosphere and perhaps differentially rotating concentric cylinders whose axes are parallel to the rotation axis of Saturn.

6. As at Jupiter, Saturn has a temperature inversion in its stratosphere. The temperature rises from a minimum of approximately 80 K near the 100-mbar level to 140 K near 10 mbar.

7. Ultraviolet light is emitted from the polar regions. These auroral emissions are associated with charged particles from the solar wind spiraling down magnetic field lines and striking atoms in the upper atmosphere.

8. Ultraviolet emissions are also seen at lower latitudes. Since these are seen only in

Figure 6.29. The extremely complex radial structure within Saturn's B ring was one of the great surprises of the Saturn encounters of Voyager. (P-23946)

   sunlit portions of the atmosphere, they have been termed 'dayglow', and their precise source is still a matter of debate.
9. Large scale heights for hydrogen lead to the conclusion that temperatures in the extreme upper atmosphere reach values of 600 to 800 K.

6.2.2   Saturn's rings

1. Saturn possesses a ring system with enormously complex radial structure and few empty gaps, even at a scale of less than a few kilometers (Figure 6.29).
2. The D ring (Figure 6.30), first proposed by Earth-based observers but too faint to have been seen by them, was found to extend from the inner edge of the C ring to about 3,200 km above the cloud tops.
3. The first images of the G ring (Figure 6.31), proposed on the basis of charged particle absorptions observed by Pioneer 12, were obtained. The ring lies between the orbits of Janus and Mimas.
4. Structure which has the appearance of braiding was seen in the narrow F ring (Figure 6.32). This is probably attributable to complex interactions between the ring particles and the sheparding satellites Prometheus and Pandora (Figure 6.33).
5. Both elliptical and discontinuous rings were found within gaps in the main ring system, generally at distances associated with strong satellite orbital resonances.

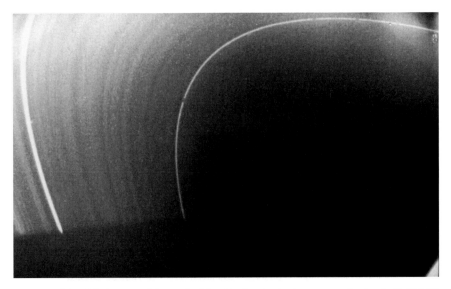

Figure 6.30. This image of Saturn's faint D ring is the best ever obtained. (P-23967)

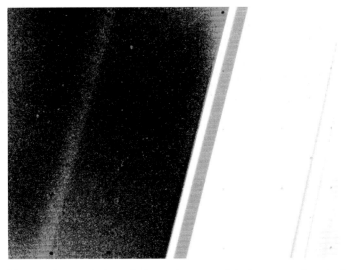

Figure 6.31. Voyager 2 passed through Saturn's equatorial plane very close to the narrow G ring. (P-23968)

6. The outer edge of the B ring is a centered ellipse (Figure 6.34) which rotates at a rate that keeps its short axis pointed approximately at Mimas.
7. Spiral density waves (Figure 6.35) and spiral bending waves also attest to gravitational interactions of the ring particles with Saturn's satellites.
8. The outer edge of the A ring is less than 10 m thick; the main rings of Saturn,

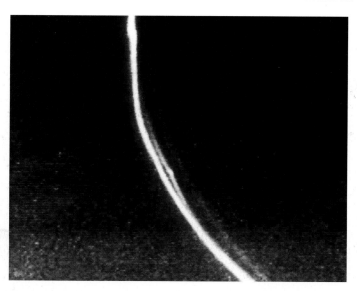

Figure 6.32. A complex structure which has the appearance of braiding was seen in Saturn's narrow F ring. (P-23099)

Figure 6.33. Prometheus and Pandora exert a 'shepherding' influence on the particles of Saturn's F ring. (260-1135A)

except for the 'corrugations' associated with the bending waves, may be uniformly thin.

9. Typical ring particle sizes range from a few micrometers to tens of meters or larger, and show radially varying size distributions.

10. Radial 'spokes' of micrometer-sized particles were seen in the outer half of the B ring (Figure 6.36). Their main periodicity is identical to the 10.657-h rotation period of Saturn's magnetic field, implying that the spokes are electrically charged and driven in part by the magnetic field.

6.2.3　Saturn's satellites

1. Four satellites (Pan, Atlas, Prometheus, and Pandora) were discovered in Voyager imaging.

2. The sizes of all 18 satellites of Saturn known prior to 2000 were first determined by Voyager imaging, leading to improved density estimates for those whose masses were known. (As of the end of 2000, Saturn has 30 known satellites.)

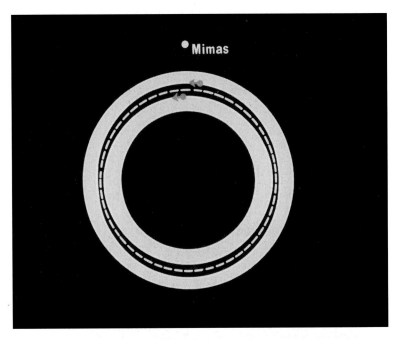

Figure 6.34. Gravitational interaction with Mimas causes particles near the outer edge of the B ring to form a 'centered' ellipse. (260-1450)

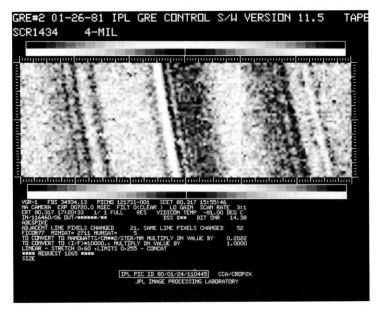

Figure 6.35. Spiral density waves within the A ring are also a result of interaction with some of the satellites of Saturn. (260-1135B)

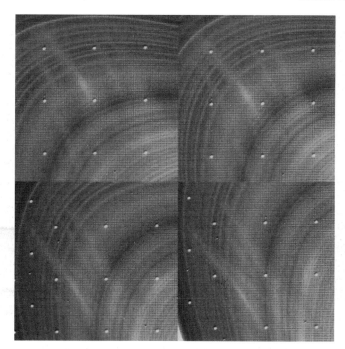

Figure 6.36. Radial 'spokes' were seen in the outer half of the B ring. Their source may be interaction of tiny ring particles whose electrostatic charges allow them to be influenced by Saturn's magnetic field. (260-1499)

3. More accurate determination of the masses of Tethys, Rhea, Titan, and Iapetus were obtained, leading in addition to a revised estimate for the masses of Mimas and Enceladus.

4. An enormous crater (named Herschel), with a diameter of about 130 km, was discovered on Mimas (Figure 6.37). Herschel, named after Sir William Herschel, is fully one-third the diameter of Mimas itself.

5. The surface of Enceladus (Figure 6.38) has undergone more recent geologic alterations than any other Saturn satellite surface seen by Voyager.

6. Tethys (Figure 6.39) also has a large 400-km diameter crater, Odysseus, and a canyon, Ithaca Chasma, which girds two-thirds of the satellite's circumference.

7. Wispy terrain and sharp albedo contrasts are characteristic of the surface of Dione (Figure 6.40). Extensive surface fracturing and horizontally variable crater size distributions are also seen.

8. Bright, wispy terrain is imbedded in an otherwise dark trailing hemisphere of Rhea (Figure 6.41). Fractures and variations in crater size distribution are also seen here.

9. Titan's diameter is 5,150 ± 4 km, making it second in size to Ganymede among Solar System satellites.

Figure 6.37. Mimas may have been very nearly broken apart by the impact that caused Herschel Crater. (P-23210)

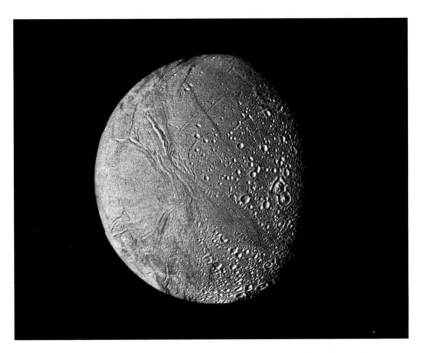

Figure 6.38. Flow lines, fractures, and reduced numbers of craters on Enceladus testify to geologically recent surface alteration on this small Saturnian satellite. (P-23956)

Figure 6.39. Ithaca Chasma very nearly encircles Tethys. It may be causally related to the 400-km diameter Odysseus Crater. (P-23948)

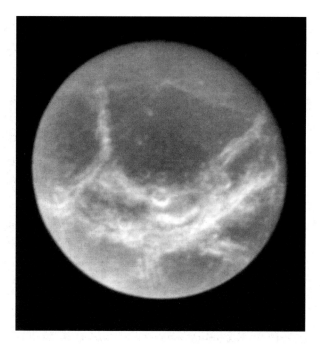

Figure 6.40. Dione, although close in size to Tethys, has wispy markings on its surface unlike anything seen on Tethys. (P-23269)

Figure 6.41. Voyager 1 passed relatively close to Rhea and obtained this mosaic of its north polar region. (P-23177)

10. Titan has a near-surface atmospheric pressure and temperature of 1.6 bars and $95 \pm 1\,K$, respectively. Figure 6.42 shows a comparison of the atmospheres of Titan and Earth.

11. Titan's atmosphere is at least 90% nitrogen (N_2); nitrogen and methane (CH_4) are the main constituents. Trace amounts of acetylene (C_2H_2), ethylene (C_2H_4), ethane (C_2H_6), methyl acetylene (C_3H_4), propane (C_3H_8), hydrogen cyanide (HCN), cyanoacetylene (HC_3N), cyanogen (C_2N_2), carbon monoxide (CO), and carbon dioxide (CO_2) were also detected.

12. There are strong indications that large portions of Titan's surface may be covered with a liquid ethane (C_2H_6) ocean, combined with 25% liquid methane (CH_4), and 5% liquid nitrogen (N_2).

13. The main haze layer (Figure 6.43) extends to 200 km above Titan's surface. Detached haze layers extend upward at least another 500 km. These haze layers made it impossible for Voyager to obtain images of Titan's surface.

14. Particles in Titan's main haze layer have a mean diameter of 1 µm, but they are probably non-spherical or consist of layers of particles with different sizes.

15. Hyperion (Figure 6.44) possesses a highly irregular, ancient surface. Its surface is relatively dark, and gravitational interaction with Titan causes Hyperion to tumble chaotically.

16. Iapetus (Figure 6.45) has an extremely dark surface on its leading hemisphere. Its edges are sharply defined and no craters are visible in the interior of the dark region. The source of this dark material is still a matter of debate.

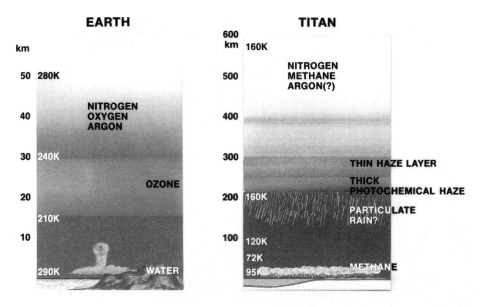

Figure 6.42. Many similarities exist between the atmospheres of Titan and Earth. Because of Titan's lower gravity, its atmosphere is much more extended than Earth's. (260-1526A)

Figure 6.43. A main haze layer obscures the surface of Titan. Detached haze layers above the main haze layer are also apparent. (P-23107)

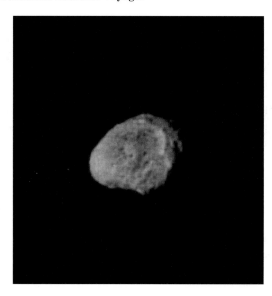

Figure 6.44. Hyperion is remarkably non-spherical for a satellite comparable in size to Mimas. Its elongated shape interacts gravitationally with Titan to cause Hyperion's spin rate and orientation to change dramatically and chaotically. (P-23936)

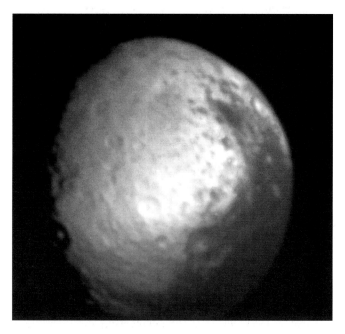

Figure 6.45. Earth-based observations of Iapetus showed that one side of this strange satellite is six times as bright as the other side. (P-23961)

Figure 6.46. Voyager 2 passed within several million kilometers of Phoebe, whose darkness and unusual orbit lead astronomers to believe that it is a captured asteroid rather than a primordial Saturn satellite. (P-24137)

17. Phoebe (Figure 6.46) has a 9-h rotation period. (Its orbital period was previously known to be 550 days.) By its dark surface and inclined, retrograde orbit, Phoebe is probably a captured asteroid.

6.2.4 Saturn's magnetosphere

1. Kilometric radio wave radiation is pulsed from Saturn at intervals of 10.657 h, which is presumably the rotation period of Saturn's deep interior and its magnetic field.
2. Saturn's magnetic field (Figure 6.47) is basically dipolar, with a surface field strength of 0.21 Oe and a tilt of less than 1° from the rotation axis.
3. The sunward magnetopause extends about 22 Saturn radii outward from the center of the planet. It undergoes variations from this value of about ±40% due to the solar wind variations.
4. An inner torus (inside the orbit of Rhea), with a population of singly charged hydrogen and oxygen ions, probably originates from the sputtering of water-ice from the surfaces of Dione and Tethys.
5. A hot ion region, with a temperature of 30 to 50 keV (which correspond to a temperature of about 500×10^6 K), exists near the outer edge of the inner torus.
6. A thick plasma sheet of ions of hydrogen, helium, carbon, and oxygen extends outward nearly to the orbit of Titan.

6.3 'URANUS THE FUZZY BLUE TENNIS BALL'

Permission from NASA and funding from the United States Congress were the major political hurdles to be cleared for Voyager 2 to have an extended mission to Uranus (see Figure 6.48). By the start of the Voyager Uranus/Interstellar Mission (VUIM) on October 1, 1981, many of the engineering and scientific problems associated with such a mission had only begun to be addressed. Contrary to popular belief, Voyager 2 did not remain idle during the four-year period between the start of

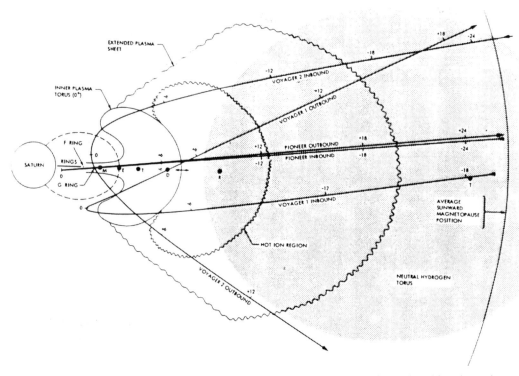

Figure 6.47. Saturn's magnetic field is closely aligned with its rotation axis. This schematic diagram shows the approximate sunward cross-sectional structure within Saturn's magnetosphere.

VUIM and the beginning of the Uranus encounter. Like Voyager 1, the spacecraft continued to transmit useful scientific and engineering data 24 hours a day. A relative dearth of other deep-space probes with competing requirements made it possible to track the two spacecraft with the DSN antennas and collect their data for an average of 12 to 16 hours a day. Some of the data was designed for post-Saturn calibrations, but the majority was either unique scientific information or the results of engineering tests being done in preparation for the Uranus encounter.

Funds for the newly approved and unplanned mission were scarce. NASA-imposed funding constraints dictated that the operating costs for the 'cruise' portion of VUIM should be substantially reduced. As a consequence, a large fraction of the Voyager Project personnel moved on to other projects. The remaining staff were kept inordinately busy conducting cruise science, long-range planning for the Uranus encounter, and engineering activities to enhance the spacecraft's capabilities needed for a successful rendezvous with Uranus.

The increasing Earth–spacecraft distance required the spacecraft's telemetry rate to be reduced to enable the DSN to decode the spacecraft's weak signal into valuable data. To compensate for this reduced rate, the FDS computers were reprogrammed to operate in a parallel mode to be able to perform onboard data processing. This

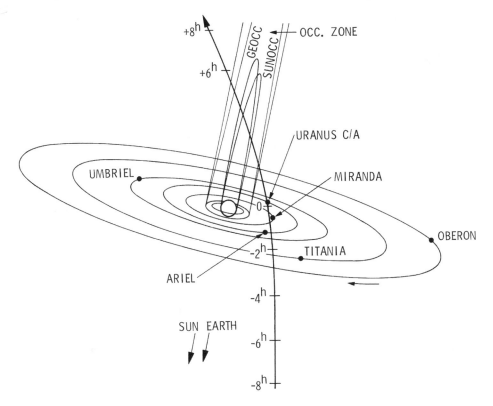

Figure 6.48. A trajectory-plane view of the path of Voyager 2 through the Uranus system. Voyager 2's closest approach to Uranus occurred on January 24, 1986. (260-1798)

allowed the data to be compressed such that fewer bits were required to be transmitted without losing any of the original information. In addition, as at Saturn, the CCS computers were used in a 'non-redundant' mode, offering the advantage of an almost doubled memory space, which could be used to carry out a more complex sequence or to operate the spacecraft for a longer period of time before reprogramming was needed. The 'cost' of this capability was that the encounter would be performed on 'single string', i.e., no backup spacecraft hardware was available in the event of a failure.

The greater distance from the Sun produced light levels that were 369 times dimmer than at Earth. The lower light levels entailed longer exposures for the images, which in turn would have greater smear due to the spacecraft's motion. To make up for this, the spacecraft was reprogrammed to provide smaller and gentler thruster firings; and the gyroscopes were modified to turn the spacecraft slowly and accurately to follow a target (known as Image Motion Compensation (IMC)). Both techniques allowed the spacecraft to keep the ISS cameras on target during the longer exposures that were now required.

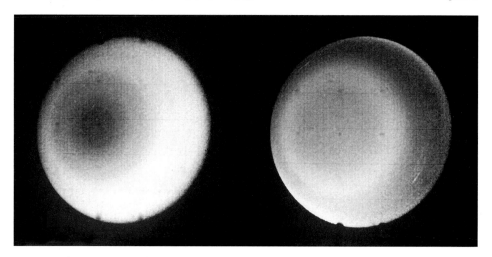

Figure 6.49. Violet, orange, and methane (red) images of Uranus. Image contrast of each has been enhanced to bring out details. The differences, although exaggerated by the contrast stretch, are nevertheless indicative of latitudinal variations in the cloud and haze structure. (P-29517)

The mission planners realized that DSN antennas could also be modified to improve their signal-capturing ability. The alteration required engineers to 'array' antennas at a given DSN complex together, which permitted the signals from Voyager 2 to be simultaneously received by two or more antennas. These signals were then electronically synchronized and combined to achieve a single, stronger signal. In effect, the two or more antennas acted as a single larger antenna.

Although the option to preserve a Uranus mission existed from the very beginning, the demands on the spacecraft were immense. Mission planners needed the spacecraft to last five more years and travel another 1.6 billion km further from the Sun. Fortunately, the Voyager 2 encounter with Uranus was a spectacular success. Information from this encounter will probably be the only close-range spacecraft data for at least the next two decades. Formal Voyager papers on the Uranus results are contained in the following dedicated issues or large groupings of scientific papers in *Science, Nature, Icarus*, and the *Journal of Geophysical Research* [7]. The University of Arizona has published *Uranus* [8], which draws heavily on Voyager results. A summary of those findings is given below.

6.3.1 Uranus's atmosphere

1. A Uranian cloud deck composed of methane was found near the 1.2-bar pressure level (Figure 6.49).
2. The atmosphere above the cloud tops of Uranus is composed of hydrogen and helium. There are 26.2 ± 4.8 g of helium for every 100 g of hydrogen.

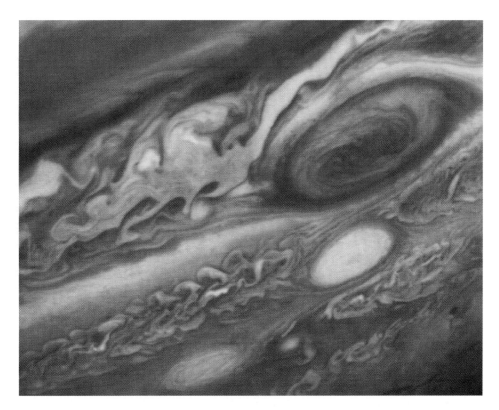

Plate 1. The Great Red Spot, an anti-cyclonic storm which has been observed in Jupiter's southern hemisphere for nearly 400 years. (P-21742)

Plate 2. Io is the innermost of the four Galilean satellites of Jupiter. (P-34590)

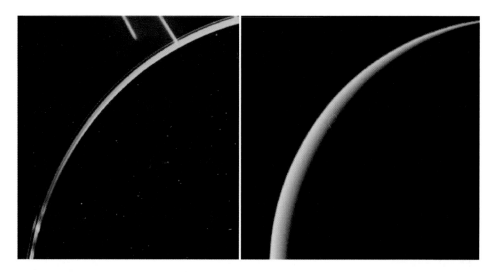

Plate 3. The image on the left was taken by Voyager 2 during the Jupiter encounter and was encoded with the Golay encoder. The image on the right was taken by Voyager 2 during the Uranus encounter with the Reed–Solomon encoder. Notice the absence of 'static' in the Uranus image.

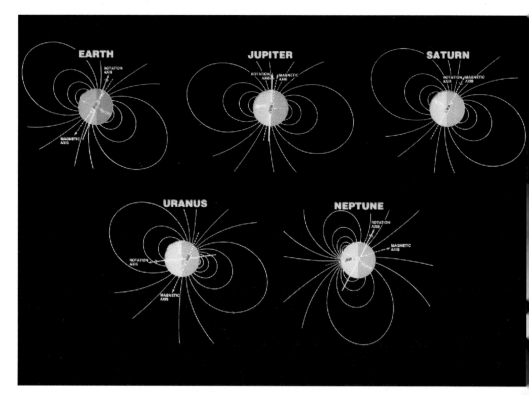

Plate 4. Schematic comparison of the magnetic field orientations and offsets for Earth and the giant planets.

Plate 5. Full-disk view of Neptune's atmosphere, including the Great Dark Spot. (P-34611)

ate 6. Members of the Voyager Science Steering Group and other Voyager Project Management personnel JPL during the Neptune encounter preparations. Pictured, from left to right, are 'Tom' Krimigis (LECP PI), hn Belcher (PLS PI), Leonard Tyler (RSS TL), Pieter de Vries (FSO Mgr), Don Gurnett (PWS PI), Norman ess (MAG PI), Lyle Broadfoot (UVS PI), Norm Haynes (Proj. Mgr), James Warwick (PRA PI), Edward one (Proj. Sci. and CRS PI), Bradford Smith (ISS TL), Rudolph Hanel (former IRIS PI), Barney Conrath RIS PI), and Herbert Bridge (former PLS PI). Absent were Lonne Lane (PPS PI) and Ellis Miner (APS).

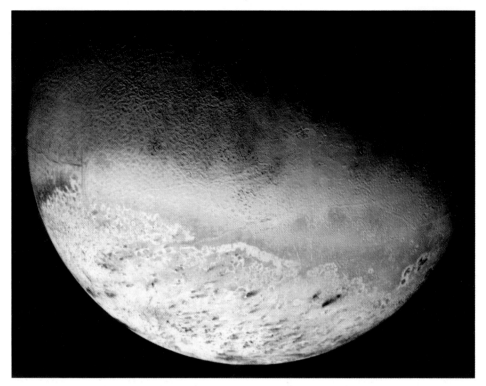

Plate 7. Photomosaic of Triton, taken during final approach to the satellite. The variety of features seen in this image is large, showing an extensive polar cap, dark wind streaks, cantaloupe terrain, large fractures, smooth terrain, sparse cratering, etc. (P-35317)

Plate 8. This artist drawing is taken from the perspective of being on Triton looking up towards Neptune. Notice that the surface was drawn with puddles of some kind of liquid, which, if they did exist, could be liquid nitrogen.

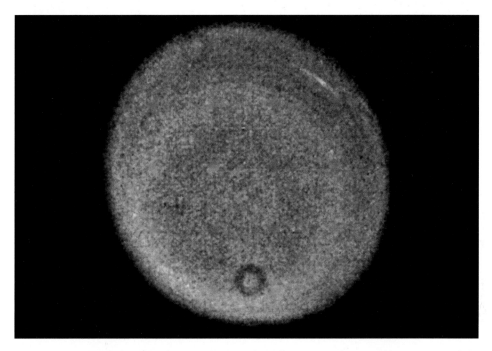

Figure 6.50. One of the eight detected discrete cloud features used to provide measures of the wind speeds is shown near the upper limb in this Voyager 2 image of Uranus. Contrast has been enhanced to bring out low-contrast details. The dark circular marks are due to dust specks in the camera optics. (P-29468)

3. A methane-to-hydrogen ratio near the cloud base was 20 times the carbon abundance seen in the Sun.
4. Unlike Jupiter and Saturn, Uranus has a near balance between internal thermal energy radiated by the planet and the solar energy received by the planet. This implies that the internal sources of heat contribute to less than 12% of the total.
5. Only a few discrete clouds were seen in the ISS images. From these, prograde zonal wind speeds were measured at 0 m/s near 20° south latitude to about 200 m/s near 60° south latitude, and back to 0 m/s at the south pole (Figure 6.50).
6. RSS data indicate that wind speeds at the equator were retrograde, with speeds near 100 m/s (Figure 6.51).
7. A temperature inversion exists in Uranus's stratosphere. The temperature rises from 52 K at the 0.1-bar level to 70 K near the 0.001-bar level.
8. Uranus has a very large hydrogen scale height in its extreme upper atmosphere. Its temperature is approximately 800 K, and the statosphere produces significant drag on the ring particles.
9. The UVS investigation revealed that the planet gives off ultraviolet emissions from the sunlit side of its extended atmosphere. This phenomenon is called 'electroglow' and is poorly understood.

VIEW FROM EARTH

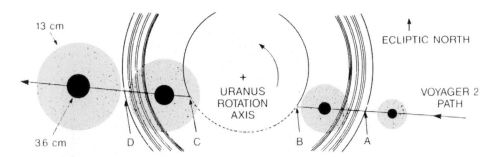

Figure 6.51. The path of Voyager 2 behind Uranus as viewed from Earth tracking stations. Radio science data for the occultation period are displayed for the same period. Abrupt dips in the intensity of the x-band signal occur as the radio signal passes through the methane cloud layer near the 1.3-bar pressure level.

Figure 6.52. This image of the ring system was taken the day before the closest approach to Uranus from a distance of $1,120 \times 10^3$ km. A previously unknown narrow ring, now known as the λ ring, is barely visible between the outer two rings (the Delta and Epsilon rings, which were discovered from Earth). Note the relatively even spacing of three inner rings (rings 4, 5, and 6). (P-29507)

6.3.2   Uranus's rings

1. During the Voyager 2 encounter, the previously known narrow rings of Uranus were imaged for the first time.
2. Two additional rings (Lambda and 1986U2R) were found (Figure 6.52). The first was narrow and located between the orbits of the Epsilon and Delta rings;

Figure 6.53. During passage through Uranus's shadow, this wide-angle 96-s exposure of the ring region was shuttered. Although the rings previously seen are identifiable, the λ ring (labeled 1986U1R) is much brighter than the rest, and most of the features are not seen in any other images or non-imaging datasets. (260-1776)

and the second was broad and closer to the planet than any other ring system around any other known planet; it has not yet received an official IAU designation.

3. Only one long-exposure image, taken at high phase angle, was able to discern the myriad of faint dust structures embedded in the rings (Figure 6.53).
4. The Epsilon ring appears to be devoid of particles in the 1-to-10-cm size range.
5. Ring particles have an extremely low albedo. This may result from methane-ice bombardment by high-energy protons found in Uranus's magnetosphere.
6. Voyager 2 recorded 40 particle hits per second as it traversed through the planet's ring plane.

6.3.3   Uranus's satellites

1. During the Voyager 2 encounter, ten new satellites were discovered (Cordelia, Ophelia, Bianca, Juliet, Desdemona, Rosiland, Portia, Cressida, Belinda, and Puck) orbiting Uranus between the rings and the orbit of Miranda (Figures 6.54 and 6.55). Their diameters ranged from 26 to 154 km.

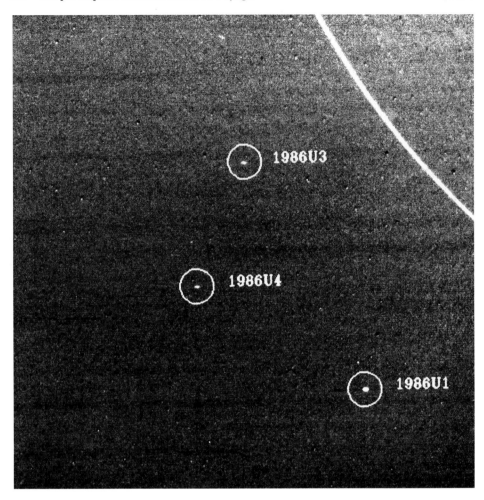

Figure 6.54. On January 18, 1986, Portia (1986U1), Juliet (1986U3), and Cressida (1986U4) were all captured in a single image which also included the outer ring of Uranus. (P-29465)

2. The disks of the previously known satellites, Cordelia and Puck, were resolved.
3. The geometric albedos for the six of the largest satellites were determined. The range was from 0.07 for Puck to 0.40 for Ariel.
4. Puck had a remarkably spherical shape for a tiny satellite, with indications of impact cratering on its surface (Figure 6.56).
5. Very little evidence of geologic activity was found on Oberon (Figure 6.57). A large mountain on its limb was seen which is possibly a central peak within an impact crater.

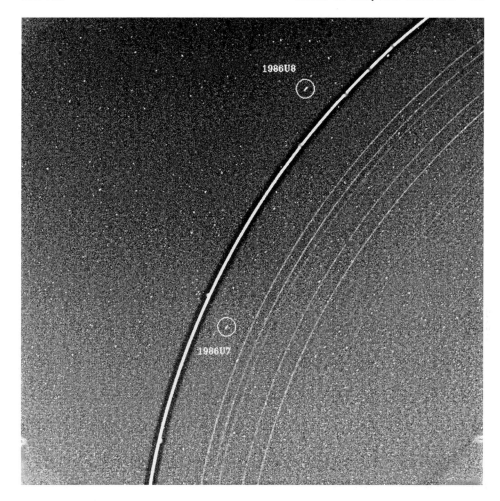

Figure 6.55. Cordelia (1986U7) and Orphelia (1986U8) are believed to be responsible for the radial confinement of Epsilon ring particles. The two were captured in a single image on January 21, 1986. The nine 'classical' rings of Uranus are also apparent. P-29466

6. Titania was verified as the largest Uranian satellite (Figure 6.58) and a surprisingly large number of fractures were found across its surface.
7. A uniformly darkened (albedo of about 0.19) surface of unknown origin for Umbriel was found, with few bright features (Figure 6.59).
8. Ariel had indications of ice flow and multiple fractures across its surface (Figure 6.60).
9. Miranda had trapezoidal regions ('coronae') unlike any seen elsewhere in the Solar System. Their origin may have resulted from incomplete differentiation in its interior (Figure 6.61).

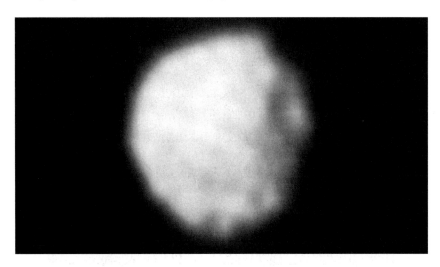

Figure 6.56. This image of Puck (1985U1) replaced a previously planned image of Miranda. Voyager 2 was 500,000 km from Puck and the resolution was about 10 km. (P-29519)

Figure 6.57. This highest resolution image of Oberon was shuttered from a distance of 660,000 km and shows features as small as 12 km. Oblique lighting enhances the vertical detail in regions near the terminator. Several craters are seen to have darkened floors. The linear rays near the bottom appear to be radiating from a common source. A mountain at the lower right edge of the disk extends 11 km above the surrounding terrain. (P-29501)

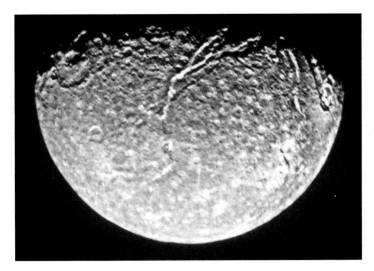

Figure 6.58. Voyager 2's highest-resolution image of Titania emphasizes the enormous fractures and a relatively low average crater diameter. Features as small as 6.8 km are resolved. Two large multi-ringed craters are seen near the terminator. Three separate concentric rings define Gertrude, the crater at the left. Ursula, on the right, appears to have two concentric structures. (P-29522)

Figure 6.59. The highest resolution coverage of Umbriel resolves features as small as 10.3 km. The bright region at the left is seen to be deposits on the floor of Wunda Crater. A dark, relatively linear feature connects the central peak of Wunda with its southern rim. To the upper right of Wunda, the central peak in Vuver Crater is also relatively bright, as are some of the scarps near the right edge of the disk. The distribution of crater sizes is very similar to that of Oberon. (P-29521)

Figure 6.60. Four images used to produce this view of Ariel, which has a resolution of about 2.4 km. The highly fractured nature of Ariel's surface is emphasized by the oblique solar illumination. (P-29520)

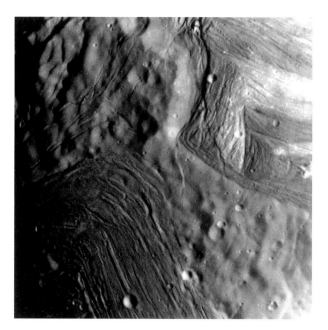

Figure 6.61. The sharpness of the coronae boundaries is emphasized in this view of Inverness Corona (upper right) and Elsinore Corona (lower left). Elsinore Corona has the same quasi-rectangular shape displayed by Inverness and Arden, but it has no apparent broad trench surrounding it, no large brightness contrasts in its interior, and consists of a band of parallel ridges surrounding an older jumbled interior. Features as small as 600 m may be seen. (P-29515)

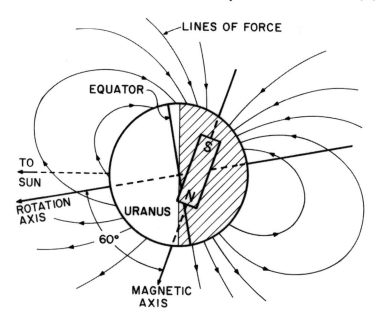

Figure 6.62. The simplest representation of the magnetic field is a dipole tilted 58.6° from the rotation axis and offset 0.3 Uranian radii toward the dark north polar region. (260-1829B)

6.3.4   Uranus's magnetosphere

1. Left-handed polarized signals with a periodicity of 17.24 ± 0.01 h were found, which is assumed to be the planet's internal rotational period.
2. The magnetic field was tilted 58.6° with respect to the planet's rotational axis and located about 0.3 radii away from its center (Figure 6.62).
3. The magnetosphere extended sunward 18 Uranus radii.
4. The magnetospheric plasma was dominated by protons, with few heavier species.
5. The most complex interaction was between the highly tilted radiation field and the larger satellites.

6.4   THE POST-URANUS MISSION OF VOYAGER

Many of the science and engineering calibrations in support of an encounter are performed during the early cruise period following an encounter. Following the Uranus encounter only a minimal amount of such activity occurred because of another Solar System object: Comet Halley. The encounter period of the European Space Agency's Giotto spacecraft with Comet Halley began less than a week after the end of the Uranus Post Encounter (PE) for Voyager. Since both spacecraft were in the same general direction in the sky there was competition for

64-m tracking stations, and in such circumstances a spacecraft in its encounter period is given precedence. To complicate matters, the Japanese Suisei and Sakigake spacecraft and the Russian Vega 1 and Vega 2 spacecraft were also nearing their encounters with Comet Halley. Voyager 1 and Pioneers 10, 11, and 12 were not in critical mission phases; they had been almost ignored during the four-month Voyager 2 encounter, and needed DSN tracking coverage as well. It was a realization of these factors that led to the attempt to complete as many of the calibrations as possible prior to the end of PE, unless their nature would permit a delay of several months.

One lesson learned from the Uranus planning experiences was that it was unwise to cut the staffing levels so low that extensive retraining would be needed when the staffing buildup occurred. Neptune preparations were planned as a more relaxed, long-term effort which would enable a relatively large fraction of the experienced personnel to be retained through the long cruise period between Uranus and Neptune encounters. The Neptune planning started very shortly after the end of the Uranus encounter.

NOTES AND REFERENCES

1. Voyager 1 Jupiter results are outlined in (1979) *Science*, **204**, 945–1008; (1979) *Nature*, **280**, 725–806; and (1980) *Geophysical Research Letters*, **7**, 1–68. Voyager 2 Jupiter results are outlined in (1979) *Science*, **206**, 925–96. Combined Voyager Jupiter results are contained in *Journal of Geophysical Research*, **85**, 8123–8841.

2. Morrison, D. and Samz, J. (1980) *Voyage to Jupiter*. NASA Special Publication #439, National Aeronautics and Space Administration, Scientific and Technical Information Branch, pp. 199.

3. Voyager 1 Saturn results are outlined in (1981) *Science*, **212**, 159–243, and (1981) *Nature*, **292**, 675–755. Voyager 2 Saturn results are outlined in (1982) *Science*, **215**, 499–594. Combined Voyager Saturn results are contained in (1983) *Journal of Geophysical Research*, **88**, 8625–9018, in (1983) *Icarus*, **53**, 165–387, and in *Icarus*, **54**, 160–360.

4. Gehrels, T. and Matthews, M. S. (eds) (1984) *Saturn*. The University of Arizona Press, Tucson, 968 pp.

5. Greenberg, R. and Brahic, A. (eds) (1984) *Planetary Rings*. The University of Arizona Press, Tucson, 784 pp.

6. Morrison, D. (1982) *Voyages to Saturn*. NASA Special Publication #451, National Aeronautics and Space Administration, Scientific and Technical Information Branch, pp. 227.

7. Voyager 2 Uranus results are outlined in (1986) *Science*, **233**, 39–109 and in (1987) *Journal of Geophysical Research*, 14,873–15,375.

8. Bergstralh, J. T., Miner, E. D. and Matthews, M. S. (eds) (1991) *Uranus*. University of Arizona Press, Tucson.

BIBLIOGRAPHY

Burns, J. A. (1986) Some background about satellites. In Burns, J. A. and Matthews, M. S. (eds) *Satellites*. The University of Arizona Press, Tucson, pp. 1–38.

Gehrels, T. and Matthews, M. S. (eds) (1984) *Saturn*. The University of Arizona Press, Tucson, 968 pp.

Kohlhase, C. (ed.) (1989) *The Voyager Neptune Travel Guide*. NASA, JPL Publication, pp. 89–24.

Miner, E. D. (1998) *Uranus: the Planet, Rings and Satellites* (2nd edn.), Wiley-Praxis, Chichester, UK.

Morrison, D. (1982) *Voyages to Saturn*. NASA Special Publication #451, National Aeronautics and Space Administration, Scientific and Technical Information Branch, 227 pp.

Morrison, D. and Samz, J. (1980) *Voyages to Jupiter*. NASA Special Publication #439, National Aeronautics and Space Administration, Scientific and Technical Information Branch, 199 pp.

Stone, E. C., Kohlhase, C. E. and the authors listed in references 9 through 19 in Chapter 5 (1977) *Space Science Reviews*, **21**, 75–376. (Contains a series of 13 papers, including a science overview, a mission description, and individual papers on each of the 11 scientific investigations.)

Caltech historical Sketch, http://www.caltech.edu/catalog/geninfo/history.html

7

The Voyager 2 encounter with Neptune

7.1 PREPARING VOYAGER 2 FOR NEPTUNE

It was called the last picture show. Of course there were ten scientific investigations other than the narrow-angle and wide-angle cameras, but, as inquisitive human beings, we use our eyes as our primary source for learning about the grandeur of the universe. Scientists, engineers, reporters, and the general public had waited three years since the Voyager 2 flyby of Uranus and 12 long years since launch for this encounter. Neptune would be the 'last stop' for Voyager 2 before it began its endless quest into the darkness of interstellar space.

Permission from NASA and funding from the US Congress to support Voyager 2 and its Uranus/Interstellar Mission were approved in 1981, after the spacecraft had completed its highly successful Saturn encounter. The additional success at Uranus raised expectations that the spacecraft could survive the three-and-a-half-year flight to Neptune. The shorter cruise duration between Uranus and Neptune, compared to that between Saturn and Uranus, meant that there would be no rest for the operations team. As soon as the Uranus encounter was over, scientists and engineers resumed their Neptune planning, which had begun prior to the Uranus encounter.

7.1.1 Detecting Voyager's weak signals

As preparation began for the Neptune encounter, engineers had to face the fact that, as Voyager journeyed further and further away from the Sun, supportable data rates with Earth were getting slower and slower. The rates had to be reduced to allow the receivers at each antenna complex to integrate for a longer period on the weaker signal in order to maintain an acceptable bit error rate. At Jupiter, Voyager could communicate with Earth at a maximum rate of 115,200 bit/s; at Saturn the rates had fallen to 44,800 bit/s; at Uranus 21,600 bit/s was the maximum sustainable rate. At Neptune the rates would be even slower had not changes in procedure been instituted. However, successively lower rates were unacceptable either to the scientists anxious to collect system-defining data or to the engineers anxious to provide capable spacecraft and telemetry link design. Something had to be done to allow Voyager to unload its trove of data at a faster rate, in spite of the greater distance from Earth.

The output power of Voyager's transmitter was limited to 22 watts at X-band and could not be increased. It was possible, on the other hand, to do something about the sensitivity of the Earth-based receiving antennas. With the approval of the Voyager Neptune mission, the Deep Space Network (DSN), composed of antennas near Goldstone (California), Canberra (Australia), and Madrid (Spain) were allocated funds to increase the size of their large 64-m antenna to 70 m (Figure 7.1). The 3-m extension allowed the DSN to begin to compensate for the natural fall-off of the received signal strength from Voyager 2.

Larger antennas were not enough. The Voyager Project once again called on the Australian government to permit the Parkes radio astronomy 64-m antenna to join in the cause. The Commonwealth Scientific and Industrial Research Organization (CSIRO), which operated this antenna, had agreed to the Parkes antenna usage during the Voyager 2 Uranus encounter and had participated in the construction of the 400-km high-quality microwave link needed to electronically combine the signal with NASA's DSN antennas near Canberra, Australia. Australian assets were ideal for Voyager, because the Neptune and Triton encounters would occur over the Australian tracking stations. The approach was to array as many antennas as possible to increase the collecting area for Voyager's extremely weak signal. At best, the 22 watts of radiated power from Neptune would drop to less than one billionth of one billionth of 1 watt collected by a 70-m antenna at Earth. To increase the received signal strength, the Parkes antenna was arrayed with the 70-m and one 34-m antenna at the Canberra DSN complex, thereby nearly doubling the signal strength over that of a lone 70-m station.

Almost 10,000 km north of Australia stood Japan's Usuda tracking antenna, and arrangements were made by the Project to use the Usuda station to supplement the radio science data collection during passage of Voyager 2 through the shadows of Neptune and Triton. Usuda was a 64-m antenna located on the island of Honshu and operated by Institute of Astronautical Science (ISAS). Combination of the Usuda data with that of the Parkes and Canberra antennas was accomplished and took place after the flyby rather than using a real-time link like that between Parkes and Canberra.

Eastward across the Pacific Ocean, three antennas – two 34 m and one 70 m – were arrayed at the DSN complex at Goldstone, California, to improve signal strength at that longitude. In addition, the array included the 27 large antennas (25 m diameter each) of the Very Large Array (VLA) of radio telescopes (Figure 7.2). The VLA is located near Socorro, New Mexico, and is operated by the National Science Foundation's National Radio Astronomy Observatory (NRAO). The VLA alone provided the collecting area of two 70-m antennas. Together, all this receiving capability on each side of the Pacific Ocean was informally thought of as listening to Voyager with the entire 'Pacific basin'.

Figure 7.1. Image of the 70-m antenna at Goldstone, California, as its structure is being increased to support the additional collecting area. (JPL-5944AC)

7.1.2   Reprogramming Voyager's computers

Each of the Voyager spacecraft was built with three pairs of computers, as discussed in Chapter 5. These are the two Computer Command Subsystem (CCS) computers, the two Flight Data Subsystem (FDS) computers, and the two Attitude and Articulation Control Subsystem (AACS) computers.

The two CCS computers issue to other spacecraft subsystems the time-sequential commands that constitute the Voyager sequence of events. At Jupiter, both CCS

Figure 7.2. These VLA antennas were enlisted by Voyager to assist in the Neptune encounter. Normally these antennas are employed to listen to galactic radio sources and were used in the late 1980s to listen for faint radio transmissions from Extra-Terrestrials, if they existed. (P-3769CC)

computers contained the same commands, one acting as prime and the other as backup. At Saturn, Uranus, and Neptune, except for a few critical commands, the CCS computers were used primarily in a non-redundant fashion. This offered the advantage of almost doubling the craft's memory space, which could be used either to carry out a more complex sequence or to operate the spacecraft for a longer period of time before reprogramming was needed. CCS 'loads' operated the spacecraft for periods from two days to six months before new sets of instructions had to be transmitted.

The two FDS computers issue timing pulses to the individual science instruments to control their routine operation modes. The changes made prior to Uranus to the operation mode of the Plasma Spectrometer (PLS) to improve its sensitivity to the lower plasma levels experienced in the outer Solar System was maintained for Neptune. The timing of the PLS modes was also changed to make the PLS less susceptible to interference caused by the stepper motor within the Low Energy Charged Particle (LECP) instrument. The FDS computers also control the formatting of data to be sent back to Earth. A number of different data formats were chosen for the Neptune encounter; data format was selected by CCS command. These new modes utilized the two FDS computers in parallel to accomplish some onboard processing of imaging data, thus reducing the number of transmitted bits without any loss of imaging information. This technique will be described in more detail below. Appropriately, the set of data formats was called the FDS 'dual-processor program'. This program also utilized the Reed–Solomon encoder hardware, first used at Uranus.

Another technique utilized during the Neptune encounter, and first employed by Voyager at Uranus, was a data scheme employing simultaneous real-time imaging and tape-recorder playback. Prior to Uranus, only real-time non-imaging data could be returned with playback data. The dual-processor program permitted complete image readout to the tape recorder in either 48 s or 4.0 min. Complete real-time image readouts could be programmed to take either 4.0 or 8.0 min. One dual-processor mode permitted reading only the top 60% of an image in 2.4 min.

The two AACS computers could not be used simultaneously, because each controls a separate system of attitude control thrusters and associated plumbing. Only the primary AACS computer on Voyager 2 had been used in flight. It had been reprogrammed on a number of occasions to correct systematic errors detected in the scan platform pointing. The reprogramming improved the spacecraft's stability for the purpose of reducing image smear, conserving attitude control gas, and permitting more flexibility in turning the spacecraft around one or more of its three principal axes.

Image data compression

Image Data Compression (IDC) in one of the FDS computers permitted the same amount of real information in the 14.4-kbit/s Neptune data stream as was contained in a 14.4-kbit/s data stream at Uranus. The Neptune 14.4-kbit/s rate, like that at Uranus, permitted transmission of real-time non-imaging science and engineering data at its maximum rate (3.6 kbit/s), plus transmission of one real-time image every 4 min. At Saturn, the full 5.12 million bits per image was transmitted. The dual-processor program for Uranus and Neptune served to inter-compare adjacent picture elements and transmit only the differences, cutting by a factor of almost 3 the total number of bits per image transmitted to Earth.

Reducing image smear

A major complication added by the remoteness of Neptune was the large reduction in light levels available for imaging. The brightness of sunlight at Neptune was a factor of 2.5 lower than that at Uranus, nearly 10 below that at Saturn, about 33 lower than Jupiter, and about 904 below that at Earth! Though Triton is relatively bright, Nereid is one of the darker satellites in the Solar System. The source of its darkness is unknown, but may be the result of the bombardment of methane (CH_4) ice by high-energy protons. Proton bombardment of methane ice results in the release of the hydrogen, leaving a residue of black carbon.

Voyager's orientation in space is maintained by keeping the Sun in its Sun sensor field of view and a reference star in its star tracker field of view. Any deviation of more than $0.05°$ from centered viewing of the Sun or the reference star in their respective sensors will trigger the firing of the appropriate attitude control thruster jet. Motion within the spacecraft attitude 'deadband' is generally relatively slow. At Saturn, for example, this gentle rocking motion averaged only about one-tenth of the angular speed of the hour hand on a clock. Even that was too rapid for the combination of high resolution and long exposures required at Uranus. A 15-s exposure required angular rates a factor of 10 slower to avoid smearing over more than two pixels of the NA camera. Spacecraft engineers again found a solution: shorten, from 10 to 5 ms, the duration of the jet pulses used to stabilize the spacecraft, effectively giving a gentler correction to detected orientation errors. This worked at Uranus, but more modifications would be needed at Neptune and longer exposures were going to be necessary. The Project embarked on a campaign to combat the lower light levels and resulting smear in the imaging. Engineers were able to reduce thruster impulses by 25% over those used during the Uranus encounter. Programmers also altered the FDS software so that the original 0- to 15-s exposures could be expanded to include many longer duration exposures up to 96 s. In addition, the software was modified to allow the Imaging Science cameras to shutter images with durations that were a multiples of 48 s. The result of all this effort was a high percentage of long-exposure images whose clarity was exceptional.

Even at Jupiter and Saturn, tape recorder motion caused smear in the images. To reduce that smear, Voyagers 1 and 2 were programmed to fire the yaw [1] thrusters to counteract the torque caused by starting or stopping the recorder. The routine was called DSSCAN, for Digital Storage Subsystem motion CANcellation. Although DSSCAN stopped the major effects of tape-recorder starts and stops, yaw-thruster firing results in small motions in the pitch axis as well. For the short exposures used at Jupiter and Saturn, this added motion was negligible, but the longer Uranus and Neptune system exposures made even second-order effects important. DSSCAN was reprogrammed prior to Uranus to include pitch-axis compensation.

At Saturn, a new technique called Image Motion Compensation (IMC) was first attempted. The spacecraft had been built to reference its orientation to internal motor-driven gyroscopes at times when celestial references could not be used (during solar occultation periods or at times when roll, pitch, or yaw maneuvers were being executed). Voyager software made it possible to compensate for slow drifts in the orientation of the gyroscopes. SCT engineers determined that the spacecraft could be turned slowly and accurately (rates up to 110° per hour in each axis) by offsetting the gyroscope drift rate factors in the software from their actual drift rates. In essence this tricked the software into believing that the gyroscopes were drifting more rapidly than normal. Such Gyro Drift Turns (GDTs) could be used to turn the entire spacecraft to pan the cameras and to compensate for apparent angular motion of a nearby satellite. During this process, images and other data would be recorded on the digital tape recorder for later playback. IMC was used for one of Voyager 1's observations of Saturn's satellite Rhea. During Voyager 2's flyby of Uranus, IMC was used for eight separate observations, one of which (Miranda closest approach imaging) utilized nine separate GDT rates.

Table 7.1. Observations which required Image Motion Compensation

Observation	Link name	Start time (enc. rel.)	Type of IMC
Ring retargetable, ISS	VRRET1	−07:16.8	NIMC
Ring retargetable, ISS	VRRET1	−06:40.8	NIMC
Ring retargetable, ISS	VRRET1	−03:12.8	NIMC
Triton longitude coverage, ISS	VTLON	−02:55.2	NIMC
High phase ring observation, ISS	VRHIPHAS	+00:40.0	MIMC
Ring plane crossing, ISS	VRING$_2$	+01:15.2	MIMC
Neptune atmosphere structure, ISS	VPHAZE	+01:38.4	IMC
Triton best color, ISS	VTCOLOR	+01:50.4	IMC
Triton map, ISS	VTMAP	+02:34.4	NIMC
Triton atmosphere, IRIS	RTATM	+04:16.8	MIMC
Triton high resolution, ISS	VTERM	+04:40.0	IMC

At Neptune, IMC was expanded to include a new technique known as Nodding Image Motion Compensation (NIMC). NIMC turned the spacecraft slowly to track the target motion during image shuttering, returning to Earth point during image readout; during NIMC the spacecraft High Gain Antenna (HGA) would remain close enough to Earth-point to avoid interruption of the data transmission to Earth. This technique, first used at Neptune, permitted IMC without requiring valuable space on the Digital Tape Recorder (DTR). The slight motion followed by a repositioning of the HGA, repeated over and over, gave the appearance that the craft was nodding back and forth. During image readout, the scan platform on which the imaging cameras were located would be re-targeted for the next frame to be shuttered.

Finally, engineers had one last trick to teach Voyager 2. For rapidly moving targets, the scan platform was commanded to move as slowly as possible in the elevation direction. This motion, although somewhat jerky, would be used to remove some of the smear from images that would otherwise have been useless. IMC of this type was called Maneuverless Image Motion Compensation, or MIMC (see Table 7.1).

Encoding the data

Although the DSN was generally able to collect almost flawless Voyager data, the quality of the received data was adversely affected by many conditions. These included poor weather conditions at the tracking stations, changes caused by Earth's ionosphere or solar wind conditions, or data transmissions which had to pass close to the Sun in transit to Earth. Encoding of the data prior to its transmission from the spacecraft helped to overcome this problem and permitted errors in the data to be detected and corrected. Such encoding adds extra bits to the data stream, often in amounts comparable to the intrinsic data prior to encoding. At Jupiter and

Saturn, for example, the encoding scheme essentially doubled the required bit rate for non-imaging data.

To conserve bits during the Neptune encounter, Voyager's Reed–Solomon data encoder hardware was used (as it had been during the Uranus encounter). Prior to Uranus, one of two redundant 'Golay' encoders had been used. As long as it functioned properly, the single Reed–Solomon encoder could cut by more than 50% the number of code bits needed for data streams of 14.4 kbit/s or less. In its initial flight testing, the Reed–Solomon hardware functioned flawlessly.

Before the Uranus dual-processor FDS program was used, imaging data had always been transmitted without either Golay or Reed–Solomon encoding. Infrequent bit errors would result in lighter or darker pixels, comparable to 'snow' in a television picture of substandard quality. Bit errors in the compressed IDC images at Neptune, as was the case at Uranus, could potentially affect entire lines of imaging and were therefore more serious than corresponding errors at Jupiter or Saturn. For this reason, whether transmitted in real time or played off the tape recorder at a later time, IDC images required the imposition of Reed–Solomon coding. The Reed–Solomon encoder could not handle data rates higher than about 14.4 kbit/s. The combined playback/imaging data format required a data rate of 21.6 kbit/s, so only a portion of the data could be encoded. Since the recorded non-imaging data was Golay encoded prior to recording, the logical choice was to Reed–Solomon encode only the real-time data. Recorded IDC imaging data could then be saved for playback at lower data rates, which would permit Reed-Solomon encoding of the entire data stream. This choice resulted in added complications for tape-recorder data management, but experienced sequence team personnel handled the problem without breaking stride. The 14.4-kbit/s limitation also explains why the Reed–Solomon encoder was not used at Jupiter and Saturn, since the prime data rates at those planets were considerably higher than 14.4 kbit/s.

The success of the encoder can be visually seen in the Uranus imaging. The Golay encoder can correct all the errors all the time in the telemetry stream unless the number of errors becomes too great. For images, as the number of errors increase, a point is reached where some of the corrupted bits cannot be reconstructed as correct pixel brightnesses. This manifests itself as a gradual increase of 'snow' in the images. The Reed–Solomon encoder also corrects all errors all the time unless the error rate got too great. When the error rate becomes too large, instead of 'gracefully' decaying, the data is completely corrupted. The ground system could not reconstruct the data into an image at all. The Project made the decision to use the lower encoding overhead to return the data to Earth, recognizing that the decision involved an increased chance of getting no data if the error rates were too high. With the use of the Reed–Solomon hardware during the Uranus encounter, 'snow' in the images was all but absent (see Figure 7.3).

Sensing and handling problems automatically

The large distances and correspondingly long communications times made it prudent to program the two Voyager spacecraft to perform periodic health checks and take

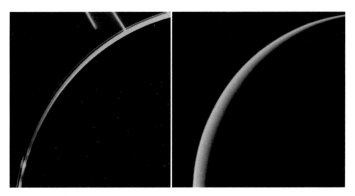

Figure 7.3. The image on the left was taken by Voyager 2 during the Jupiter encounter and was encoded with the Golay encoder. The image on the right was taken by Voyager 2 during the Uranus encounter with the Reed–Solomon encoder. Notice the absence of 'static' in the Uranus image. (See color plate 3.)

appropriate actions if faults were detected. For Voyager 2 it was also important to consider what should be done if the remaining radio receiver should fail. Since no further commanding of the spacecraft would be possible, all future actions of the spacecraft would have to be programmed into the spacecraft computers prior to receiver failure. For this reason, Voyager science and engineering personnel periodically loaded a part of the CCS computer with a skeleton sequence known as a Backup Mission Load (BML). Only during the busiest portions of an encounter, when CCS memory space is at a premium, is BML removed. If Voyager 2's radio receiver had failed more than 12 days before closest approach to Neptune, the spacecraft would have obtained a series of pictures of Neptune, collected fields and particles data, and performed a simple radio science occultation experiment as it passed through the shadow of Neptune. Concurrent IRIS, PPS, and UVS data would also have been obtained, though the optimum viewing geometries actually used during the more detailed encounter would not have been accommodated. No high-resolution satellite data would have been obtained.

A large number of other 'Failure (or Fault) Protection Algorithms' (FPAs) were programmed into Voyager 2. In most of these FPAs, several preliminary steps would be tried before drastic steps would be taken to correct the problem. All of the FPAs were thoroughly tested on the ground to be certain they worked properly, but few were triggered in flight as a result of actual problems on the spacecraft. A relatively detailed description of these FPAs is given by Jones [2]. If excessive electrical current were used by the spacecraft, the consequent drop in power-supply voltage would be detected by onboard monitors. In such a circumstance, the spacecraft would have shut down all non-essential systems, rechecked the voltage, and continued along one of a number of different courses of action, depending on the conditions detected. In one of these scenarios, for example, the imaging cameras (largest power users among the science instruments) would be shut off. This would be followed by the other three scan platform instruments being pointed in a 'safe' direction in space, the scan platform motors turned off, and Voyager would continue to transmit non-imaging data at a rate of 4,800 bit/s. Leaks in the propulsion subsystem associated with the attitude control jets (not correctable by lesser actions) would have been countered by shutting down the primary plumbing and activating an entirely separate backup plumbing system.

If a non-catastrophic collision or a passing bright particle were to cause the Sun Sensor (SS) to lose the Sun, the spacecraft would perform a series of yaw and pitch turns to search the sky systematically to relocate the Sun, switching to a backup SS if necessary. Once the Sun was located in the SS field of view, a slow roll turn would be executed until the predesignated star was sighted in the Canopus Star Tracker (CST). A backup CST is also available for usage were the primary CST is unable to detect a star. Seizure of the scan platform could also result in turning off power to the platform motors to prevent damage to these critical pieces of hardware.

Should one of the spacecraft transmitters fail, backups for both the S-band and X-band transmitters were available. The precise actions taken by the spacecraft would depend on whether or not Voyager was in a critical part of the encounter operations.

Some automatic responses to science instrument problems were also available. Since IRIS needed to be at a precisely controlled temperature to operate successfully, it has a backup heater in the event that the CCS computer senses that the primary heater is not operating properly. Occasionally the PRA erroneously sensed an overload and inserted signal attenuators. This condition is called PRA 'Power-On Reset' (POR). When Voyager sensed that a PRA POR had occurred, it incremented a counter and sent the appropriate commands to the PRA to restore normal operation. Since the PPS could be damaged or destroyed by high light levels, the CCS was programmed to 'safe' the instrument. Thus, if certain responses to other problems on the spacecraft occurred, the CCS could command the PPS to assume its smallest field of view and lowest gain state, as well as pointing the scan platform to a predetermined 'safe' position.

The CCS computer itself performed frequent self-tests. If a command was transmitted from Earth that the CCS was not programmed to receive, or if partial memory failure occurred, the response could result in termination of the executing sequence of events until corrective commands were received from Earth.

These are just a few samples of the Voyager's preprogrammed 'thought' processes. It is comforting to know that these silent software code sentinels stood ready to guard against such a wide variety of potentially life-threatening conditions on Voyager. Though few have actually been called to service, such safeguards have undoubtedly contributed to Voyager's lifetime extending far beyond early expectations.

7.1.3 Planning a sequence of events

Those who have driven a vintage automobile for a lengthy period of time are well aware of the idiosyncrasies and strong points of these 'old' machines. As they age and develop their own personalities, their owners and operators become intimately familiar with such 'personalities'. Love and tender care can often elicit responses from these machines far beyond those available to 'strangers' and so it has been with Voyager. Years of experience have taught the Voyager science and engineering personnel how to utilize the capabilities of the spacecraft more and more efficiently to collect unique science data about these distant giant worlds. This was enhanced by the fact that Voyager's computers were reprogrammable. By modifying its 'operating systems' the Voyager 2 that passed Uranus was more capable than the Voyager 2 that passed Jupiter and Saturn. Similarly, the Voyager 2 that passed Neptune was a much better spacecraft than the Voyager that encountered Uranus.

For those closely involved with the long-term deliberations and planning, Voyager has offered an intellectually stimulating challenge to find new and innovative methods and techniques for investigation of the outer Solar System.

Assessing what was already known

Because of the great distance of Neptune and the fact that no spacecraft had blazed a trail for Voyager 2, much less was known about this giant planet than was known about Jupiter and Saturn prior to their encounters. A group of over 100 scientists gathered in Pasadena, California, on February 4 through February 6, 1984, to discuss the state of knowledge of the Uranus and Neptune systems. This conference resulted in the publication of *Uranus and Neptune* [3], which was referenced extensively in Chapter 4. In preparation for the upcoming encounters, the Uranus/Neptune Science Working Groups were formed in the early spring of 1984. The working groups each handled one of three disciplines: atmospheres, rings, and satellites/magnetosphere. Satellites and magnetosphere disciplines were combined due to the limited amount of information known about the targets and the relatively easier planning of observations for those disciplines. A report entitled *Science Objectives and Preliminary Sequence Designs for the Voyager Uranus and Neptune Encounters* [4], produced on January 28, 1985, from those workshops, summarized the most current information available to guide the Project in preparing the encounters.

Selecting a trajectory through the Neptune System

The process of selection of a trajectory for a planetary encounter is bound by two guiding principles. The first is the safety of the spacecraft (and its ability to perform the many activities planned for it); the second is the scientific desire to answer as many of the fundamental questions about the body being encountered.

Previously, Voyager 2's trajectory had been dictated primarily by the need for a gravitational assist at each planet to facilitate the spacecraft's rendezvous with the next planet on its grand tour of the Solar System. This time was different. There were no further worlds to explore. Voyager 2 was free to be targeted to the trajectory that would maximize the science return, as long as that trajectory was one that would not unduly jeopardize the safety of the craft. One of the main science objectives was to obtain Earth and Sun occultations of both Neptune and Triton. An Earth occultation by Neptune would allow the mission controllers to point the spacecraft antenna toward Earth to transmit its signal through the atmosphere, thereby providing a temperature and pressure profile at the particular latitude being probed. Performing the same experiment on Triton would allow atmospheric scientists the same opportunity to characterize the satellite's rarefied atmosphere, if it had one.

A Sun occultation of Neptune would allow the UVS to investigate the hydrogen content of the planet's upper atmosphere. The combination of radio and solar occultations could go a long way toward providing an understanding of the atmospheric temperatures and structures of Neptune and Triton. The primary question remaining was whether the spacecraft could safely pass through the Neptunian system and have occultations at both Neptune and Triton.

The required geometry included a 'polar crown' passage to accomplish these objectives. The spacecraft would be targeted to swing over the north pole of Neptune and then swing southward, passing Triton 5 h 14 min later. Achieving the desired gravitational bending at Neptune was a potentially risky factor, and would require the spacecraft to come very close to the planet's atmosphere. The spacecraft had to stay high enough to avoid Neptune's atmosphere and rings (if, indeed, Neptune did have rings or ring arcs). Traveling along a polar crown trajectory would bring Voyager 2 within about 5,000 km of these environmental hazard zones (see Figures 7.4 and 7.5).

Selecting major scientific goals

Armed with the latest available Neptune Earth-based data and a sparse understanding of Voyager's capabilities, the four sections of the Neptune Science Working Group (NSWG) held a series of meetings between August 1986 and July 1987. Their primary purpose was to recommend to the Voyager Science Steering Group a set of time-sequential observations to address as many as possible of the significant scientific questions at Neptune. To address these questions, four working groups were organized to address, respectively, the atmosphere, the potential rings, the satellites, and the magnetosphere. With two years remaining before the Neptune encounter, it was almost embarrassing how little was actually known about the planet and its system. The rotation period of Neptune was little more than an estimate, and there was no direct knowledge on the nature of the magnetic and associated radiation fields. There were only tantalizing indications that Neptune possessed rings, and only the barest of information existed on its two known satellites. On July 15, 1987, the NSWG's final report was published [5], documenting the recommendations of the atmospheres, rings, satellites, and magnetospheres working groups.

The atmospheric goals, arranged in descending order of priority, included:

- Bulk atmospheric composition: hydrogen (H_2), helium (He), and methane (CH_4) abundances in Neptune's upper atmosphere and gravitational harmonics.
- Global energy budget: thermal emission and albedo of Neptune to more accurately determine the amount of heat coming from the interior of the planet.
- Vertical structure: troposphere to exosphere with some latitude structure. Composition and variable constituents: methane (CH_4), ethylene (C_2H_4), acetylene (C_2H_2), atomic hydrogen (H), *ortho/para* [6] hydrogen (H_2) ratio, etc.

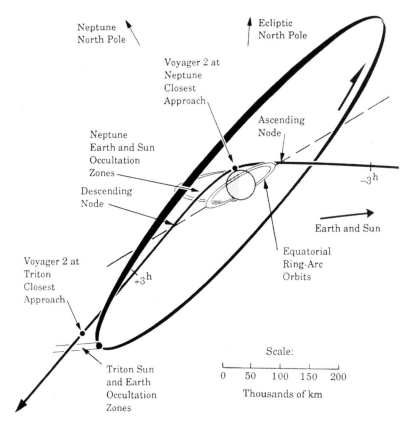

Figure 7.4. An edge-on view of Voyager 2's trajectory through the Neptune system.

- Clouds and haze vertical structure and particle properties.
- Horizontal cloud and temperature structure, composition, physical processes, and rotation rate: meteorology.
- Auroral structure and appearance at both poles.

Goals for the Voyager observations of the proposed Neptune rings (if they existed), again in order of priority, included:

- Ring-arc structure: both radial and azimuthal profiles and their changes with time.
- Ring satellite search: moonlets in or near the rings.
- Search for additional ring material.
- Orbital kinematics: improve orbit models, compare ring motions to satellite motions.
- Ring particle motions: importance of self-gravity, vertical thickness of rings, collisions between particles.
- Ring particle properties: size, shape, reflectivity, temperature, and composition.

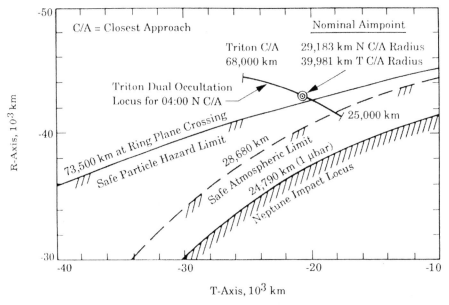

Figure 7.5. The Neptune northern polar aiming space. The aim-point satisfied all major science objectives, but introduced some environmental hazards that needed to be avoided.

- Ring environment: interaction with magnetospheric plasma or with an extended neutral atmosphere.

Prioritized observational objectives for Voyager studies of Neptune satellites included:

- Global properties: sizes, shapes, masses, densities, variations with longitude and time.
- Surface characteristics: types of surface structures, characterization of surface processes, variations with solar phase angle.
- Satellite system: searches for satellites both interior and exterior to the orbit of Triton.

Prioritized objectives for Voyager studies of Neptune's magnetosphere, again in order of priority, included:

- Study Neptune's interaction with the solar wind.
- Characterize the magnetic field of Neptune with the maximum accuracy.
- Determine the plasma physics in the polar region of Neptune.
- Measure the (internal) rotation rate of Neptune.
- Investigate the physics of the Neptunian magnetosphere.

Matching scientific goals with spacecraft capabilities

With the assistance of Voyager Project personnel at JPL, the NSWG members then proceeded to describe the instruments, instrumental configurations, and times when Voyager 2 observations contributing to the selected science objectives could be accomplished. This task initially was done independently within the four separate working groups. Each working group was encouraged to select observation designs that would simultaneously contribute to more than a single science objective. A particular design for an observation was known as a 'link' (representative of a link in a chain of events). An individual link might be used several times, but always with either the same design or the same science objective. Details of the link designs were not included in this early stage of planning, but were developed later.

Once the sets of links and execution times were defined, individual link execution times were incorporated in a computerized Time-Ordered Listing (TOL) of events and laid out on graphical timelines. The TOLs and the graphical timelines also included the estimated capabilities of the communications link between Voyager and the DSN tracking stations. Assuming full station arraying, the maximum data rate available while Voyager was over the Goldstone and Canberra tracking stations was 21.6 kbit/s; the Madrid tracking stations could receive data at rates up to 14.4 kbit/s. Canberra had the longest tracking passes (12 h). Available rates were lower near the beginning and end of each pass. Early guidelines suggested that, for all but a two-week period surrounding the closest approach to Neptune, only a single 70-m station should track Voyager 2. Real-time imaging was only possible for rates of 8.4 kbit/s or greater. Where the data rates required for a particular link exceeded the stated DSN capabilities, decisions had to be made about whether to shift the link execution times, delete particular link executions, delete the link entirely, or recommend augmentation of the DSN station coverage.

In an iterative fashion, each of the four working groups eventually put together a recommended sequence of atmospheric (or magnetospheric, or ring, or satellite) observations. Each also categorized their recommended links as first-priority, second-priority, or third-priority science. Although these relative priorities existed only within a given discipline, they nevertheless served a very useful purpose in the later resolution of conflicts between intergroup recommendation. Although most NSWG members were investigators on one of the science teams, it is fortunate indeed that all were objective enough to consider the relative merits of scientific observations from all disciplines. Otherwise, the processes required to optimize the overall science return would have been enormously more difficult.

Resolving conflicts

Conflicts between the requested observations of the four working groups became immediately apparent once their requests were merged into a single timeline. For all but a three-week period from 14 days before to seven days after closest approach to Neptune, these conflicts were mainly resolved by relatively minor time shifts of one or more of the observations. Occasionally, routine periodic observations would have to be stopped prior to a unique, long-period observation and thereafter resumed.

The highest-quality observations were concentrated within a few days of close approach to Neptune, especially on the approaching leg where the planet, suspected rings-arcs, and satellites were nearly fully illuminated. Here conflict resolution was considerably more painful because many good observations had to be discarded to retain the highest-priority science and to provide a relatively uniform mix of atmospheric, ring-arc, and satellite science. Some observations proved to be impossible due to the combination of geometrical factors and other spacecraft and sequence constraints.

A multitude of conflicts, followed by resolution via compromises, characterized the design of the Neptune encounter sequence of events. The sense of pride in a job done well was felt by most of those intimately involved in the preparation of Voyager 2 for the encounter. The feelings bordered on euphoric during those dramatic few days when the major discoveries about Neptune and Triton were being received, analyzed, and disseminated to the waiting press and public. For many it was difficult to go home for a much-needed rest each day or night in case the receipt of some new item of information might be missed.

Translating commands to Voyager-readable language

The end product of the working group activities was a 'strawman' time-ordered listing of events for the entire encounter period. Starting from this listing, Voyager Flight Science Office personnel, assisted by Flight Engineering Office and Mission Planning Office personnel, began the development of a 'Scoping Product' which provided the first quantitative estimate of spacecraft resources needed to accomplish the recommended observations. The first result of scoping was the selection of boundaries between encounter phases and between the CCS computer 'loads' within those phases (see Table 7.2). As shown in the table, the encounter phases were specified as Observatory, Far Encounter, Near Encounter, and Post Encounter. The nature of the science observations in each phase is discussed below. Scoping of the encounter was completed during the period from October 15, 1984, to January 15, 1988. As the dates indicate, the first Neptune Observatory load was developed during the Uranus encounter sequence development process to obtain a head start on Neptune encounter planning and to utilize the experienced personnel from the Uranus encounter.

Each of the successive levels of sequence preparation contained several opportunities for new inputs, incorporation of those inputs, generation of revised sequence products, and critical review of those products. It was this iterative process which enabled Voyager Project personnel to ferret out both major and minor problems, in addition to improving the scientific and engineering efficiency of the final sequence of events.

Following Scoping, the next level of complexity was introduced. During this 'Integrated Timeline' (IT) Product development period, supporting engineering

Table 7.2. CCS computer load boundaries for the Neptune encounter

Phase/load	Phase/load start time [date], [h:min (GMT)], [day/h:min (from C/A)]			Phase/load duration
Observatory	[1989 Jun 05],	[06:42.4],	[−80/21:17.6]	62d 02h
B901	[1989 Jun 05],	[06:42.4],	[−80/21:17.6]	18d 00h
B902	[1989 Jun 23],	[06:42.4],	[−62/21:17.6]	34d 22h
B903	[1989 Jul 28],	[04:42.4],	[−27/23:17.6]	9d 04h
Far Encounter	[1989 Aug 06],	[08:42.4],	[−18/19:17.6]	18d 07h
B921	[1989 Aug 06],	[08:42.4],	[−18/19:17.6]	7d 14h
B922	[1989 Aug 13],	[22:42.4],	[−11/05:17.6]	5d 23h
B923	[1989 Aug 19],	[21:42.4],	[−05/06:17.6]	4d 18h
Near Encounter	[1989 Aug 24],	[15:42.4],	[−00/12:17.6]	5d 06h
B951	[1989 Aug 24],	[15:42.4],	[−00/12:17.6]	2d 06h
B952	[1989 Aug 26],	[20:42.4],	[+01/17:42.4]	3d 00h
Post-Encounter	[1989 Aug 29],	[20:42.4],	[+04/17:42.4]	33d 15h
B971	[1989 Aug 29],	[20:42.4],	[+04/17:42.4]	12d 23h
B972	[1989 Sept 11],	[19:42.4],	[+17/16:42.4]	20d 16h
end	[1989 Oct 02],	[11:42.4],	[+38/08:42.4]	

spacecraft configuration, and science calibration and configuration commands were added. DSN station coverage needed to support the observations was also negotiated and, for the first time, link designs were specified in full-blown detail. The software used to store and manipulate the computer files was known as ASSET (for Automated Science/Sequence Encounter Timeline), and utilized a specially programmed version of dBASE II. Each IT Product was critically reviewed by all elements of the Voyager Project. Once the IT Product was formally approved, often with a short list of matters still needing resolution, the sequence development baton was passed from the Science Investigations Support Team (SIS) to the Sequence Team (SEQ). Though SEQ took the lead, the SIS Team, the Spacecraft Team, the Navigation Team, and the Mission Planning Office all had major supporting roles. The IT Product development for the Voyager 2 Neptune encounter took place in the time interval from April 7, 1986, through March 27, 1987.

From IT Product approval forward, sequence development was dependent on several Voyager computer programs which existed prior to Voyager launch (ASSET was initially developed in 1984 especially for the Uranus and Neptune encounters). Some of these programs have been inherited from other planetary exploration projects, but all were adapted to Voyager, and all have since been updated on numerous occasions to correct problems, improve their efficiency, or add new capabilities. Each change required approval by a Voyager Change Board (VCB) consisting of the Mission Director and managers of each of the Voyager Offices. The set of approved software was known as the 'Mission Build'.

The first in this series of programs was POINTER, which (a) translated sequence inputs into detailed science instrument commands, (b) determined, in scan platform coordinates, the appropriate pointing for each target, and (c) performed simple verification of the sequence validity. The output of POINTER included pictorial representations of the scenes to be viewed by ISS, IRIS, PPS, or UVS (output by the TARGET module of POINTER). It also included a detailed time-ordered printout of the sequence of commands, generated by a module of POINTER known as OPSGEN (Observation Pointing Sequence GENerator). Both OPSGEN printout and the TARGET plots were reviewed thoroughly, and errors in implementation were corrected. If the TARGET plots or OPSGEN printout showed the need for a change of strategy or other modification, written sequence change requests first had to be approved by the VCB. Approved changes were incorporated into the SERF (Science Events Request File), a computer-readable POINTER output file which was used as input for the next stage of sequence development. The SERF was therefore representative of the 'Final Timeline' (FT) Product.

Another module of POINTER (known as VERIFY) was used near the conclusion of sequence preparation to provide verification that the intent of each science observation was preserved through the end-to-end series of program runs. Observatory Phase (OB) OPSGEN review started on February 2, 1987; the Post-Encounter (PE) OPSGEN review was started on June 26, 1987. OB and PE were done first to allow any new information about Neptune to be incorporated into the later FE and NE development products.

The SERF and several other files (the moment-by-moment orientation of the spacecraft, initial conditions, and other information needed to properly configure the spacecraft systems) were processed by a program called SEQGEN (SEQuence GENerator). In addition to the sequence integration task performed by SEQGEN, one of its main purposes was to verify that none of the constraints on sequence structure imposed by spacecraft limitations or accepted usage rules were violated. The primary outputs of SEQGEN were the ESF (Event Sequence File) and the SRF (Sequence Request File). Computer printouts of each of these files were then thoroughly reviewed to ascertain their correctness. An additional output of SEQGEN was a file specifying scan platform pointing as an input to the POINTER VERIFY process mentioned earlier.

The corrected SRF served as input to the program SEQTRAN (SEQuence TRANslator), which put the instructions into a language that the spacecraft could store, understand, and execute. The output of SEQTRAN was the DMWF (Desired Memory Word File), which was then processed through the COMSIM (COMmand SIMulation) program to make a final verification that all commands were correctly formulated and in accord with the understood operation modes of the spacecraft. The GCMD (Ground CoMmanD) file that was generated by COMSIM could be transmitted to the spacecraft. COMSIM also generated an EVTSDR (EVenT System Data Record). Together with the corrected ESF from SEQGEN, the EVTSDR permitted the generation of an SOE (Sequence Of Events) listing, which detailed the spacecraft events that would occur and was a reminder of ground event timing in support of Voyager. The SEQGEN, SEQTRAN, and COMSIM programs together generated what was called the 'Advanced Planning' (AP) Uplink Product. AP Product development started with the Observatory Phase and then progressed to Post-Encounter, Far Encounter, and ended with the Near Encounter sequence development. The time period for this work spanned from June 30, 1987, through February 15, 1989.

The POINTER VERIFY module was used near the end of the final Uplink Product (UP) development as a verification that instrument pointing was as desired. VERIFY used inputs from SEQGEN and COMSIM to specify scan platform pointing, spacecraft attitude, and ISS operation timing. Since IRIS, PPS, and UVS essentially operated continuously, times for individual instrument-pointing verification were critical and had to be specified by the respective Experiment Representatives.

To accommodate new information on Neptune received during the sequence generation, or to correct or improve the observations, a limited update of each of the UP Products was scheduled. This update process occupied most of the final six weeks prior to transmission of each CCS computer load to Voyager 2. The update included automatic pointing adjustments to account for any detected changes in the spacecraft trajectory.

One other fact had to be realized in generating command loads. The Trajectory Correction Maneuvers (TCMs) were critical for targeting the spacecraft to its aim-point. Unfortunately, not enough information about the system physical constants (e.g., planet mass, satellite masses, planet spin axis, satellite positions, etc.) would be known at the time the computer sequences were built. Even more serious, the Near Encounter sequence would be built, validated, and transmitted to the craft prior to learning enough about Neptune to navigate Voyager through it accurately. To combat this, a technique known as Late Ephemeris Update (LEU), which was used at Uranus, was employed at Neptune. Here a computer sequence which had already been built and validated was rerun through the final set of programs with the most current information. This rerun, using the most current spacecraft, planet, and satellite ephemerides, produced the best that sequence engineers could produce prior to transmitting the sequence to the spacecraft. An LEU was performed for two Far Encounter sequences (B922 and B923), both Near Encounter sequences (B951 and B952), and the first Post-Encounter sequence (B971).

In some cases, even the LEU technique was inadequate. The most sensitive measures of mass and timing came from tracking data obtained only hours before the relevant observations. Voyager 2 required a Late Stored Update (LSU) to preserve the targeting of the closest approach remote-sensing observations. Engineers would run the Near Encounter sequence again using the updated ephemerides, even after the sequence was already stored in the spacecraft CCS. The LSU was then radiated to the spacecraft to rewrite some of the parameters for the computer sequence that it had already begun to execute! (see Table 7.3).

But pointing updates would not correct an early or late arrival at Neptune. With the planet 4,500 million km from Earth, Voyager would probably not arrive at its aim-point at precisely the planned time. The problem was further complicated by the fact that if Voyager 2 missed its desired Neptune aim-point and timing, the aim-point

Table 7.3. Observations which required LSUs

Observation	Link name	Start time (enc. rel.)	Required update
Neptune movable block	NMB	−03:20.0	Timing
Retargetable ring, ISS	VRRETIN3	−03:12.8	Retarget, pointing
Triton longitude coverage, ISS	VTLON	−02:55.2	Pointing
Neptune photometry, PPS	PPVPHOT3	−00:39.2	Pointing
Limbtrack maneuver	X.LMBTRK	+00:06.6	Maneuver
Neptune atmosphere structure, ISS	VPHAZE	+01:38.4	Pointing
Triton movable block	TMB	+01:50.4	Timing
Triton best color, ISS	VTCOLOR	+02:06.4	Pointing
Triton map, ISS	VTMAP	+02:34.4	Pointing
Triton map, NIMC maneuver	V.NIMC	+02:34.4	Maneuver
Triton atmosphere, IRIS	RTATM	+04:16.8	Pointing
Triton IMC maneuver	V.IMC	+04:39.2	Maneuver
Triton high resolution, ISS	VTERM	+04:40.0	Pointing
PPS configuration	PTBETCMA	+05:12.0	Timing
Triton darkside map, IRIS	RTMAPOUT	+06:48.0	Pointing
Triton crescent drift, UVS	UTCRDRFT	+07:48.0	Pointing
Triton darkside map, IRIS	RTMAPOUT	+08:50.4	Pointing
Neptune heat flux, IRIS	RPNORTH	+10:04.8	Pointing
Neptune photometry, PPS	PPVPHOT5	+11:46.4	Pointing
Triton heat flux, IRIS	RTDISK	+12:40.8	Retarget, pointing
Triton heat flux, IRIS	RTDISK	+15:40.8	Pointing
Triton heat flux, IRIS	RTDISK	+17:42.4	Retarget, pointing
Triton hydrogen blowoff, UVS	UTHYDRGN	+18:07.2	Pointing

and timing at Triton would also be in error. The approach adopted by the Project to correct for timing errors was the use of a 'movable block' of commands. Commanding a change in the start time of the block would correspondingly shift all CCS commands internal to the block by the appropriate amount. Unfortunately, the shifted start time for observations at Neptune closest approach was different from the time shift needed for the CCS commands at Triton. It therefore became necessary to design two separate movable blocks. The Neptune movable block started 3 h 20 min before Neptune closest approach and continued until 1 h 46.4 min after closest approach. The Triton movable block started 1 h 50.4 min and ended 8 h 38.4 min after closest approach. The start time of both blocks could be altered independently during a Late Stored Update and could be shifted by ±9.6 min.

The entire, complex process was completed on schedule and Voyager 2 performed all its encounter observations exactly as expected. A single exception was a minor problem with one frame in the high-resolution mosaic of Triton (VTERM) where small timing and IMC rate errors resulted in excessive imaging smear in one of the imaging frames.

7.2 CHARACTERISTICS OF THE NEPTUNE ENCOUNTER

The actual Neptune encounter spanned 119 days (see Table 7.2). A number of encounter-related science activities occurred during periods immediately preceding and following the encounter. Closest approach to Neptune occurred at 04:00 Greenwich Mean Time (21:00 Pacific Daylight Time) on August 25, 1989. The one-way light time at Neptune closest approach was 4 h 6 min. Radio signals transmitted from Voyager 2 when it was closest to Neptune actually arrived at the Canberra Deep Space Communications Complex at 08:06 GMT (01:06 PDT) and were relayed within fractions of a second to JPL in Pasadena, California. The Voyager science activities associated with each of the encounter phases, the geometry of the Neptune encounter, and a brief outline of the handling of the data from Voyager are described below.

7.2.1 Encounter science activities

Generally, all Voyager spacecraft activities between the beginning of June 1989 and the end of October 1989 were science activities, since all were aimed at providing the best possible science return from Neptune, its magnetosphere, its ring-arcs, and its satellites. An extensive network of supporting science activities and observations from Earth was also an important factor in the success of the encounter. Some of those observations were discussed in Chapter 4. Here we will concentrate on the encounter activities of the Voyager spacecraft itself, including engineering activities in support of science observations, science calibrations, and the science observations themselves. Even these activities were so extensive that only the barest of details can be given here. For convenience in discussing the activities, they have been grouped by mission phase, and the phases are discussed in chronological order.

Uranus–Neptune cruise

A series of six Trajectory Correction Maneuvers (TCMs) was planned for Voyager 2 between the close approaches of Uranus and Neptune. Initially, these TCMs were used to target the spacecraft to arrive at Neptune at 23:12 GMT on August 24, 1989. Following the planetary rendezvous, the gravitational field of Neptune would divert the spacecraft such that it would come within 44,000 km of Triton. Science Investigators wanted Voyager 2 to come as close as possible to this satellite to accomplish the critical science observations, many of which were range dependent.

The first post-Uranus TCM (B15) occurred just 21 days after the Uranus encounter. Its accuracy was sufficient that the scheduled TCM B16 was cancelled. During 1985, the science community requested that the Triton miss-distance be reduced and the arrival at Neptune be slightly delayed. The change in Triton miss-distance would decrease image smear; and the later Neptune arrival time would improve the Radio Science Earth occultation experiment by centering the event over the Australia tracking stations. The Canberra DSN antennas would thereby observe Voyager 2 higher in the sky, and attenuation of the radio occultation signal by Earth's atmosphere would be lessened. The Neptune closest approach timing was delayed to 04:00 GMT on August 25, 1989, and the Triton miss-distance was reduced to 10,000 km.

In late 1985, values for several Neptune constants were updated on the basis of better Earth-based observations. These changes forced the Project to move the Neptune aim-point further away from the planet to avoid what were now estimated to be excessive drag forces from the upper atmosphere, and to avoid potential impact with the predicted location of Neptune ring arcs. On March 13, 1987, TCM B17 adjusted the speed and direction of Voyager 2 to achieve the newly defined rendezvous parameters.

To ensure that the Voyager 2 encounter was a success, the Project embarked on a series of test and training activities. These activities involved Project personnel and sometimes included the spacecraft; the first of these was performed in October 1987, and the testing continued into the encounter time period. One test and training activity was known as an Operational Readiness Test (ORT). This test was used to demonstrate that all components of the Project (operational personnel, proce-dures, hardware, software, and ground communication facilities) where ready for the encounter. In addition, Capability Demonstration Tests (CDTs) were done using the improved flight software onboard the spacecraft. Though it may sound risky in the cautious NASA of today, this seemed the most straightforward means of assuring that the software did indeed work as planned. In essence, Voyager 1 was a 'flying' testbed whose employ was critical to the success of the Voyager 2 Neptune encounter.

In April 1989, a CDT was performed to validate that a TCM could be designed to use only the spacecraft roll thrusters. A TCM of this type would maintain the spacecraft's attitude (except for roll) while gently changing its trajectory. This enabled the spacecraft to continue its communications with Earth while altering its direction. Maintaining communications would be critical during the encounter. The test confirmed the viability of roll-turn TCMs; TCM B20 could therefore use this technique during the Neptune close-approach period. The test itself also slightly changed the spacecraft's velocity and became known as TCM B17C.

Occasional imaging of the planet was performed during the period between the end of the Uranus encounter and the beginning of the Neptune encounter. The frequency of such imaging increased as Voyager 2 approached Neptune. By May 1988 Voyager's pictures were of higher resolution than any previously obtained from Earth-based telescopes. On May 9, 1988, at a distance from Neptune of 684 million km, Voyager 2 obtained the image shown in Figure 7.6. Neptune is featureless in this view, other than the general darkening toward the limb (the edge of the planet's disk). Triton, Neptune's largest satellite, can be seen even from this distance.

Spacecraft navigation at Neptune involved some special challenges. A technique successfully used at the other planets was known as optical navigation, or OPNAV for short. The technique required shuttering long-exposure images of the satellites in the foreground of a known star-field, and these observations served to improve knowledge of the satellite ephemerides, the location of Neptune's center of gravity, and the position of Voyager 2 relative to Neptune and its satellites. The

Figure 7.6. This image of Neptune and Triton was taken on May 9, 1988. Triton has a reddish-yellow hue that is caused by methane-derived organic material on its surface or atmosphere. (P-33154)

challenge at Neptune was that the planet had only one known moon by which the spacecraft could be optically navigated. Triton's orbital period of 5.9 days was short enough that useful OPNAVs could be obtained over many orbits of the satellite around Neptune. Tiny Nereid, the other known satellite, has an orbital period of 365 days; but the relatively small size and long period rendered it of very little use for anything other than its own ephemeris.

Although optical navigation was widely recognized as the superior navigation data type, OPNAVs limited solely to Triton seemed unacceptable. With its usual inventiveness, the Navigation Team had a novel idea. Why not place OPNAV observations in the command sequences without specifying their targets; then as new satellites are discovered, the OPNAVs could be pointed at one of these new satellites? The amount of time set aside for pointing the cameras at such satellites would be indeterminate during the planning stages, but if sufficient time were set aside for a wide variety of possibilities, the suggestion was thought to be viable.

The idea of planning to navigate Voyager 2 with OPNAVs of as-yet-unknown satellites sounded risky. What if none was discovered? While this outcome was possible, it was unlikely. Voyager had discovered three satellites at Jupiter, three at Saturn, and ten at Uranus; it was not only possible, but probable that many new satellites would be discovered during the Neptune encounter. The Project agreed to use 'retargetable OPNAV' observations to help Voyager to meet its planetary rendezvous. The main question was not whether Voyager would discover new satellites but rather whether they would be discovered early enough to enable the determination of their orbits before critical OPNAV targeting was finalized. There seemed to

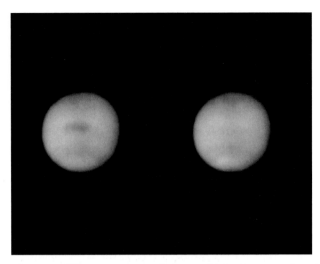

Figure 7.7. These images of Neptune were made by combining images taken through a violet, clear, and orange filter and have a resolution of 3,256 km per line pair. The picture on the right was taken 5 h after the image on the left. During that period the planet had rotated 100 degrees. (P-34255)

be no alternative. Following TCM B17, the first of the Neptune OPNAV observations began.

Seven months prior to closest approach a dark spot appeared in Neptune's southern hemisphere. The dark spot (later named the Great Dark Spot, or GDS, after Jupiter's Great Red Spot) became clearer and more pronounced as Voyager's range from the planet decreased. Few explanations were found for this atmospheric disturbance, but its discovery contributed significantly to the expected excitement the summer would bring. Four months before closest approach, and a scant five weeks before the start of the Observatory Phase, numerous atmospheric features began to appear in the atmospheric images. On April 26, 1989, from a distance of 176 million km, Voyager 2 obtained the two images of Neptune shown in Figure 7.7. The GDS is prominent in the image on the left. There is also the hint of a dark band in the southern hemisphere (below the GDS) that appears to encircle the planet. A white feature, also in the southern hemisphere, is visible in the image on the right. These features were quite surprising because sunlight at Neptune is barely one-tenth of 1% of the solar energy density received at Earth.

One of the more important activities just prior to the start of the Observatory Phase was the 'Near Encounter Test' (NET). Because of the complexity of the sequence of activities surrounding Neptune closest approach, the most critical of these activities was tested on Voyager 2 during the NET. Maneuvers to provide image motion compensation and to track the limb of the planet during the radio occultation were tested. The power system was also checked to better determine the power margins in circumstances comparable to those near closest approach to Neptune. There was also an attempt to determine how the spacecraft radio receiver would drift in frequency during the series of commands needed to perform the radio science occultation experiment adequately. The simulated time period covered by the NET was from 3.5 h before Neptune closest approach to approximately 3.5 h after Triton closest approach (N−3 h 30 min to N+8 h 40 min). Wide-angle camera images of the star background served to check the accuracy of the various maneuvers during their execution. Because the spacecraft scan platform had not been moved at its medium rate (0.33°/s) since the Uranus encounter, the eight medium-rate slews (seven two-axis slews and one one-axis slew) to be used during Near Encounter were also tested. Only minor flaws were revealed by the NET, and appropriate corrections were made for the actual Near Encounter. In addition to testing the spacecraft Near Encounter sequence of events, the NET also served as a good test of the readiness of both the DSN tracking stations and personnel and of the Ground Data System at JPL.

One other major type of activity performed during the latter portions of Uranus–Neptune cruise was the series of instrument calibrations performed in preparation for the encounter. Measurements were made to determine more precisely the sensitivity, pointing direction, and relative response across the fields of view of the PPS. Canting the spacecraft to one side served to illuminate with sunlight a target plate attached to the spacecraft. Because the target plate had precisely known reflective properties, and because the distance and brightness of the Sun were also well known, ISS imaging and IRIS radiometer responses were calibrated by viewing the target plate. The MAG was calibrated several times during the Uranus–Neptune cruise period by rotating the spacecraft through several complete turns, first in the roll direction, and then in the yaw direction. The response of the magnetometers themselves was not expected to change appreciably, but differing power loads and distribution on Voyager give rise to spacecraft magnetic fields that must be measured and subtracted from the total field in order to determine the pre-existing ambient field. Most of the instruments had calibration or conditioning modes, which were used to accomplish routine periodic checks of their responses and health during cruise and the encounter.

The primary data type during cruise was the 12 to 16 h per day of low-rate (160 bit/s) science data. The UVS was used to observe stars; CRS, LECP, MAG, and PLS were used to study magnetic fields and charged particles between the planets; and the PRA and PWS investigations were assisting in the study of the interplanetary environment as well as searching for radio signals which might indicate the impending presence of a Neptune magnetic field.

Observatory Phase

The actual encounter period began with the Observatory Phase (OB), extending from June 5, 1989, through August 6, 1989. OB was divided into three CCS computer loads: B901, B902, and B903. Although Voyager imaging of Neptune exceeded Earth-based resolution as early as May 1988, OB offered the first opportunity for nearly continuous observations of the planet and its system. The beginning of the command sequence saw the start of the first VPZOOM observations (V = Visual,

P = Planet, ZOOM = Approach movie). Each VPZOOM observation consisted of five-color narrow-angle (NA) camera imaging of the planet every 72° of longitude (i.e., five times each rotation period). These imaging frames would be combined to make an approach movie to look for atmospheric features that moved across the face of Neptune. Five such movies done in OB were spaced in such a way that resolution had improved by a factor of 1.4 ($= \sqrt{2}$) in each successive series. These movies were designed to look for cloud features that might be followed as the planet rotated to determine prevailing wind speeds at various latitudes.

The start of OB also saw the initiation of the Ultraviolet Spectrometer (UVS) scans of the Neptune system to search for ultraviolet emissions from Neptune's atmosphere and from the space between Neptune and its satellites. These were designed to detect hydrogen and other gases escaping from the planet or its satellites. The chemical composition of those gases provided clues about the chemical composition of Neptune's atmosphere or the surfaces of its satellites.

Voyager 2 continued its search for evidence that radio signals were being generated by solar wind interaction with Neptune's magnetic field. To accomplish this, quick bursts of high-rate Planetary Radio Astronomy (PRA) and Plasma Wave Subsystem (PWS) data were recorded once a day throughout the Observatory Phase. In addition, as stated in Section 4.1.2, the most definitive technique for determining the planet's rotation rate was to measure the periodicity of long-wavelength radio-waves from charged particles 'frozen' into the planet's rotating magnetosphere. These observations were not possible prior to OB, because they could only be done when the instruments were in the proximity of the planet's magnetosphere.

The Sun and other stars to be used in Near Encounter observations were measured twice by PPS and UVS during OB (B902) as a calibration of the star brightnesses. Similar calibrations would later be done in Post-Encounter, so that the critical observations would occur midway between two high-quality calibration points. The IRIS Flash-Off Heater (FOH) was turned off and the IRIS instrument was turned on in B902 to allow time for cool-down of the instrument for critical Far Encounter and Near Encounter observations.

Periodic calibrations of the fields and particles instruments continued, including an abbreviated version of the roll and yaw maneuvering to calibrate MAG. RSS also performed 'occultation-like' tests in B902 to exercise both the spacecraft radio system and the ground systems and personnel in preparation for the Neptune occultation experiment. This ORT, known as ORT-3.5, was a scaled-down version of ORT-3 performed during the May 1989 NET, just prior to the start of OB. ORT-3.5 included the use of the Usuda site in Japan for the first time, and turned on the S-band transmitter. Usuda's data would be combined with data from the DSN to increase Voyager's received signal strength. The S-band transmitter was normally off to conserve power and to increase the device's lifetime, but was turned on to duplicate accurately the spacecraft configuration during the Earth occultation experiment.

As Voyager 2 raced toward Neptune, images were continuously shuttered on either side of the planet to look for ring material and new satellites. As previously discussed, the Navigation Team was counting on these searches to discover at least

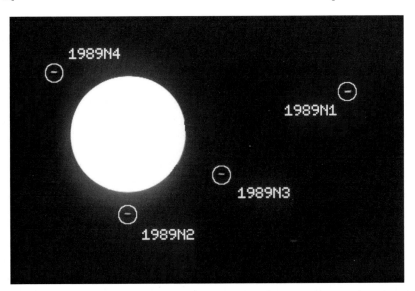

Figure 7.8. This 46-s Voyager 2 exposure was taken with the narrow-angle camera using the clear filter. Neptune appears as a white globe in this overexposed image while the new satellites appear as streaks due to movement by the spacecraft. (P-34540)

one new satellite to guide Voyager 2 more accurately to its aim-point. In early July, during B902, it happened: the first new satellite was discovered. It was called 1989N1, was irregularly shaped, and had a radius of about 200 km. The Voyager Project began determining the orbit for the new satellite and then started calculating scan platform coordinates for the retargetable OPNAV frames. With more than a month before Neptune closest approach, 1989N1 had indeed been found early enough for it to be used to navigate Voyager it to its desired aim-point.

When it rains it pours. Throughout July more and more satellites were discovered. Though none was as large as 1989N1, they appeared to be satellites rather than large ring particles. On July 30, 1989, Voyager 2 took the image shown in Figure 7.8 of four newly discovered satellites.

Engineering activities in OB included a measurement called a 'Torque Margin Test' (TMT). A TMT was devised by spacecraft engineers prior to the Uranus encounter to measure how freely the scan platform was turning. Normally, the stepping motors on the scan platform were driven by electronic pulses of approximately 0.04 s duration. By reprogramming the AACS computer the motor-drive pulses could be made as short as 0.001 s. Post-Saturn testing showed that a healthy motor will still drive at full rate with 0.006-s pulses, but will slow perceptibly with 0.005-s pulses. The TMT in OB utilized 0.006-s pulsing to ascertain whether the spacecraft scan platform motors would still operate at full speed. No slowing was noted. Other OB engineering activities included conditioning of the spacecraft gyroscopes, calibration of the gyroscope drift rates, and a check to ascertain whether the FDS and CCS computer clocks were synchronized.

During the middle of B903 and 23 days prior to the Neptune encounter, the 444-s TCM B18 maneuver was performed. This maneuver was designed to remove all known errors, including targeting errors and timing errors, from the trajectory. The placement of the spacecraft at its aim-point at the right time was critical to the successful execution of the Near Encounter sequence. But at such a distance from Neptune, Navigators could only promise a delivery of the spacecraft to within a 300-km radius circle about the aim-point and a timing error of 165 s. Once again, the accuracy of the maneuver was done so skillfully that TCM B19, scheduled during the Far Encounter Phase, was deemed unnecessary and was deleted.

Another nuance with this maneuver was to ensure that the extra power needed to perform the maneuver did not force reduction in power usage by the science instruments and other important spacecraft subsystems. To provide more power margin, the radio transmitters were reconfigured from their normal state of X-band high power/S-band off to X-band low power/S-band low power. Near the end of the Observatory Phase the first observations designed specifically to search for satellites were performed.

Far Encounter Phase

By August 6, 1989, the apparent size of Neptune was increasing rapidly and soon would no longer fit reliably within a single NA field of view. A four-frame (or larger) NA imaging mosaic was needed during the start of the Far Encounter (FE) Phase to reliably capture the planet. The activities of FE were split into three CCS computer loads: B921, B922, and B923. The more complex observations in FE produced data at an ever-increasing rate. The use of additional antennas was needed to maintain the relatively high communication rates and to relieve Voyager of its data. To increase the downlink capability, the Very Large Array was added to the array at Goldstone on a daily basis.

Less than one day into the start of Far Encounter, Radio Science ORT-4 was performed. The test lasted approximately 10 h, involved a large portion of the Voyager Project, and included personnel at DSN complexes located around the world. It was the last full dress rehearsal for the upcoming near encounter, and, like all tests, it revealed areas that needed to be improved, but nothing major.

An additional movie of the type done in OB was inserted early in FE. The movie sequence was specially designed to observe faint details in the planet's atmosphere. Most of the atmospheric imaging emphasis centered on WA coverage of the entire disk plus scattered higher resolution NA mosaics targeted at atmospheric disturbances. Detailed imaging of the rings-arcs also began in FE.

Voyager 2 continued in earnest its satellite search campaign that it had started near the end of OB. Each VSATSRCH (V = Visual, SATSRCH = SATellite SeaRCH imaging) observation shuttered images on either side of the planet in its quest to discover more satellites. Imaging coverage of Triton also began in FE.

On August 11, five days into B921, two ring-arcs, which appeared to be 50,000 and 10,000 km long, were discovered (Figure 7.9). This confirmed what astronomers had suspected for years: Neptune had rings, or at least arcs of material. The objective

Figure 7.9. A 155-s exposure taken with the narrow-angle camera captured this image of one of the two ring-arcs and the newly discovered satellite 1989N4. (P-34578)

now was to determine their composition, structure, age, and the mechanism by which they were created and maintained. One leading candidate for their origin was the imbedded (but unseen) satellites. Possible debris from collisions with other objects could have ejected material that eventually became partially distributed around the planet.

At this point in the encounter, the gravitational pull of Neptune on Voyager 2 was beginning to grow. Measurements of Voyager's acceleration toward the planet now permitted refinement of the planet's mass. Prior to FE, the value of Neptune's mass had been estimated purely on the basis of Earth-based measurements of the satellite orbits.

The temperature of IRIS, which had been turned on in OB, stabilized sufficiently to begin its planetary observations by the middle of FE. Its most important measurements in FE were the B922 temperature measurements of the disk of Neptune about nine days before planetary closest approach. At that time the apparent angular size of Neptune matched the 4° diameter IRIS field of view. Observing the disk of the planet in this way allowed atmospheric scientists to gain half of the information needed to determine how much heat the planet was radiating into space. The other half would have to wait for Voyager's outbound leg when the huge dark disk of Neptune had shrunk once again to the size of the IRIS field of view.

UVS observations similar to those done in OB continued during FE, but at a reduced frequency. Although the PPS had to be used sparingly to preserve its useful lifetime, it also began definitive measurements of the planet, the satellites, and the rings-arcs during FE. Daily high-rate samples of PRA and PWS data supplemented the lower-rate data from these two investigations and from the four fields and particles investigations.

Two days prior to the start of B923 (August 17 at 06:10 SCET), at 470 Neptune radii away, PRA and PWS first began to sense radio emissions from Neptune's magnetic field. Although actual penetration of the field would not occur until the Near Encounter Phase, radio waves generated by interaction of the solar wind with the planetary magnetic field were seen during the last seven days of FE. It was estimated that Voyager 2 would enter the planet's magnetosphere somewhere between 27 and 9 hours before Neptune closest approach.

Observations of Neptune continued to increase in frequency and intensity. The B923 sequence focused on atmospheric features, the ring-arc system and both previously known satellites. The fields, particles and waves instruments continued to study the radio noise from the planet. Engineers at JPL began work to 'tweak' some of the parameters for TCM B20, the 4 roll-turn pair TCM. They also prepared for the flurry of activity that was associated with both the critical LEU and LSU for Neptune closest approach. The stage was now set for the all-important NE observations.

Near Encounter Phase

The Near Encounter Phase included almost 90 Neptune optical remote-sensing science observations rated by the NSWG members as highest priority. NE extended from August 24 through August 29, 1989. CCS computer load B951 lasted 53 h; the B952 load lasted 72 h. The plan for Near Encounter would be to use Canopus (the brightest star in the southern constellation Carina) as the roll reference star. Prior to closest approach, the spacecraft would be rolled to +61° with respect to Canopus to align PLS and LECP for measurements of precipitating electrons over Neptune's north pole. After Neptune and just prior to Triton, the spacecraft was commanded to roll to the star Alkaid, which forms the end of the handle of the 'Big Dipper' (the star constellation Ursa Major), to align the PLS detectors for measurements of Neptune's co-rotating plasmas. Voyager 2 traversed Neptune's bow shock at 14:38 UT on August 24 [7].

By the start of the first Near Encounter sequence, Neptune was filling a WA frame and Triton, which had been just a few pixels across for most of the encounter, now spanned halfway across a NA field of view. At this point Voyager 2 was moving at approximately 61,000 km/h, and continuous data collection was typical of all the Voyager investigations. Comparison of the investigation findings yielded far more information about the Neptune system than would otherwise have been possible.

Table 7.4. Neptune encounter characteristics

Voyager 2 event	Time from Neptune C/A ($\pm$h:min)	Spacecraft event time (GMT h:min)	Distance (km)
Nereid C/A	−03:49.1	Aug 25, 1989 00:06.6	4,652,880
Ascending node	−01:02.6	Aug 25, 1989 02:53.0	85,290
Neptune C/A	00:00.0	Aug 25, 1989 03:55.7	29,240
Enter 63K Ring Sun occ.	+00:04.1	Aug 25, 1989 03:59.8	29,770
Enter 63K Ring Earth occ.	+00:05.1	Aug 25, 1989 04:00.8	30,060
Enter Neptune Sun occ.	+00:05.9	Aug 25, 1989 04:01.6	30,330
Enter Neptune Earth occ.	+00:06.4	Aug 25, 1989 04:02.1	30,500
Exit Neptune Earth occ.	+00:55.2	Aug 25, 1989 04:50.8	76,880
Exit Neptune Sun occ.	+00:56.0	Aug 25, 1989 04:51.6	77,780
Exit 63K Ring Sun occ.	+01:00.4	Aug 25, 1989 04:56.0	82,730
Exit 63K Ring Earth occ.	+01:00.5	Aug 25, 1989 04:56.2	82,880
Descending node	+01:19.2	Aug 25, 1989 05:14.8	103,950
Triton C/A	+05:14.4	Aug 25, 1989 09:10.1	39,790
Enter Triton Sun occ.	+05:43.7	Aug 25, 1989 09:39.4	50,930
Enter Triton Earth occ.	+05:43.8	Aug 25, 1989 09:39.5	50,960
Exit Triton Earth occ.	+05:46.4	Aug 25, 1989 09:42.1	52,740
Exit Triton Sun occ.	+05:46.6	Aug 25, 1989 09:42.3	53,010

Each observation link had a designated leading investigation; other investigations were 'riders' whose observation objectives also had to be considered. In the discussion below, the observations will be grouped by target (atmosphere, rings, satellites, or magnetosphere) and by lead investigation (RSS, IRIS, UVS, ISS, etc.), so the interested reader needs to remember that useful data was received by more than just the leading investigation. The Neptune encounter characteristics are presented in Table 7.4.

The most important (and most complex) of the atmospheric observations was the RSS radio occultation experiment. When telemetry data was being transmitted to Earth, the data consisted of 'dots and dashes' in a binary code, which were superimposed on the basic X-band radio frequency. Just before the spacecraft passed behind Neptune as viewed from Earth (Figure 7.10), the telemetry stream was turned off so that more power could be concentrated in the main X-band and S-band 'carrier frequencies' used by RSS. During passage behind the planet, the spacecraft was continuously maneuvered to keep the antenna pointed at the closest point along the planetary limb to the spacecraft–Earth line. The atmosphere of the planet refracted (i.e., deflected) the radio beam toward Earth, and radio signals were successfully received at the Australian tracking stations during the entire occultation period. The experiment provided pressure, temperature, and composition information about the deep atmosphere of Neptune.

Atmospheric temperature, pressure, and composition information was also provided by a series of IRIS observations. IRIS measurements near the latitude and longitude of the RSS occultation exit point enabled the two experiments to be

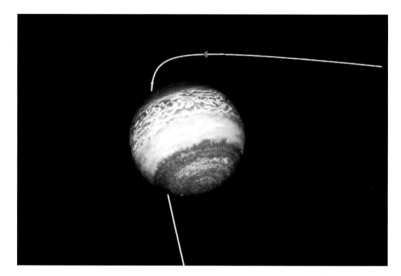

Figure 7.10. The path of Voyager 2 behind Neptune as viewed from Earth is depicted. The RS atmospheric and ring occultation experiments occurred during this period of time.

coupled to provide more accurate atmospheric composition data (especially, the helium abundance) than could be provided by either RSS or IRIS alone. Composition of the equatorial regions was obtained during long periods of staring at the planet during Far Encounter. Horizontal temperature variations were measured at closer range by scanning the IRIS field of view in latitude or in raster patterns across the illuminated and dark hemispheres.

PPS and ISS were used to obtain a series of seven combined measurements of the atmospheric brightness under a variety of solar illumination and viewing conditions. These would later be used in conjunction with the IRIS whole-disk temperature measurements almost seven days after closest approach to determine the planet's heat budget: the fraction of the observed thermal energy that comes from solar heating, and the fraction that comes from internal heat sources. ISS also executed one inbound mosaic of six WA images designed to provide relatively high-resolution views of the illuminated atmosphere and its cloud structure.

UVS observations of Neptune's atmosphere in NE included a series of mosaic and fixed recordings of auroral emissions, one bright limb profile, and two observations of occultations of the Sun. The auroral emission studies helped to verify the location of the magnetic pole and to characterize interactions of the atmosphere and the magnetic field. The limb profiles and occultation measurements provided information on the vertical distribution of hydrogen and other gases in the extreme upper atmosphere.

NE was also the prime period for the collection of CRS, LECP, MAG, PLS, PRA, and PWS data on the magnetosphere of Neptune. Frequent high-rate samples of PRA and PWS data were recorded for later playback. The LECP telescope systems were commanded to step to a new position in a specific pattern to avoid

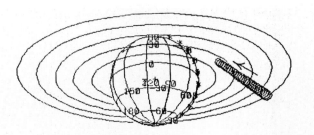

Figure 7.11. The apparent path of the star Nunki (sigma Sagittarii) due to the motion of Voyager 2 provided information on the high-resolution radial structure of the rings of Neptune.

interference with the PLS instrument. The stepping cyclic began at N−12 h and continued until N+24 h. Approximately 35 Neptune radii away, Voyager 2 crossed a well-defined Bow Shock.

Ring-arc observations were primarily the purview of ISS. Three inbound mosaics, which were shuttered during approach to the planet, had been designed to be retargeted to newly discovered ring material. Six hours prior to closest approach Voyager imaged a single WA frame above the planet's north pole in search of a polar ring. Two WA frames were shuttered near the time Voyager passed through the equatorial plane of the planet to provide a relatively high-resolution view of the entire radial extent of the rings. Three WA frames were shuttered while the spacecraft was in the shadow of Neptune. These provided viewing of the backlighted rings without the normal concurrent problem of glare associated with sunlight on the camera optics. It was also an ideal viewing geometry to look for fine dust material in the rings. Since these images were shuttered during spacecraft maneuvering in support of the RSS atmospheric occultation experiment, the timing and pointing of the images had to be carefully chosen to minimize image smear due to the combined effects of maneuvering, spacecraft motion, and ring motion. A full ring system mosaic was obtained during the outbound portion of the encounter. Two ring-arc movies were executed almost four days after closest approach.

Another ring observation was of the occultation variety. PPS and UVS combined to watch the star Nunki (sigma Sagittarii) as the motion of the spacecraft caused the rings to appear to move in front of it. The apparent path of Nunki (Figure 7.11) cut almost completely through the proposed location of the Neptune ring system.

At 03:56 on August 25, 1989, Voyager 2 arrived at the targeted aim-point for its Neptune encounter. Voyager passed the planet at a distance of 29,240 km from the planet's center. After 12 years of travel Voyager 2 had passed closest approach to its final planetary target.

A little more than one hour later the spacecraft radio signal, as viewed from Earth, passed behind the rings after the planetary occultation, enabling RSS to obtain a high-resolution radial profile of the rings. The RSS, PPS, and UVS data

Figure 7.12. This12-frame mosaic is of Triton's south polar cap. Notice that one frame on the limb has slightly poorer resolution due to a small error in the programmed IMC rates. (P-34754)

was used to disclose radial variations in the optical thickness of the rings at three different wavelengths. As Voyager 2 crossed the plane of the rings about 60 min before and 85 min after Neptune closest approach, PWS obtained high-rate samples of its data. Tiny particles striking the spacecraft at high relative velocities will be instantly vaporized and ionized (electrically charged) by the impact. The ionization process creates radiowaves in the frequency range sampled by PWS; the bursty radio signals detected by PWS during ring-plane crossing are thus a record of particle impacts on the spacecraft. ISS also shuttered three WA frames just prior to the outbound ring-plane crossing.

One hundred and ten minutes after closest approach Voyager changed its attention from the dark disk of Neptune to the brightly lit disk of Triton, still 3.5 h distant. Once again ISS was called upon to obtain high-resolution color imaging of Triton. Observations to study surface color and map its surface appearance yielded surface cartography, photometry, and inferred planetochemistry. The highest resolution imaging of Triton included 12 NA frames and required the use of IMC (Figure 7.12). Twelve different pitch rates and two average yaw rates were needed to obtain this mosaic.

Two minutes prior to closest approach, PPS and UVS took advantage of a Triton stellar occultation of Gomeisa (beta Canis Majoris). This occultation was used to detect a Triton atmosphere if other observations had failed to detect one or, if previously detected, determine its extent, optical depth, and structure. Prior to the end of this occultation, Triton occulted the Sun. RSS and UVS used this opportunity to probe the atmosphere all the way to the surface, obtain atmospheric composition and temperature structure, determine an electron density in its ionosphere, and accurately determine Triton's diameter and density.

Figure 7.13. Voyager was 4.86 million km from Neptune as it took this dramatic image of Neptune and Triton. Triton's orbit is about to take the satellite behind the planet as seen from Voyager. (P-34761)

Closest approach to Triton occurred at 09:10 UTC on August 25, 1989, at a range of 39,800 km from the satellite's center. Voyager 2 exited Neptune's magnetosphere 38 h after first entering it. More than two days after closest approach Voyager 2 experienced multiple Bow Shock crossings. At the time of these measurements, the estimated distance of the craft was more than 160 Neptune radii away from the planet. With the spacecraft diving southward at an angle of 48° to the ecliptic, all four remote-sensing instruments observed the illuminated crescent of both Triton and Neptune. Three days and 6.5 h after Neptune closest approach, Voyager imaged the crescents of both Neptune and Triton at the same time (Figure 7.13).

Deviations from expected Voyager and DSN performance did occur during NE, but the deviations were few in number and minor in their consequences. The Voyager Team and its aging hero had succeeded once again in producing a stellar performance.

Post-Encounter Phase

The Post-Encounter (PE) Phase was in some ways a little like the cleaning crew that does its work the morning after an all-night party. While the peak of exciting discoveries had passed, there was still much in the way of important observations and calibrations that remained to be done. PE began on August 29 and ended on October 2, 1989. It was divided into CCS computer loads B971 and B972. Among the more important activities of PE was the completion of Digital Tape Recorder (DTR) playback of the important recorded NE events. The highest priority events recorded on the DTR were transmitted to Earth twice to improve both the probability of successful return and to decrease data noise and data gaps often inherent in single playbacks.

PRA continued to monitor the rotation of Neptune's magnetic field for an additional few days until the pulsed radio waves were too weak to be detected. A combination of PRA data and MAG data was used to determine the rotation period of Neptune's magnetic field (and presumably its interior) to an accuracy of less than a minute.

During PE, IRIS continued its observations of the dark hemisphere of Neptune, basically repetitions in reverse time order of the IRIS observations performed in FE. After completion of the disk temperature measurements about 11 days after closest approach, IRIS was turned off for the last time. IRIS had indeed made it to Neptune and had acquired data that was critical to understanding planet's atmosphere.

UVS repeated its scans of the system and its auroral emission searches using observation designs similar to those done in OB and early FE. The Sun and stars used for the NE occultation experiments by UVS and PPS were recalibrated. Periodic calibrations of the fields and particles instrumentation also resumed, including the series of roll and yaw turns used to calibrate the MAG experiment.

Voyager Interstellar Mission

At the end of the Neptune encounter, both spacecraft began their last mission, the Voyager Interstellar Mission (VIM). Its objective was to extend NASA's exploration beyond the Solar System. The spacecraft would characterize the environment of the outer Solar System, search for the boundary between the supersonic flow of particles from the Sun and that region beyond (e.g., the heliopause), and characterize the interstellar medium. The trajectories of Voyagers 1 and 2 and Pioneers 1 and 2 can be seen in Figure 7.14.

VIM can be divided into three phases. The first is the Terminal Shock Phase. During this time the region of space where the supersonic solar wind slows from a supersonic flow to a subsonic flow will be studied. Changes in the magnetic field direction and changes in the plasma flow direction characterize this environment. Though the exact location of the Terminal Shock is not known, it is currently estimated to be reached by Voyager 1 sometime between 2001 and 2003; and Voyager 2 will reach that boundary a few years later. The Terminal Shock is expected to be between 80 and 90 AU from the Sun. Passage through the Terminal Shock will end this phase and begin the second phase, the Heliosheath Exploration Phase.

It is believed that the Heliosheath is dominated with solar wind particles and is influenced by the magnetic field from the Sun. No one knows the extent of the Heliosheath but it is estimated to be 'tens of AU' wide. Traveling at their current speeds, the twin spacecraft will take years to cross this region. The phase ends when the spacecraft pass through the Heliopause, that boundary between interstellar space and the region dominated by our Sun. Once passed, the Voyagers will begin to collect the ultimate VIM dataset: measurements of the interstellar medium.

After achieving a successful encounter of Saturn and its large moon Titan, Voyager 1 is heading northward at about 35° above the ecliptic. It is moving at approximately 3.5 AU per year in the direction of the Solar Apex (the direction that the Sun is moving relative to nearby stars). Voyager 2 is heading southward at about 48° below the ecliptic. It is moving at approximately 3.1 AU per year and also in the

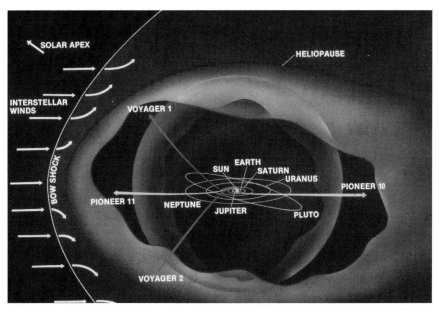

Figure 7.14. Diagram of the trajectories of Voyagers 1 and 2 and Pioneers 10 and 11. These spacecraft are the only craft made by humans that, through the use of gravitational assists, have enough energy to escape the Solar System. (JPL-12396)

general direction of the Solar Apex. Voyager 2 is moving slower than Voyager 1 as a consequence of the slower flyby of Saturn needed to continue to Uranus and Neptune and because its polar crown trajectory was slightly in front of Neptune as the planet orbits the Sun. In such a passage, rather than obtaining a gravitational assist and increasing its speed, Voyager 2 gave up energy to Neptune and, relative to the Sun, actually slowed down.

Currently there are five functioning instruments on Voyager 1 and six on Voyager 2. Both spacecraft have operational CRS, LECP, MAG, PRA, and PWS instruments. In addition Voyager 2's PLS instrument is still returning valuable data. The limiting factor for returning instrument data to Earth, assuming that no catastrophic failure occurs, is electrical power and fuel for attitude control. At launch, Voyager's Radioisotope Thermoelectric Generators (RTGs) were producing approximately 470 watts of power. With the natural decay of the plutonium dioxide fuel source, Voyagers 1 and 2 were each producing about 315 watts by the beginning of March 2001. As power levels decrease, more and more of the spacecraft's subsystems will have to be turned off. Except for UVS, the remote-sensing instruments on Voyagers 1 and 2 (IRIS, ISS, and PPS) were powered off soon after the Neptune encounter. The UVS instruments were left on to help to determine the distribution of hydrogen within the outer Heliosphere. By the year 2000 both UVS instrument were powered down to reduce power demands and keep electrical current flowing in the major spacecraft components. Next, supplemental heaters will be turned off. Power management will become more and more difficult as fewer

Table 7.5. The approximate year Voyager capability is reduced to preserve power

Event	Voyager 1	Voyager 2
Terminate scan platform and UVS observations	2000	1998
Termination of gyro operations	2011	2010
Termination of DTR operations	2010	2012
Initiate instrument power sharing	2018	2016
Can no longer support one instrument	2020	2020

Table 7.6. Nearby star encounters by Voyager 1 and 2 [8, 9]

Spacecraft	Star	Year of closest approach	Encounter range (light years)
Voyager 1	AC+79 3888	40,272	1.64
	DM+25 3719	146,193	2.35
	Kruger 60	172,867	2.85
	DM+21 652	497,482	2.08
Voyager 2	Barnard's Star	8,571	4.03
	Proxima Centauri	20,319	3.21
	Alpha Centauri	20,629	3.47
	Lalande 21185	23,274	4.65
	Ross 248	40,176	1.65
	DM+15 3364	129,704	3.44
	Sirius	296,036	4.32

non-critical operational items are left. The spacecraft RTGs will lose about 7 watts every year until they can no longer support minimum operations (see Table 7.5).

The hydrazine fuel for the attitude control system is also a spacecraft consumable. The average spacecraft use is about 6 to 8 g of fuel every week, and the consumption increases if a MAGROL calibration maneuver is required. MAGROLs are performed six times a year, consist of ten 360° roll turns, and are used to calibrate the magnetometers. Each of the Voyager spacecraft began its journey with 100 kg of fuel. At the beginning of March 2001, Voyager 1 had 31.6 kg remaining, while Voyager 2 had 33.5 kg.

Voyagers 1 and 2 do not need their thrusters to maintain their velocity. The Jupiter gravitational assist back in 1979 gave them enough energy to forever leave the influence of the Sun. When the race between fuel consumption, power output, and hardware lifetime is finally lost, the Voyager spacecraft will continue to glide silently through interstellar space (see Table 7.6).

NOTES AND REFERENCES

1. Rotational motion of a spacecraft can be described in terms of rotations around one or more of three axes: roll, pitch, and yaw. Roll motion turns the spacecraft around the axis

of the 3.7-m antenna, very nearly preserving antenna-pointing in space. Pitch and yaw motions are perpendicular to roll and to each other and change the antenna-pointing direction.

2. Jones, C. P. (1979) Automatic fault protection in the Voyager spacecraft. American Institute of Aeronautics and Astronautics, Paper No. 79-1919, p. 11.

3. Bergstralh, J. T. (1984) *Uranus and Neptune*. NASA Conference Publication 2330. National Aeronautics and Space Administration, Scientific and Technical Information Branch.

4. Miner, E. D., Ingersoll, A., Esposito, L., Johnson, T. and Wessen, R. (1985) *Science Objectives and Preliminary Sequence Designs for the Voyager Uranus and Neptune Encounters*. JPL Publication PD 1618-57. Jet Propulsion Laboratory, California, Pasadena.

5. Miner, E. D., Ingersoll, A., Kurth, W., Esposito, L. and Johnson, T. (1987) *Science Objectives and Preliminary Sequence Designs for the Voyager Neptune Encounter: Report of the Neptune Science Working Group*. JPL Publication D-4607/PD 1618-66. National Aeronautics and Space Administration, Jet Propulsion Laboratory, California Institute of Technology, Pasadena.

6. The hydrogen molecule, consisting of two hydrogen atoms, has two different structures. In one of these (*ortho*-hydrogen), the nuclear spins of the two atoms are parallel. In the other (*para*-hydrogen), the nuclear spins are anti-parallel. *Ortho*-hydrogen preferentially occupies odd-numbered molecular energy levels; *para*-hydrogen occupies even-numbered levels. The fraction of *para*-hydrogen in hydrogen gas is dependent on the temperature of the gas; for a temperature of 64 K, the equilibrium fraction of *para*-hydrogen is about 62%.

7. Schulz, M., McNab, M. C. and Lepping, R. P. (1995) Magnetospheric Configuration of Neptune. In Cruikshank, D. P. (ed.) *Neptune and Triton*. The University of Arizona Press, Tucson, pp. 233–77.

8. Kohlhase, C. (1989) *The Voyager Neptune Travel Guide*. JPL Publication 89-24. National Aeronautics and Space Administration, Jet Propulsion Laboratory, California Institute of Technology, Pasadena, p. 163.

9. Cesarone, R. J., Sergeyevsky, A. B. and Kerridge, S. J. (1983) Prospects for the Voyager Extra-Planetary and Interstellar Mission. *AAS/AIAA Astrodynamics Specialist Conference*, Lake Placid, New York. Paper 83-308, p. 24.

BIBLIOGRAPHY

Dodd, S. R. (1988) *Voyager 2 Neptune Encounter Near Encounter Phase Delivery Package*. JPL Internal Document. National Aeronautics and Space Administration, Jet Propulsion Laboratory, California Institute of Technology, Pasadena.

Kohlhase, C. (1987) *Voyager Mission Design Guidelines and Constraints Document: Uranus-to-Neptune Cruise/Neptune Encounter*. JPL Publication PD 618-123, Volume IV. National Aeronautics and Space Administration, Jet Propulsion Laboratory, California Institute of Technology, Pasadena.

Kohlhase, C. (1989) *The Voyager Neptune Travel Guide*. JPL Publication 89-24. National Aeronautics and Space Administration, Jet Propulsion Laboratory, California Institute of Technology, Pasadena.

Miner, E.D. (1987) *Voyager Neptune/Interstellar Mission: Flight Science Office Science and Mission Systems Handbook*. JPL Publication D-498/PD 618-128. National Aeronautics and Space Administration, Jet Propulsion Laboratory, California Institute of Technology, Pasadena.

Miner, E.D. (1998) *Uranus: the Planet, Rings and Satellites*. (2nd edn.) Wiley-Praxis, Chichester, UK.

Voyager Interstellar Mission facts: http://vraptor.jpl.nasa.gov/voyager/vimdesc.html

8

The interior of Neptune

8.1 CONSTRAINTS IMPOSED BY VOYAGER RESULTS

The planets of the inner Solar System (Mercury, Venus, Earth, and Mars) all possess solid surfaces, as does tiny Pluto in the outer reaches of the Solar System. The giant planets (Jupiter, Saturn, Uranus, and Neptune) have no such solid surfaces, but are instead enormous balls of mainly gaseous matter. Deep in their interiors, pressure from overlying materials, combined with heat-producing processes within each of these giant planets, create high temperatures. These interior temperatures are probably high enough to liquefy any normally solid materials in the deep interior or core of the planet.

Jupiter and Saturn differ from their smaller neighbors, Uranus and Neptune, in a number of significant ways. In addition to their much larger size and smaller distance from the Sun, Jupiter and Saturn have a different outward appearance at both large and small scales and both rotate once every 10 hours (approximately). Uranus and Neptune also have shorter rotations than Earth, but still require more than 16 hours for a single rotation. Jupiter and Saturn have 'surface' (cloud-top) temperatures of 165 and 134 K, respectively; those of Uranus and Neptune are 76 and 73 K, respectively. Jupiter and Saturn have a predominantly yellow color; Uranus and Neptune are distinctly blue. All four planets have internal magnets (as does Earth). (The nature and contents of the magnetic 'field' of Neptune, as seen by Voyager 2, will be discussed in detail in Chapter 10.) The magnetic poles of Jupiter and Saturn are more or less aligned with the rotation poles; but this is far from true for Uranus and Neptune.

The differences between the giant planets must also be reflected in the nature of their interiors. Although we cannot observe the interiors directly, a number of relevant clues are available, and if we consider these clues carefully, we may, like Sherlock Holmes, be able to deduce the nature of the things we cannot observe. This is essentially the task of the scientists who model the interiors of the planets. As with most aspects of planetary science, the same physical laws that govern processes in a scientific laboratory can be assumed to apply within the planet Neptune. If these laws are well enough understood, they can be formulated as mathematical expressions which describe the relationship between temperature, pressure, density, and chemical composition at various depths. The mathematical expressions that describe the conditions in the interior of Neptune then constitute the planet's interior model. The relevant laws of physics are not always well understood (primarily because we cannot fully reproduce within any laboratory on Earth the conditions of high temperature, high pressure, and precise chemical composition found in Neptune's interior), and as we continue our studies of the planets we also continue our schooling in the nature of the laws that govern processes within Neptune and other giant planets.

As Voyager 2 flew by Neptune, the precise motion of the spacecraft was tracked from Earth. This was accomplished primarily with two types of radio-wave data: Doppler tracking and ranging. The tracking data provided accurate measurements of Voyager 2's speed away from the tracking station; the ranging data provided accurate distances from Voyager 2 to the tracking station on Earth. By carefully watching how each of these measurements changed during the Neptune encounter and subtracting (a) the known motion of the tracking station around the center of the Earth, (b) the motion of the Earth around the gravitational center of the Earth–Moon system, and (c) the motion of the Earth–Moon system around the Sun, the residuals would be due to the gravity of Neptune. In fact, these measurements are so accurate that information can also be gleaned about how Neptune's mass is distributed within the planet's interior. Radio astronomers express this distribution as the mass of the planet (or more precisely, the mass, M, times the universal gravitational constant, G) and the gravity harmonics of the planet, J_2 and J_4, which describe the planet's deviation from a purely spherical mass distribution. Two other measurements were needed before the analysis could begin: the rotation period of the planet (i.e., the length of its day), and the oblateness (a measure of the equatorial bulge) of the planetary disk.

The rotation period of Neptune could normally be determined by time-lapse imaging. However, because no surface can be seen, and because high-speed winds, which vary with latitude, hide the motion of the bulk of Neptune, the best way to measure the length of a Neptune day is to monitor the interaction of electrically charged particles trapped in Neptune's magnetic field with electrically charged particles streaming out from the Sun in the so-called solar wind. The solar wind is relatively constant over short periods of time, and the magnetic field of Neptune is assumed to rotate with the bulk of Neptune's interior. Although the precise nature of the interaction between the solar wind and the trapped particles is poorly understood, the low-frequency radio signals generated by the interaction are believed to fluctuate with a period equal to the rotation period of the planet's interior.

The oblateness of Neptune in visible light was measurable from images of the planet's disk. The radio oblateness was roughly determined by an occultation experiment in which Voyager 2 passed behind Neptune as viewed from tracking stations on Earth. The path of the spacecraft, whose position relative to the center of Neptune was precisely determined from the Doppler tracking and ranging signals described earlier, passed behind Neptune at a relatively high latitude ($+61°$) and reappeared at

Figure 8.1. Geometry of the radio science occultation of Neptune as viewed from Earth. Ingress (entry) was at latitude +61°; egress (exit) was at latitude −42°. (P-33721AC)

Table 8.1. Neptune mass and other parameters from Voyager 2

Quantity	Units	Value
GM of Neptune	km^3/s^2	$6,835,427 \pm 27$
Mass of Neptune	kg	10.246×10^{25}
J_2 gravitational harmonic	–	$(3.538 \pm 0.009) \times 10^{-3}$
J_4 gravitational harmonic	–	$(-3.8 \pm 1.0) \times 10^{-5}$
1–bar equatorial radius of Neptune (A)	km	$24,764 \pm 20$
1–bar polar radius of Neptune (C)	km	$24,340 \pm 30$
Oblateness of Neptune disk ($1-C/A$)	–	0.0171 ± 0.0015
Mean density of Neptune	g/cm^3	1.64 ± 0.01
Rotation period of Neptune	h	16.11 ± 0.05

a latitude closer to the equator (−42°). The geometry of the occultation is illustrated in Figure 8.1. The measure of the radius of the planet at those two latitudes therefore provided an estimate of the difference between the polar and equatorial diameters of the planet at radio wavelengths.

The mass of Neptune and its gravity harmonics are given in Table 8.1. The oblateness of the Neptune disk, both at visible and at radio wavelengths, the rotation period of Neptune's interior, and the planet's average density, are also given in Table 8.1 [1].

The theoretical relationship between the rotation period of the planet, the planetary oblateness, and the gravitational harmonic, J_2, are shown in Figure 8.2.

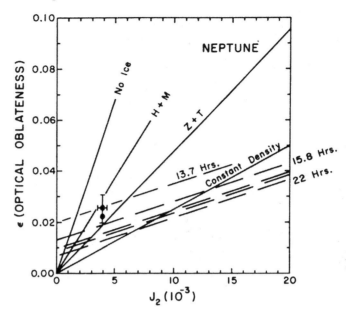

Figure 8.2. Theoretical relationship between rotation period, planetary oblateness, the radial mass distribution inside the planet, and the gravitational harmonic, J_2.

For a planet with a dense core, the rotation period would be expected, on the basis of this figure, to be much shorter. The implications of the actual, longer period of rotation are that Neptune is not likely to have a separate dense core, but that molten icy and rocky materials (if any molten rocky materials exist) may still be mixed within the interior.

The chemical composition and pressure/temperature structure of the detectable atmosphere of Neptune were determined from a combination of infrared spectrometer data and the radio occultation experiment during passage behind Neptune as viewed from Earth [2]. Compositions at levels higher in the atmosphere were determined by the ultraviolet spectrometer [3]. (The derived atmospheric structure and chemical composition of Neptune are discussed in Chapter 9.) The variations of temperature, pressure, and composition within Neptune must coincide near the sensible atmosphere with the observed characteristics, which therefore provide additional constraints which must be met by any proposed interior models.

Earth-based measurements of the microwave brightness of Neptune at many wavelengths also serve to constrain interior models. Microwaves generally originate in deep layers of the atmosphere, and the microwave brightness provides a measure of the temperature at those levels, except as modified by absorption of that radiation in the overlying atmosphere. Detailed interior models should also be consistent with the large amounts of heat escaping from the interior as measured by Voyager 2's infrared spectrometer [4].

Neptune interior models based on the Voyager 2 data are very similar to those derived for Uranus. These models are not unique: none fits the Voyager data perfectly, and none is even close enough to the Voyager data to provide some certainty about the nature of the interior of the planet.

8.2 MODELS CONSISTENT WITH VOYAGER DATA

The mathematical details associated with modeling the interior of Neptune are presented elsewhere [5]. They include specification of the relationships between mass, density, pressure, temperature, composition, and physical state at various depths within the planet. It is not the purpose of this chapter to duplicate those mathematical derivations, but rather to summarize the salient features of the resulting models.

While the composition of the atmosphere of Neptune above its clouds is known from Voyager 2 measurements, the composition of the deeper layers of the planet cannot be determined precisely . Consequently, in models of the interior the composition is best described in terms of three general types of material, namely, 'gas', 'ice', and 'rock'. These descriptions are compositional only, and not an indication of temperature or physical state. In this context, gas includes primarily hydrogen and helium; ice includes methane, ammonia, and water; rock includes heavier materials such as silicon and iron. Because of the high temperatures and pressures in the interior of Neptune, the gas, ice, and rock components are all expected to behave as liquids.

Post-Voyager models of the interior of Neptune by different authors [6] all possess some characteristics in common. In each of the models the rock core is relatively small (or non-existent), constituting less than 2% of the total mass of the planet. Those models that have rock cores estimate that the density of the core lies between 10 and 12 g/cm^3.

If a central core of molten rocky materials exists, it is overlain by a thick layer of predominantly icy materials. Close to 90% of the total mass of Neptune is contained in this intermediate region, which varies in density from about 1 g/cm^3 near its outer boundary to about 5 g/cm^3 near its inner boundary. If there is a rocky core, the density at the boundary, which is about 0.1 to 0.2 of the total radius, is slightly less than 5 g/cm^3; if there is no rocky core, the central density of this inner portion of the interior is slightly greater than 5 g/cm^3.

The outermost region of the planet is predominantly hydrogen and helium, but with concentrations of water, methane, and ammonia enhanced by some factor, A, over solar abundances. The radio occultation experiment (measurements of the changes in the Voyager radio signal at Earth-based receiving stations after passage of that signal through the atmosphere of Neptune) indicates that A is between 30 and 60 for methane at pressures greater than 1 bar. Although this is a higher value than that given for Uranus, the uncertainties in the measurements mean that the methane abundance in each of these planets is essentially indistinguishable. This upper portion of Neptune's interior is predominantly gaseous and occupies the outer 10 to 20% of Neptune's radius. It is unclear from the evidence whether there is a

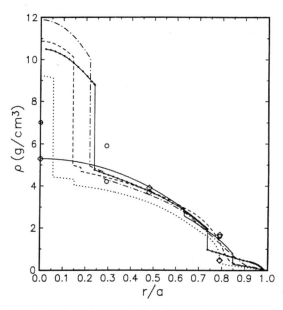

Figure 8.3. Variation of density with depth in the interior of Neptune for a number of different models, including one model of Uranus (dotted line). The Uranus model has lower densities at essentially all depths because Uranus has an almost identical radius and a smaller total mass than Neptune.

gradual or a sharp transition between this layer and the deeper layer(s). This outermost layer contributes less than 15% of Neptune's total mass.

Although the details of the models presented by different authors [6] vary, their predictions of density variation with depth are remarkably similar (see Figure 8.3). This similarity in density profiles occurs because the distribution of density with depth is reasonably well constrained by the mass and the J_2 of Neptune. The resulting models of Neptune's interior are not unique; each author has a certain amount of freedom to insert his or her own personal biases regarding the internal composition of Neptune, provided those models continue to be consistent with the mass and the J_2 gravitational harmonic of Neptune.

The variation of temperature with depth (and to a lesser extent, the variation of pressure with depth) is similarly poorly constrained. The pressure at the centre of Neptune may be close to 8 million bar. Central temperatures are generally calculated assuming a temperature increase with depth which is adiabatic [7]. With such a 'lapse rate', the central temperature of Neptune probably lies between 6,000 and 10,000 K, essentially identical with similar estimates for Uranus.

Although Neptune is half again as far from the Sun as Uranus and therefore receives only about $4/9$ [$= (1/1.5)^2$] as much solar radiation as Uranus, the cloud-top temperatures of the two planets are remarkably similar. Neptune's extra thermal energy comes from the interior of the planet. Although the amount of energy per square centimeter is less than that for Jupiter or Saturn, the *ratio* of the internal heat from Neptune to the heat due to sunlight is larger than for any other planet in our Solar System. This strange blue planet emits nearly 2.6 times as much heat as it received from the Sun [4]. Uranus, by way of contrast, emits less than 1.13 times the energy it receives from the Sun, and measurements of Uranus are consistent with no internal heat whatsoever (i.e., the Uranus heat balance may be 1.00) [8]. Although Uranus and Neptune are similar in size and mass, there are clearly some major differences in their present thermal behavior. Because the absence of any appreciable internal heat from the interior of Uranus is basically unexplained, it is not surprising that internal models do not predict this major difference between these two giant planets.

8.3 SUMMARY

In summary, our present understanding of the interior of Neptune is one in which there are either two or three main regions (see Figure 8.4). The upper region is primarily gaseous hydrogen and helium, with a methane abundance enhanced over solar values by a factor of 30 to 60 and occupying the outer 10 to 20% of the planet. Beneath the upper region is a region of icy materials, possibly mixed with rocky materials, ranging in density from almost $1 \, g/cm^3$ at its outer edge to about $5 \, g/cm^3$ at its inner edge. That inner edge may be the center of Neptune, or there may be a denser core of rocky materials inside the icy region. If such a central rocky core exists, its size is much less than the size of the Earth.

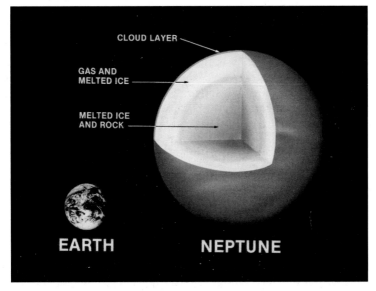

Figure 8.4. Representation of the interior structure of Neptune, depicting the extent of the three main regions. It is possible that no central core of molten rock exists. (JPL-12266BC)

NOTES AND REFERENCES

1. Adapted from Table I on page 112 of Hubbard *et al.* (1995) The interior of Neptune. In Cruikshank, D. P. (ed.) *Neptune and Triton*. University of Arizona Press, Tucson.

2. Radio science measurements determine the temperature and mean molecular weight of the atmospheric gases; the primary constituent gases of the atmosphere above the cloud tops is determined from the infrared spectra. By combining the results of the two experiments, it is possible to determine the gases that are present at various altitudes above the cloud tops and their relative amounts. The Voyager 2 radio science preliminary results were reported in Tyler, G. L., Sweetnam, D. N., Anderson, J. D., Borutzki, S. E., Campbell, J. K., Eshleman, V. R., Gresh, D. L., Gurrola, E. M., Hinson, D. P., Kawashima, N., Kursinski, E. R., Levy, G. S., Lindal, G. F., Lyons, J. R., Marouf, E. A., Rosen, P. A., Simpson, R. A. and Wood, G. E. (1989) Voyager radio science observations of Neptune and Triton. *Science*, **246**, 1466–73; the infrared spectrometer preliminary results were reported in Conrath, B., Flasar, F. M., Hanel, R., Kunde, V., Maguire, W., Pearl, J., Pirraglia, J., Samuelson, R., Gierasch, P., Weir, A., Bézard, B., Gautier, D., Cruikshank, D., Horn, L., Springer, R. and Shaffer, E. (1989) Infrared observations of the Neptunian system. *Science*, **246**, 1454–9.

3. Ultraviolet spectra, obtained both during normal viewing of Neptune and by viewing the Sun through the atmosphere of Neptune, provide data on the gaseous and ionic composition of the upper atmosphere. The Voyager 2 ultraviolet preliminary results were reported in Broadfoot, A. L., Atreya, S. K., Bertaux, J. L., Blamont, J. E., Dessler, A. J., Donahue, T. M., Forrester, W. T., Hall, D. T., Herbert, F., Holberg, J. B., Hunten, D. M., Krasnopolsky, V. A., Linick, S., Lunine, J. I., McConnell, J. C., Moos, H. W., Sandel, B. R., Schneider, N. M., Shemansky, D. E., Smith, G. R., Strobel, D. F. and Yelle, R. V. (1989) Ultraviolet spectrometer observations of Neptune and Triton. *Science*, **246**, 1459–66.

4. Information on the details of the calculations used to determine the Neptune internal heat flux are found in Pearl, J. C. and Conrath, B. J. (1991) The albedo, effective temperature, and energy balance of Neptune as determined from Voyager data. *Journal of Geophysical Research*, **96**, 18,921–30.

5. Hubbard, W. B., Podolak, M. and Stevenson, D. J. (1995) The interior of Neptune. In Cruikshank, D. P. (ed.) *Neptune and Triton*. University of Arizona Press, Tucson, pp. 109–38.

6. Podolak, M., Reynolds, R. T. and Young, R. (1990) Post Voyager comparisons of the interiors of Uranus and Neptune. *Geophysical Research Letters*, **17**, 1737–40. Hubbard, W. B., Nellis, W. J., Mitchell, A. C., Holmes, N. C., Limaye, S. S. and McCandless, P. C. (1991) Interior structure of Neptune: Comparison with Uranus. *Science*, **253**, 648–51. Podolak, M. and Marley, M. (1991) Interior model constraints on superabundances of volatiles in the atmosphere of Neptune. *Bulletin of the American Astronomical Society*, **23**, 1164 (abstract); Zharkov, V. N. and Gudkova, T. V. (1991) Models of giant planets with a variable ratio of ice to rock. *Annale Geophysicae*, **9**, 357–66.

7. If the volume of gas confined in an insulated container (one which allows no heat flow through its walls) is compressed, the absolute temperature of the gas will rise by an amount that is determined by the changes in the volume and the pressure of the gas. Furthermore, for a gas composed of 90% H_2 and 10% He, the increase, in pressure is related to the decrease in volume by the relationship, $pV^{1.44} = $ constant. For an ideal gas, $p_iV_iT_f = p_fV_fT_i$, where i and f, respectively, denote the initial and final values of the

pressure (P), volume (V), and absolute temperature (T). An adiabatic lapse rate is one for which the temperature of the gas increases with pressure by precisely the same amount as it would if it were confined in such an insulated container. Mathematical combination of the two above relationships yields $T_f = T_i(p_f/p_i)^{0.30}$. Thus, for an adiabatic lapse rate in a gas which is 90% H_2 and 10% He, the temperature approximately doubles for each decade increase in pressure.

8. Pearl, J. C., Conrath, B. J., Hanel, R. A., Pirraglia, J. A. and Coustenis, A. (1990) The albedo, effective temperature, and energy balance of Uranus, as determined from Voyager IRIS data. *Icarus*, **84**, 12–28.

BIBLIOGRAPHY

Bergstralh, J. T., Miner, E. D. and Matthews, M. S. (eds) (1991) *Uranus*. University of Arizona Press, Tucson. (See chapter on pp. 29–61 by Podolak, Hubbard and Stevenson, entitled 'Models of Uranus' interior and magnetic field' and the references cited therein.)

Cruikshank, D. P. (ed.) (1995) *Neptune and Triton*. University of Arizona Press, Tucson. (See chapter on pp. 109–38 by Hubbard, Podolak and Stevenson, entitled 'The interior of Neptune', and the references cited therein.)

Hubbard, W. B. (1999) Interiors of the giant planets. In Beatty, J. K., Petersen, C. C. and Chaikin, A. (eds) *The New Solar System (Fourth Edition)*. Sky Publishing Corporation, Cambridge, MA, pp. 193–200.

9

The atmosphere of Neptune

9.1 ATMOSPHERIC FEATURES

Neptune is the most distant of the giant planets of the Solar System. The primary gas detectable in its atmosphere by remote-sensing techniques is hydrogen, the simplest of the elements. As with the other giants, this hydrogen exists mainly as diatomic molecular hydrogen, H_2. Hydrogen is the most abundant element in the Sun, in the Solar System as a whole, in the Milky Way Galaxy, and in the universe. It has therefore become common practice to specify the abundance of other elements and gases in terms of their amounts relative to hydrogen.

Helium (He), whose atomic nucleus contains two protons and two neutrons, is the second most abundant gas in the Sun and in the atmosphere of Neptune and the other giant planets. Helium gas is an inert, non-flammable gas that interacts chemically with few other elements. Both hydrogen and helium gases are essentially transparent to visible light. Although hydrogen has a few narrow absorption lines within the visible spectrum of light, most of the strong absorption lines and bands in the spectrum of these two gases occur in the ultraviolet part of the spectrum at wavelengths invisible to the human eye.

The dominant bluish tint of Neptune is due to the third most abundant gas in the planet's upper atmosphere, namely methane (CH_4). Unlike the colorless hydrogen and helium, methane gas and ice readily absorb sunlight at the red end of the spectrum. The resulting subtraction of red from sunlight incident on the atmosphere and cloud tops of Neptune results in the characteristic blue tint of the planet. This contrasts markedly from the browns and yellows of the atmospheres of Jupiter and Saturn, where methane absorption of the redder parts of the spectrum is insignificant.

Water (H_2O) and ammonia (NH_3) are probable constituents of the deep atmosphere of Neptune. The temperature of Neptune's upper atmosphere reaches a minimum at a pressure level of 100 millibars, or about one-tenth of the pressure of the Earth's atmosphere at sea level. At that pressure level, the atmospheric temperature is about 55 K. The temperature increases with depth in the atmosphere, reaching about 70 K near the 1-bar level (1 bar is very nearly equivalent to Earth's sea-level atmospheric pressure). Methane gas rising from the interior of Neptune remains in a gaseous state until it reaches the atmospheric temperature of about 80 K, where it freezes to form methane ice clouds. The base of the methane clouds on Neptune is therefore just below the 1-bar level, but their tops may extend to, or above, the 1-bar level. These clouds apparently do not form an opaque, continuous layer but permit visibility down to a lower, more continuous cloud layer, whose composition remains unknown. By analogy with Jupiter and Saturn, the lower cloud layer was thought to be ammonia, which freezes at a temperature of approximately 125 K, where the atmospheric pressure is between 3 and 4 bars. It is perhaps more likely that the cloud layer is composed of hydrogen sulfide (H_2S), which freezes at a temperature of about 145 K, near the 5-bar pressure level. Water freezes at much deeper layers in the atmosphere; its freezing temperature of roughly 275 K may not be reached until the pressure is nearly 50 bars! Upward motions in the atmosphere may carry small amounts of ammonia and water to much higher levels in the atmosphere, but not nearly high enough or in sufficient quantities to be detected by Voyager or Earth-based observations.

9.1.1   Helium abundance

One of Voyager 2's main scientific objectives in the atmospheric studies of the giant planets was to determine the helium abundance in each planet. This abundance is generally expressed as the ratio of the mass of helium to the total mass of the atmosphere, but may also be expressed as a ratio of the numbers of helium molecules to total molecules, or the helium mole fraction. Initially, the giant planets were expected to be relatively unaltered reservoirs of hydrogen and helium from the primitive solar nebula [1]. It was assumed that the giant planets would have the same helium-to-hydrogen ratio, and that the measured value would be representative of the helium-to-hydrogen ratio in the material from which the Solar System was formed. The actual measurements by Voyager have shown that assumption to be flawed.

Measurements of the helium abundance in the outer atmosphere of the Sun [2] yield a helium mass fraction near 0.28 (corresponding to a helium mole fraction near 0.16). Thermonuclear reactions within the Sun's core are constantly creating helium out of hydrogen, so the solar helium abundance was expected to represent an upper limit for the helium abundance in the primordial Solar System. Initial analysis of the Voyager results at Jupiter [3] yielded a helium mass fraction of 0.18 ± 0.04 (helium mole fraction of 0.10 ± 0.02); similar analysis for Saturn [4] was interpreted as a helium mass fraction of 0.06 ± 0.05 (helium mole fraction of 0.03 ± 0.02). However, direct measurements by the atmospheric probe carried by the Galileo Mission to Jupiter showed that the actual helium mass fraction for Jupiter was 0.234 ± 0.005 (helium mole fraction of 0.157 ± 0.003), much closer to the solar value. The error in the Voyager measurement apparently lay in incorrect correlation of radio science and infrared data. Using infrared data only [5], the Voyager result for Saturn is

0.22 ± 0.04 (mole fraction 0.14 ± 0.03), which, while lower than the Sun's value of 0.28 ± 0.01, is not nearly as extreme as previously believed.

The high (greater than 3 million bars) pressures in the interiors of Jupiter and Saturn transform hydrogen into a metallic state [6]. Temperatures in the interiors of Jupiter and Saturn are low enough to allow the helium to separate from the hydrogen and form liquid droplets. Because of their higher density, these droplets sink slowly through the metallic hydrogen, resulting in a depletion of helium at higher levels, which eventually manifests itself in lower helium abundance in the outer atmosphere. The high pressures needed to transform hydrogen into its metallic form occur at lower temperatures inside Saturn than inside Jupiter. Helium depletion should therefore be more pronounced in the outer atmosphere of Saturn than in the outer atmosphere of Jupiter, consistent with the helium abundance numbers cited above.

Pressures within the 'gas' regions of Uranus and Neptune are never high enough to create metallic hydrogen. As a consequence, helium depletion by the same mechanism invoked for Saturn is ineffective, and the helium abundance of Uranus and Neptune may be representative of the primordial solar nebula. If Uranus or Neptune were formed with an abundance of elemental carbon (C) or nitrogen (N), combination with hydrogen to form methane (CH_4) or ammonia (NH_3) would reduce the amounts of hydrogen and increase the mole fraction of helium [7]. Other authors [8] suggest that at high temperatures (above 2,000 K) and high pressures (above 200,000 bar) methane has a tendency to break apart into its carbon and hydrogen components. This would lead to an increase of hydrogen in the outer atmosphere, effectively reducing the helium mole fraction. Neither of these mechanisms is likely to have caused substantial changes to the helium mass fractions of either Uranus or Neptune. The best estimates of the helium mass fractions for the Sun and the four giant planets of our Solar System are shown in Figure 9.1.

9.1.2   Methane abundance

Methane is the only 'ice' component of Neptune whose abundance can be determined directly from Voyager 2 data. While water and ammonia are believed to be major components of the interior of Neptune, they would be detectable only at depths beyond those probed by Voyager 2 instrumentation (see Figure 9.2).

The abundance of methane was determined from radio occultation measurements during the period when Voyager 2 flew behind Neptune as viewed from Earth. Figure 9.3 depicts the path of Voyager 2 and the raw radio science data for that period. The intensity of the signal received at Earth dropped precipitously as the spacecraft all but disappeared behind Neptune. Signals were detected continuously during the entire 49-min period of the occultation. During two relatively brief periods during the occultation the signal strength decreased even further before returning to prior levels. These two features correspond to those times when the radio signals reached depths corresponding to a methane cloud deck. Between the times of these two events, part of the radio beam penetrated the Neptune atmosphere

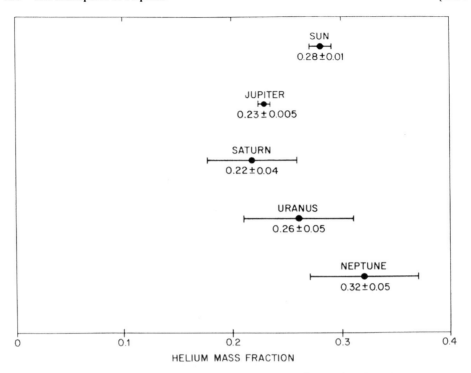

Figure 9.1. The helium mass fractions for the outer atmospheres of Jupiter, Saturn, Uranus, Neptune, and the Sun. Uncertainties are shown by the superimposed error bars.

to depths below the methane cloud deck and part of the beam was scattered by the cloud itself.

Analysis by Gautier et al. [9] of Voyager 2 and Earth-based measurements led to the conclusion that the methane mole fraction was between 0.02 and 0.04 (corresponding to a methane mass fraction of 0.17 ± 0.06). Assuming that the vast majority of carbon (C) in Neptune's atmosphere is tied up in methane, that amount of methane leads to a carbon abundance (C/H) between 30 and 60 times the solar value! Within the uncertainties of the respective measurements, the methane abundance on Uranus and Neptune are the same.

9.1.3   Minor constituents

No chemical constituents other than methane (CH_4), helium (He), and molecular and atomic hydrogen (H_2 and H) were detected by Voyager in the lower atmosphere of Neptune. Acetylene (C_2H_2) was positively detected in the upper atmosphere, and there is also evidence for ethane (C_2H_6) [10]. Both of these gases are byproducts of methane photochemistry. The estimated mixing ratios (amounts relative to hydrogen) in the stratosphere near the millibar pressure level are 5×10^{-8} for acetylene and 10^{-6} for ethane [11]. Other expected byproducts of methane photo-

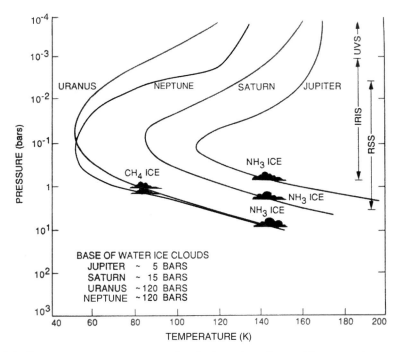

Figure 9.2. Temperature–pressure curves for the four giant planets. The approximate pressure levels probed by ultraviolet, infrared and radio science investigations are shown at the right of the figure. The approximate condensation temperatures and corresponding pressure levels for methane, ammonia, and water ice clouds are indicated.

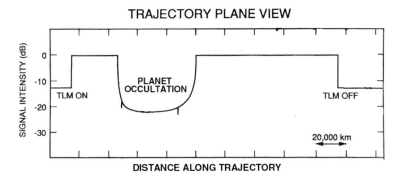

Figure 9.3. Radio science data for the occultation period are displayed for the period that Voyager 2 was behind Neptune as viewed from Earth. Abrupt dips in the X-band signal intensity occur as the radio signal passes through a 2- to 3-km thick methane cloud layer whose base lies near the 1.9-bar pressure level.

chemistry at Neptune include the polyacetylenes ($C_{2n}H_2$ with $n = 2, 3$, etc.), the most important of which is diacetylene (C_4H_2) [12]. Although these minor constituents of the Neptune atmosphere contribute little to the gaseous composition of the atmosphere, they condense in the upper atmosphere to form hazes which affect the atmosphere's reflective characteristics, raise the temperature of the upper atmosphere and lower the temperature of the underlying atmosphere.

Earth-based observations at submillimeter wavelengths have identified hydrogen cyanide (HCN) in the stratosphere and carbon monoxide (CO) in the stratosphere and troposphere of Neptune [13]. The estimated mixing ratios are $\sim 10^{-9}$ for hydrogen cyanide and $\sim 10^{-6}$ for carbon monoxide [13]. Nearly simultaneous measurements of Uranus with the same observing equipment failed to detect any evidence of either of these gases. The presence of either of these gases in the stratosphere of Neptune is somewhat mystifying, since atmospheric scientists had supposed that nearly all the carbon was contained in methane and nearly all the nitrogen was contained in ammonia. It is possible that the carbon monoxide seen in the stratosphere is produced in the troposphere; the minimum temperature of 55 K is not cold enough to freeze out and hence trap carbon monoxide below the 100-mbar pressure level. Hydrogen cyanide, on the other hand, freezes out at temperatures well above the 55 K minimum in Neptune's atmosphere. Any hydrogen cyanide produced below the 100-mbar level in the atmosphere is effectively cold-trapped well below the stratosphere. The presence of hydrogen cyanide in the stratosphere implies that the nitrogen may have a source outside the atmosphere (nitrogen-rich meteorites or neutral atoms of nitrogen from the surface of Neptune's largest moon, Triton) or that vigorous upward motions of nitrogen (N_2) gas from the deep troposphere might be sufficient to carry the nitrogen across the minimum temperature of 55 K at the top of the troposphere. Extensive discussions of the hydrogen cyanide problem are given by Bishop *et al.* [14] and by Gautier *et al.* [9].

9.1.4   Isotopic ratios

In addition to the abundances of different elements in the atmospheres of the giant planets, it is instructive to look at the relative abundances of different isotopes of the same element. Isotopes differ only in the number of neutrons in their nuclei. So, for example, deuterium (D) is the same element as hydrogen, but contains both a proton and a neutron in the atomic nucleus. That extra neutron changes slightly the absorption spectrum and is relatively easily detectable. Comparison of the relative amounts of methane (CH_4) and deuterated methane (CH_3D) [15] yield a D/H ratio of order 10^{-4}, a factor of 2 to 3 higher than for the Sun. The enhancement is comparable to that for Uranus and may be due to a larger proportion of meteoritic and cometary material in these planets than in Jupiter or Saturn.

Another interesting isotopic ratio is that of carbon-12 (^{12}C) to carbon-13 (^{13}C). As in the case of D/H, the $^{12}C/^{13}C$ ratio is thought to be higher for cometary material that may be more abundant in Uranus and Neptune than in Jupiter and Saturn. The $^{12}C/^{13}C$ ratio for Neptune was measured for ethane (C_2H_6) near

$820\,\text{cm}^{-1}$ ($\sim 1.2\,\mu\text{m}$) in the infrared [11] and found to be 78 ± 26, consistent with the common Solar System value of 90 [16].

9.2   GLOBAL ENERGY DISTRIBUTION

Although Neptune is half again as far from the Sun as Uranus (and therefore less than half as much solar energy falls on its atmosphere), temperatures in the lower atmospheres of the two planets are almost the same. This unusual circumstance is a consequence of differences in the amount of internal heat escaping from the two planets. Whereas Uranus appears to have little or no heat escaping from its interior (at least during the time of the 1986 Voyager 2 encounter), Neptune radiates more than twice the thermal energy it absorbs from the Sun. In other words, the heat escaping from the interior of Neptune is greater than the amount of heating due to the rays of the Sun falling on the planet.

One of the principal scientific objectives of the Voyager 2 encounter of Neptune was to determine the nature and distribution of the escaping heat from Neptune's interior. To accomplish that goal, Voyager 2 had first to make accurate measurements of the temperature and reflectivity of Neptune under a variety of illumination conditions. This data, combined with the knowledge that Neptune's distance from the Sun (at the time of the Voyager 2 encounter) diminished the intensity of sunlight illuminating the planet by a factor of 908 relative to Earth, would enable scientists to calculate the fraction of radiated heat that was generated in the planet's interior. Precise determination of the internal heat helps to differentiate between model predictions for Neptune.

9.2.1   Amount of sunlight absorbed

An accurate estimate of the amount of sunlight absorbed by Neptune requires (1) knowledge of the amount of sunlight incident on the planet and (2) determination of the amount of sunlight reflected back into space. The difference between these two quantities represents the solar energy absorbed by the planet.

The total flux of solar radiation per unit area received outside the Earth's atmosphere at the average Sun–Earth distance (1.000 AU) is known as the solar constant. Its value is slightly variable, but seems to range between 1,365 and 1,369 W/m². At greater distances from the Sun, the flux of solar energy per unit area drops by a factor of r^{-2}, where r is the distance from the Sun in AU. The mean Sun–Neptune distance is 30.110 AU; the total flux of solar radiation at the distance of Neptune is therefore 1.508 ± 0.002 W/m². For reference, the Sun–Neptune distance at the time of the Voyager 2 encounter on August 25, 1989, was 30.214 AU. The corresponding incident solar flux was 1.497 ± 0.002 W/m².

The brightness of sunlight scattered from the planet was carefully measured by Voyager 2 for a variety of illumination and viewing conditions. From these and earlier Earth-based measurements it was concluded [17] that, when averaged over the 164-year orbital period of Neptune, the ratio of solar energy scattered away to space relative to that incident upon Neptune was 0.29 ± 0.07. In other words, 0.71 ± 0.07 of the solar radiation incident upon Neptune during its passage around the Sun is being absorbed by the planet.

Sunlight illuminates the planet from one direction only. The heat escapes primarily as infrared radiation which is emitted in all directions from both the illuminated and dark parts of the planet. One must therefore effectively reduce the solar flux per unit area by an additional factor of 4 (the ratio of the area of a sphere to that of a circle of the same radius) before comparing it to the heat energy being radiated by the planet. The effective planet-wide solar flux input, averaged over a Neptune year, is thus $(1.508)(0.71)/4 = 0.27 \pm 0.03 \, \text{W/m}^2$. From the Stefan–Boltzmann law [18], the solar heating alone would result in a planet-wide average equilibrium temperature [17] of $46.6 \pm 1.1 \, \text{K}$.

9.2.2   Amount of re-radiated energy

Neptune has a relatively normal polar tilt of about $29.6°$ to its orbital plane (Earth's tilt is $23.44°$). Over the 164-year orbital period, Neptune receives more total solar radiation near its equator than near its poles by more than a factor of 2. The solar input should give rise to an equator-to-pole difference in temperatures of at least 3 K. However, within the troposphere, the poles are less than 1 K cooler than the equator. It is apparent that some mechanism for transfer of heat from the equator toward the poles must be occurring within Neptune. The most likely mechanism is a convective interior that is somehow in dynamic contact with the portions of the atmosphere where solar radiation is absorbed. The circulation patterns then effectively block internal radiation from escaping at Neptune's equator, forcing most of it toward higher latitudes.

Variation in temperature with longitude is even less than that from equator to pole, especially near the cloud tops. While temperatures may vary over relatively short periods of time in areas where storms are present, the average zonal temperatures appear to change over timescales comparable to or greater than the 164-year orbital period, rather than with the 16.11-hour rotation period.

If the atmosphere did not selectively absorb some of the emitted radiation, the brightness of Neptune at any thermal infrared wavelength would provide a measure of the temperature of the planet. The actual situation requires modeling of the atmosphere and the infrared spectrum to provide a measure of the total thermal flux escaping the planet. The errors introduced by such modeling are not large (generally on the order or 0.1 K or so). The Neptune effective temperature as derived from Voyager measurements is $59.3 \pm 0.8 \, \text{K}$. This number is derived from an estimated average thermal flux emitted from the planet of $0.70 \pm 0.04 \, \text{W/m}^2$ [19].

9.2.3   How large is the internal heat source within Neptune?

A simple comparison of the thermal flux emitted by Neptune to the absorbed solar flux makes it apparent that internal heat is the primary source of heat in the Neptune atmosphere. Without the internal heat source, Neptune's emitted flux would be more

Table 9.1. Energy balance for Jupiter, Saturn, Uranus, and Neptune

	Jupiter	Saturn	Uranus	Neptune
Mean solar distance (AU)	5.203	9.555	19.218	30.110
Incident flux (W/m^2)	50.50	14.97	3.70	1.51
Total reflectivity	0.34 ± 0.03	0.34 ± 0.03	0.30 ± 0.05	0.31 ± 0.07
Reflected flux (W/m^2)	17.32 ± 1.62	5.12 ± 0.45	1.10 ± 0.18	0.47 ± 0.10
Absorbed flux/4 (W/m^2)	8.29 ± 0.40	2.46 ± 0.11	0.65 ± 0.05	0.26 ± 0.03
Equilibrium temperature (K)	109.5 ± 1.4	82.4 ± 0.9	58.2 ± 1.0	46.6 ± 1.1
Effective temperature (K)	124.4 ± 0.3	95.0 ± 0.4	59.1 ± 0.3	59.3 ± 0.8
Emitted flux (W/m^2)	13.73 ± 0.13	4.47 ± 0.08	0.68 ± 0.01	0.70 ± 0.04
Energy balance	1.66 ± 0.09	1.82 ± 0.09	1.05 ± 0.08	2.69 ± 0.28
Internal flux (W/m^2)	5.46 ± 0.43	2.01 ± 0.14	0.03 ± 0.05	0.44 ± 0.05

than a factor of 2.5 smaller than was actually measured. The ratio of these fluxes, sometimes known as the 'energy balance,' is 2.61 ± 0.28. Inspection of Table 9.1, adapted in part from Pearl and Conrath [17], reveals that Neptune has the largest energy balance of the giant planets, although the actual amount of internal flux is much smaller than that of either Jupiter or Saturn.

It is apparent from Table 9.1 that the amount of internal heat escaping from their interiors varies greatly from planet to planet, and that each (with the notable exception of Uranus) has a substantial contribution from internal heat. While we cannot be certain what causes these differences, it is highly likely that vertical motions within the visible atmosphere accompany the escaping heat. As will be shown later in this chapter, Neptune's atmosphere apparently has much more storm activity than is true for Uranus, and it is possible that the apparently shorter lifetimes of Neptune atmospheric storms (relative to storms on any of the other three giant planets) may also be related to the higher energy balance of Neptune.

9.3 VERTICAL STRUCTURE WITHIN THE ATMOSPHERE

A 'black-body' has uniform temperature, absorbs all sunlight incident upon it, and re-radiates the same amount of energy in a mathematically precise fashion. The amount of thermal radiation at any given wavelength (color) of light is defined by a relationship known as the Planck function, after German physicist Max Planck (1858–1947). It shows that a very hot black-body (several tens of thousands of degrees Kelvin) emits most of its energy at blue, violet, and ultraviolet wavelengths. A cool black-body (comparable in temperature to Neptune's effective temperature of 59.3 K) radiates most of its energy at red and infrared wavelengths. The Planck function also predicts the T^4 dependence (where T is the temperature) of total emitted energy. Typical black-body curves for temperatures of 10,000, 1,000, and 100 K are shown in Figure 9.4.

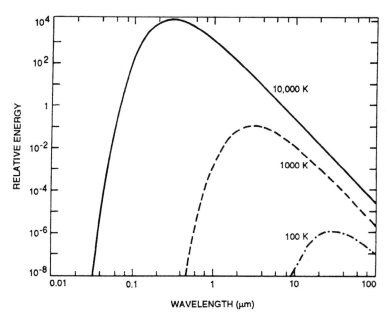

Figure 9.4. Typical black-body curves for temperatures of 10,000, 1,000, and 100 K. Note the decrease in energy and increase in the wavelength of maximum energy as temperature decreases.

The Neptune atmosphere departs substantially from black-body behavior, as do most real bodies. The presence of hydrogen gas, helium gas, and methane gas and ices causes selective absorption of light at predictable wavelengths. The measured thermal emission escaping the planet is affected by, and in turn affects, the relative chemical composition and temperature in the overlying atmosphere. As a result, the net thermal energy measured at any given wavelength is the sum of contributions (both emission and absorption) from many levels in the atmosphere. The relative amounts of energy from each level (at a given wavelength) are often referred to as the 'weighting function' for that wavelength. At wavelengths where the atmosphere is strongly absorbing, the weighting function peaks at high altitudes in the atmosphere; where little or no absorption occurs the weighting function has its maximum much deeper in the atmosphere. The nature of the weighting function at each wavelength is determined by a combination of theoretical and empirical methods and must include instrument response. Once determined, these weighting functions can be used to derive the temperature and pressure of the atmosphere at levels near each weighting function peak. By selecting the appropriate wavelengths, one can in principle reconstruct vertical profiles of temperature, pressure, and composition within the atmosphere. This was one of the primary goals of the infrared and ultraviolet investigations on board Voyager.

The above procedures provide atmospheric vertical profiles for altitudes where the pressures are less than 0.5 bar. The radio science occultation experiment probed the Neptune atmosphere to at least 6.3 bars and therefore extended to greater depths the vertical structure information obtained by the ultraviolet and infrared investigations.

9.3.1 Temperatures and pressures at different altitudes

For a planet with a solid surface the atmospheric altitude scale is usually referenced to the level of the average surface, or in the case of Earth to mean sea level. A useful reference point for a giant planet is the level in the (equatorial) atmosphere where the pressure is 1.0 bar. For Neptune this falls at a distance of $24,764 \pm 20$ km from Neptune's center. The temperature a few kilometres lower in the atmosphere (at a pressure level of 1.2 bars) is 78 K. Table 9.2 presents the pressure, temperature, and methane relative abundance (as a percentage of the total number of molecules) at intervals of ~ 10 km above and below the 1-bar reference level [20]. The ratio of helium to hydrogen molecules is assumed to be constant at 19/81. The data was

Table 9.2. Atmospheric data for Neptune

Altitude (km)	Temperature (K)	Pressure (bar)	Number of molecules ($10^{17}/cm^3$)	CH_4 (% molecules)
200.0	129.8	0.0005	0.5	0.0
190.0	127.1	0.0007	0.7	0.0
180.0	123.9	0.0010	1.0	0.0
170.0	121.5	0.0013	1.3	0.0
160.0	117.1	0.0016	1.6	0.0
150.0	112.3	0.0020	2	0.0
140.0	106.9	0.0025	2.5	0.0
130.0	100.4	0.0032	3	0.0
120.0	94.4	0.0040	4	0.0
110.0	87.4	0.0050	5	0.0
100.0	77.7	0.0071	7	0.0
90.0	67.6	0.0112	11	0.0
80.0	60.4	0.0200	20	0.0
70.0	55.3	0.0398	40	0.0
60.0	53.6	0.0562	56	0.0
50.0	52.5	0.0794	80	0.0
40.0	52.7	0.141	140	0.0
30.0	52.6	0.224	220	0.0
20.0	56.0	0.398	400	0.0
0.0	71.5	1.000	1,000	0.1
−10.0	81.4	1.41	1,400	1.4
−20.0	98.4	2.24	2,200	2.3
−30.0	110.4	3.16	3,200	2.3
−40.0	127.2	5.01	5,000	2.3

derived on the basis of the two radio occultation measurements: an ingress near latitude $+61°$ and an egress near latitude $-44°$.

9.3.2   Temperature variations with latitude

The radio science occultation profiles were limited to mid-latitudes by the geometry of the encounter, which required that Voyager 2 pass near Neptune's north pole in order to have close encounters of both Neptune and Triton. The atmospheric observations of Neptune by the infrared and ultraviolet investigations have no such limitation, but cover all latitudes except those poleward of $+70°$. The latitudinal variations of atmospheric temperature for two different pressure levels as derived by the infrared investigation [21] are shown in Figure 9.5. The upper curve corresponds to an altitude of about 60 km above the 1-bar reference level. The lower curve corresponds to an altitude of about 15 km. The scatter in the data is a consequence of both longitudinal variations and data noise.

At altitudes near 15 km the temperatures are quite uniform, varying by less than 2 K from the mean of 61 K. Temperature excursions at the 60-km altitude level are

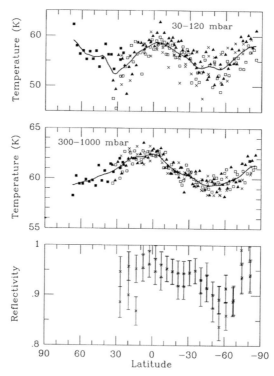

Figure 9.5. Latitude scans from several infrared observations. Similar symbols are used for data from the same observation (i.e., approximately the same longitude). The solid curves are constructed from the averages of $10°$ latitude bins. (P-35413A)

somewhat larger. Note the tendency in Figure 9.5 toward higher temperatures at high latitudes relative to mid-latitudes, contrary to predictions.

The relatively rapid rotation of Neptune also changes the shape of the planet, causing oblateness (polar flattening). The polar radius out to the 1-bar reference level is 424 km smaller than the equatorial radius, measuring only $24,340 \pm 30$ km [22]. As with Uranus, the oblateness is larger than would have been predicted by the measured rotation rate, a fact that is reflected in post-Voyager interior models of Neptune (see Chapter 8).

9.3.3   Discrete cloud and haze layers

Prior to the Voyager 2 encounters, very little was known about the haze and cloud layers within the atmospheres of Uranus and Neptune. The possibility of a thin layer of haze near the 200-km altitude had been suggested for both planets. A relatively transparent cloud deck of methane ice crystals was also suggested somewhere near the 2-bar level. An opaque cloud of unknown composition near the 3-bar level formed the primary reflecting layer for incoming solar radiation.

The post-Voyager understanding is different primarily in the details of cloud composition and altitude. Beginning with the deepest hypothesized cloud layers, a thick layer of clouds involving water and/or water ice may extend downward to as much as 400 bars pressure, far beyond direct observation. The deepest portions of this cloud probably include mixtures of water, ammonia, and hydrogen sulfide (H_2S). The top of this cloud may extend upward to the 40-bar level. Above this deepest cloud layer, with a base near 37 bars (perhaps 150 km below the 1-bar reference level), a cloud of ammonium hydrosulfide (NH_4SH) ice may exist, again well below Voyager or Earth-based detection.

The deepest cloud detected in Voyager 2 radio occultation measurements was an optically thick cloud layer between ~ 3.3 bars (~ -30 km; ~ 110 K) and ~ 5 bars (~ -50 km; ~ 140 K). This cloud has an unknown composition, but may be hydrogen sulfide (H_2S) ice. Another possibility, somewhat less favored, is that the cloud is ammonia (NH_3) ice. Significant amounts of sunlight reach this main cloud deck, and, for reasons not clearly understood, much of the red part of sunlight is absorbed (not reflected) by this cloud deck. Hence, this cloud is sometimes referred to as the 'blue-tinted' cloud. As a consequence of the red absorption of overlying methane gas and ice crystals and the red absorption of the blue-tinted cloud, Neptune has a deeper blue color than Uranus.

A physically thick but usually optically thin haze layer of methane (CH_4) ice has a base near 1.5 bars (~ -10 km, ~ 85 K) and extends upward to ~ 0.4 bar ($\sim +20$ km, ~ 60 K). Occasionally, strong upward welling in local areas may produce more opaque clouds of methane ice, and under the appropriate lighting conditions these methane clouds can cast shadows on the deeper blue-tinted cloud, permitting direct measurement of altitude differences from images of these areas (see Figure 9.6).

Above the tropopause (minimum temperature, ~ 55 K near 0.1 bar) in the stratosphere are one or more haze layers of much smaller optical depth. In order

Figure 9.6. Voyager 2 narrow-angle camera image of the south polar region, showing what are believed to be dark shadows adjacent to bright cloud streaks. They are believed to be shadows because they lie on the anti-sunward sides of the bright clouds and appear to be lengthened in regions where the Sun is closer to the horizon. (P-34709)

of decreasing optical thickness these are methane (CH_4) or ethane (C_2H_6) ice haze (near 0.02 bar, $+70\,m$, $\sim 60\,K$), a possible acetylene (C_2H_2) ice haze (near 0.01 bar, $+80\,km$, $\sim 70\,K$), and a possible diacetylene (C_4H_2) ice haze (near 0.006 bar, $+90\,km$, $\sim 80\,K$).

Figure 9.7 depicts in cartoon form the various haze and cloud layers discussed in this section.

9.3.4 The upper atmosphere

The very limited dataset makes the structure of the extreme upper atmosphere much less certain than that of the lower atmosphere, but some estimates are possible on the basis of ultraviolet spectrometer data. Table 9.3 lists temperatures, pressures, and number densities to altitudes of 5,000 km. Above 500 km the relative abundance of helium decreases rapidly. At 1,000 km, helium constitutes about 1% of the atmosphere and it has dropped to less than 0.1% at 2,000 km altitude. Hydrogen – mainly in its diatomic molecular form – is the sole detectable gas at higher altitudes. The fraction of hydrogen in its atomic form continues to increase with altitude. At 7,000 km about one-quarter of the hydrogen is atomic (H) and three-quarters is molecular (H_2).

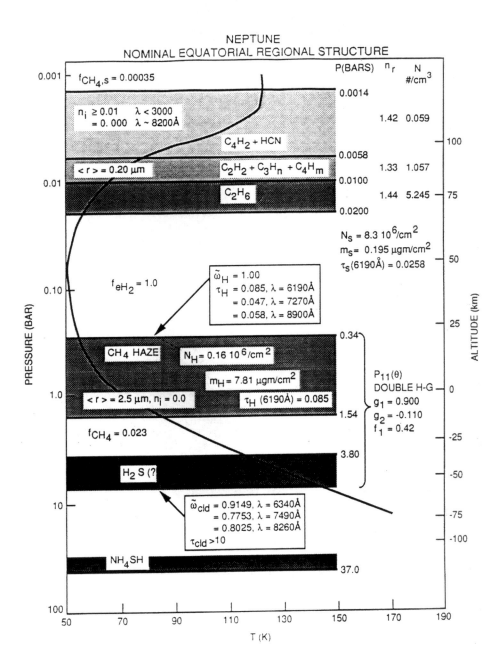

Figure 9.7. Possible atmospheric structure near Neptune's equator, depicting the measured or theorized haze and cloud layers. Adapted from Figure 15, Baines *et al.* [23].

Table 9.3. Upper atmospheric data for Neptune

Altitude* (km)	Temperature (K)	Pressure (bar)	Total (molecules/cm^3)
5,000	750	5×10^{-14}	1×10^4
4,000	750	3×10^{-13}	2×10^5
3,000	750	2×10^{-12}	5×10^6
2,000	700	1×10^{-11}	1×10^8
1,000	400	1×10^{-9}	1×10^{10}
500	175	1×10^{-6}	1×10^{12}

* Above the 1-bar level in the atmosphere.

At even higher altitudes in the 'exosphere', atomic hydrogen becomes the dominant constituent. The 750 K temperature above 2,000 km [24] is representative of a very slow decrease in the number of atoms of hydrogen per volume as the altitude increases. Extrapolation to the altitudes of the Neptunian rings (10,000 to 32,000 km altitude) shows that enough atomic hydrogen must be present to cause the ring particles to experience orbital drag. This force and its implications for the lifetimes of the rings will be discussed further in Chapter 11.

As the atomic hydrogen is exposed to ultraviolet light from the Sun, much of it is ionized (the electrons absorb enough energy to escape from the hydrogen atoms), resulting in a large population of free electrons. The resulting 'ionosphere' was detected during the radio occultation experiment. Both ingress and egress measurements yielded electron concentrations of 1,000 to 2,000 per cubic centimeter at an altitude of 1,400 km. Above that altitude, no well-defined concentrations were seen, but a steady decrease in electron density was observed up to an altitude of 5,000 km, where the density was ~ 100 per cubic centimeter. The egress profile showed a number of sharp layers below an altitude of 1,400 km, particularly one at 1,200 km, where the electron density reached 7,000 per cubic centimeter. Several other large-scale fluctuations occurred below 1,000 km, some of which may be ionospheric layers with concentrations of up to 10,000 electrons/cm^3 and some of which may be 'aliases' in the data due to radio science data complications [25].

9.4   HORIZONTAL CLOUD AND TEMPERATURE STRUCTURE

At the time of the Voyager 2 encounter, the atmosphere of Uranus displayed few discrete clouds. Although Earth-based and Hubble Space Telescope imaging of Uranus have since shown that the conditions within the atmosphere of Uranus may have been anomalously quiet during the Voyager 2 encounter, Uranus nevertheless continues to be the most pacific of the giant planets in its outward appearance.

Neptune, on the other hand, perhaps as a result of its large internal heat source, is, in comparison to its close twin, a boiling cauldron of atmospheric activity. The

Figure 9.8. Full-disk view of Neptune's atmosphere, including the Great Dark Spot. (P-34611) (See color plate 5.)

largest of the apparent storms in Neptune's atmosphere was an Earth-sized spinning oval comparable to Jupiter's Great Red Spot (GRS). In honor of its Jupiter counterpart, the Neptune oval was called the Great Dark Spot (GDS). Both the GDS and the GRS are located near 20° south latitude on their respective planets. Both are anticyclones; i.e., they each rotate in a counterclockwise direction. They are of comparable size relative to the planets in whose atmospheres they exist: the GDS had an average extent of 38° in longitude by 15° in latitude, and the GRS is about 30° by 20°. The GDS time-lapse movies give the impression that it rolls around its center with a period of about 16 days; a more objective observation is an oscillation in shape with a periodicity of 8 days; Jupiter's GRS rotates once every 7 days. However, while the GRS appears to dominate its latitude on Jupiter, creating enormous turbulence in its wake, the GDS appears to have very little influence outside its immediate neighborhood. Also, the GRS moves westward at a rate of only 3 m/s relative to the internal rotation of Jupiter; the GDS drifts westward at speeds in excess of 300 m/s relative to Neptune's internal rotation!

The best full-disk image of Neptune, including the GDS, is shown in Figure 9.8. The longevity of the GDS is not well determined. The bright companion on its southern flank was first observed by Voyager 2 in January 1989, seven months before the Neptune encounter [26], and was also seen in Earth-based images [27].

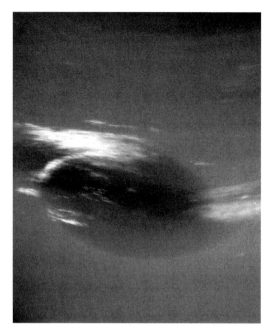

Figure 9.9. High-resolution view of the Great Dark Spot and its bright companion. (P-34672)

This companion is probably similar to orographic clouds on Earth, which form as a result of rapid cooling as winds are forced to higher altitudes by the presence of mountains. In the case of Neptune, the GDS is apparently the 'mountain' over which the Neptune winds are forced to flow. During the entire set of GDS observations by Voyager 2 the bright companion remained intimately associated with the larger companion. A high-resolution view of the two is seen in Figure 9.9. Additional HST observations failed to see the bright companion during a following apparition of Neptune [28], a likely indication that the GDS had disappeared. Jupiter's GRS has been telescopically observed for nearly 400 years!

9.4.1   Other cloud features

Another much smaller dark spot with a bright core is shown in Figure 9.10(a). It was called 'D2', a *double entendre* denoting both the fact that it was the second dark spot discovered and in honor of the popular little robot, R2D2, of *Star Wars* fame (*Star Wars* was still a popular movie at the time of the Voyager 2 Neptune encounter). D2 did not appear to have appreciable rotational motion, but upward vertical motion near its center may have been responsible for the formation of methane ice clouds. D2 was located at 55° south latitude.

A line of bright features near 71° south latitude extends across nearly 90° of longitude. It appears to be an active zone in the atmosphere rather than a discrete

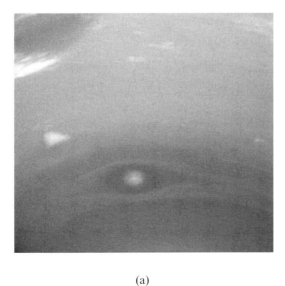

(a)

(b)

Figure 9.10. (a) A small dark spot ('D2'), seen in southern latitudes in the top image, was observed during most of the encounter period near 55° south latitude. The bright core appeared later. (b) The bright spot called 'Scooter' is shown in the bottom image at high resolution. Its latitude is near 42° south. As in the case of the core of D2, the high-resolution images show the bright feature to be composed of several linear clouds aligned in latitude. (P-35025)

long-lived feature. The intensity and distribution of brightness along this arc varied significantly, even in a single rotation of the planet.

Numerous other discrete features were seen in the atmosphere and were useful in providing a sense of the winds at various latitudes. However, few if any of them were long-lived enough to be given specific names or other designations.

A bright feature named 'Scooter' [Figure 9.10(b)] was also observed at 42° south latitude during the entire 80-day duration of the encounter. It was named for its rapid westward motion relative to many other observed cloud features, but it was not the fastest moving of these spots. Although most of the other bright features appear to be methane ice clouds, Scooter appears to reside much lower in the atmosphere and all but disappears in ultraviolet light and in the red methane filter. This is interpreted as meaning that the feature is below the methane cloud/haze deck near 1 bar.

9.4.2   Zonal wind profiles

As seems to be the case with all the giant planets, the dominant wind patterns are almost exclusively east–west in direction. They vary in latitude and generally show a pattern of wind speeds that has a degree of symmetry around the equator, often reaching maximum magnitude near the equator. Because the poles are single points, the latitudinal wind patterns must include zero wind speeds at 90° north or south latitude.

Neptune's wind patterns also fit this general description. Winds that blow in the same direction as the rotation of the interior of the planet are called *prograde* winds. Atmospheric features at latitudes with prograde winds circle the planet in less time than the rotation period of the planet's interior. Winds that blow in a direction opposite that of the planet's rotation are known as *retrograde* winds. Features carried by retrograde winds require more time to circle the planet than the rotation period of the interior.

Jupiter and Saturn have strong prograde winds at their equators. Uranus and Neptune have retrograde equatorial winds. In fact, Neptune's retrograde equatorial winds are by far the strongest retrograde winds seen anywhere in the Solar System, reaching speeds of nearly 400 m/s. Neither Jupiter nor Saturn have retrograde (westward-blowing) winds in excess of 50 m/s. Near-equatorial winds on Uranus reach about 110 m/s, still far shy of those measured for Neptune. Furthermore, prograde winds on Neptune reach speeds in excess of 200 m/s near 70° north and south latitude. Hence, while Saturn's near-equatorial winds of about 450 m/s appear to be the strongest among the planets, the total *contrast* in wind speed of about 600 m/s exhibited by Neptune is greater than for any other planet in the Solar System. When one considers the great distance of Neptune from the Sun and the much smaller energy input than that which drives winds on Jupiter and Saturn (see Table 9.1), the high wind speeds on Neptune are startling. The 'rotation periods' of atmospheric features in Neptune's atmosphere vary from approximately 11 hours to more than 20 hours! The wind speed profile and individual data points for Neptune are shown in Figure 9.11.

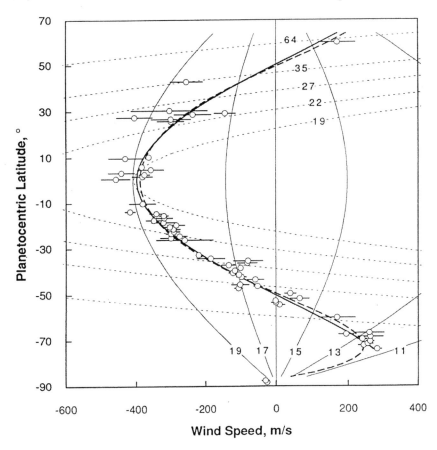

Figure 9.11. Variation of wind speed with latitude near the 1-bar pressure level in the atmosphere of Neptune. A single determination of wind speed was made at 60° north latitude during the radio occultation experiment. That data point is also shown at 60° south latitude.

9.4.3 The color of Neptune

As with Uranus, Neptune is deficient in its reflectivity at red wavelengths. For Uranus, this characteristic is mainly due to methane gas and ice in the atmosphere. While this same mechanism works for Neptune, there is an additional factor which removes even more red light from Neptune. This is the blue-tinted (hydrogen sulfide?) cloud below the 3-bar level in Neptune's deep troposphere. Hence, Neptune has a deep blue color rather than the blue-green typical of Uranus.

Much more in the way of banding appears in the atmosphere of Neptune than in the atmosphere of Uranus. Although the banding has neither the distinct boundaries nor as many bands as Jupiter, it is easily seen in the images of the planet without major stretches in contrast or color. Part of this banding may be due to non-uniform distribution in latitude of the (methane) cloud and haze layers. What causes this non-uniform distribution is unknown, as is the source of the color of the blue-tinted cloud. As in most scientific endeavors, our studies of Neptune's atmosphere have answered many of our a priori questions while raising many new ones.

NOTES AND REFERENCES

1. Cameron, A. G. W. (1973) Abundance of the elements in the solar system. *Space Science Reviews*, **15**, 121–46.
2. Anders, E. and Grevesse, N. (1989) Abundances of the elements: Meteoritic and solar. *Geochimiche Cosmochimiche Acta*, **53**, 197–214.
3. Gautier, D., Conrath, B., Flasar, M., Hanel, R., Kunde, V., Chedin, A. and Scott, N. (1981) The helium abundance of Jupiter from Voyager. *Journal of Geophysical Research*, **86**, 8713–20.
4. Conrath, B. J., Gautier, D., Hanel, R. A. and Hornstein, J. S. (1984) The helium abundance of Saturn from Voyager measurements. *Astrophysical Journal*, **282**, 807–15.
5. Conrath, B. J. and Gautier, D. (2000) Saturn helium abundance: A reanalysis of Voyager measurements. *Icarus*, **144**, 124–34.
6. Nellis, W. J., Ross, M., Mitchell, A. C., van Thiel, M., Young, D. A., Ree, F. H. and Trainor, R. J. (1983) Equation of state of molecular hydrogen and deuterium from shock-wave experiments to 700 kbar. *Physical Review*, **A27**, 608–11.
7. This mechanism for helium 'enrichment' was suggested by Fegly, B. and Prinn, R. G. (1986) Chemical models of the deep atmosphere of Uranus. *Astrophysical Journal*, **307**, 852–65; and by Pollack, J. B., Podolak, M., Bodenheimer, P. and Christofferson, B. (1986) Planetesimal dissolution in the envelopes of the forming giant planets. *Icarus*, **67**, 409–43.
8. MacFarlane, J. J. and Hubbard, W. B. (1982) Internal structure of Uranus. In Hunt, G. (ed.) *Uranus and the Outer Planets*. Cambridge University Press, Cambridge, pp. 111–24.
9. Gautier, D., Conrath, B. J., Owen, T., de Pater, I. and Atreya, S. K. (1995) The troposphere of Neptune. In Cruikshank, D. P. (ed.) *Neptune and Triton*. University of Arizona Press, Tucson, pp. 547–611.
10. Bézard, B., Romani, P. N., Conrath, B. J. and Maguire, W. C. (1991) Hydrocarbons in Neptune's stratosphere from Voyager infrared observations. *Journal of Geophysical Research*, **96**, 18961–75.
11. Orton, G. S., Lacy, J. H., Achtermann, J. M., Parmar, P. and Blass, W. E. (1992) Thermal spectroscopy of Neptune: The stratospheric temperature, hydrocarbon abundances, and isotopic ratios. *Icarus*, **100**, 541–5.
12. Atreya, S. K., Sandel, B. R. and Romani, P. N. (1991) Photochemistry and vertical mixing. In Bergstralh, J. T., Miner, E. D. and Matthews, M. S. (eds) *Uranus*. University of Arizona Press, Tucson, 110–46.
13. Marten, A., Gautier, D., Owen, T., Sanders, D. B., Matthews, H. E., Atreya, S. K., Tilanus, R. P. J. and Deane, J. R. (1993) First observations of CO and HCN on Neptune and Uranus at millimeter wavelengths and their implications for atmospheric chemistry. *Astrophysical Journal*, **406**, 285–97.
14. Bishop, J., Atreya, S. K., Romani, P. N., Orton, G. S., Sandel, B. R. and Yelle, R. V. (1995) The middle and upper atmosphere of Neptune (especially IV.E. on pp. 469–72). In

Cruikshank, D. P. (ed.) *Neptune and Triton*. University of Arizona Press, Tucson, pp. 427–87.

15. de Bergh, C. Lutz, B. L., Owen, T. and Maillard, J. P. (1990) Monodeuterated methane in the outer solar system. IV. Its detection and abundance on Neptune. *Astrophysical Journal*, **355**, 661–6.

16. Anders, E. and Grevesse, N. (1989) Abundances of the elements: Meteoritic and solar. *Geochimiche Cosmochimiche Acta*, **53**, 197–214.

17. Pearl, J. C. and Conrath, B. J. (1991) The albedo, effective temperature, and energy balance of Neptune, as determined from Voyager data. *Journal of Geophysical Research*, **96**, 18921–30.

18. The Stefan–Boltzmann law describes the relationship for a black-body between the temperature of that body and the amount of energy radiated from that body per unit area of that body. Numerically, this law states that the amount of energy radiated by a square meter of a black-body, $B = \sigma T^4$, where B is in units of W/m^2, T is the black-body absolute temperature in kelvins (K), and σ is a constant equal to $5.67 \times 10^{-8}\,W/m^2\,K^4$.

19. Using the emitted flux of $0.70\,W/m^2$ and the Stefan–Boltzmann law described in note [18], one must then take the fourth root of $(0.70/5.67 \times 10^{-8})$, which yields 59.3 K.

20. Adapted from Lindal, G. F., Lyons, J. R., Sweetnam, D. N., Eshleman, V. R., Hinson, D. P. and Tyler, G. L. (1990) The atmosphere of Neptune: Results of radio occultation measurements with the Voyager 2 spacecraft. *Geophysical Research Letters*, **17**, 1733–6.

21. Conrath, B., Flasar, F. M., Hanel, R., Kunde, V., Maguire, W., Pearl, J., Pirraglia, J., Samuelson, R., Gierasch, P., Weir, A., Bézard, B., Gautier, D., Cruikshank, D., Horn, L., Springer, R. and Shaffer, W. (1989) Infrared observations of the Neptunian system. *Science*, **246**, 1454–8.

22. Tyler, G. L., Sweetnam, D. N., Anderson, J. D., Borutzki, S. E., Campbell, J. K., Eshleman, V. R., Gresh, D. L., Gurrola, E. M., Hinson, D. P., Kawashima, N., Kursinski, E. R., Levy, G. S., Lindal, G. F., Lyons, J. R., Marouf, E. A., Rosen, P. A., Simpson, R. A. and Wood, G. E. (1989) Voyager radio science observations of Neptune and Triton. *Science*, **246**, 1466–73

23. Baines, K. H., Hammel, H. B., Rages, K. A., Romani, P. N. and Samuelson, R. E. (1995) Clouds and hazes in the atmosphere of Neptune (especially Figure 15 on p. 539), in Cruikshank, D.P. (ed.) *Neptune and Triton*, University of Arizona Press, Tucson, pp. 489–546.

24. Broadfoot, A. L., Atreya, S. K., Bertaux, J. L., Blamont, J. E., Dessler, A. J., Donahue, T. M., Forrester, W. T., Hall, D. T., Herbert, F., Holberg, J. B., Hunten, D. M., Krasnopolsky, V. A., Linick, S., Lunine, J. I., McConnell, J. C., Moos, H. W., Sandel, B. R., Schneider, N. M., Shemansky, D. E., Smith, G. R., Strobel, D. F. and Yelle, R. V. (1989) Ultraviolet spectrometer observations of Neptune and Triton. *Science*, **246**, 1459–66.

25. Eshleman, V. R., Tyler, G. L., Anderson, J. D., Fjeldbo, G., Levy, G. S., Wood, G. E. and Croft, T. A. (1977) Radio science investigation with Voyager. *Space Science Reviews*, **21**, 207–32.

26. Smith, B. A., Soderblom, L. A., Banfield, D., Barnet, C., Basilevsky, A. T., Beebe, R. F., Bollinger, K., Boyce, J. M., Brahic, A., Briggs, G. A., Brown, R. H., Chyba, C., Collins, S. A., Colvin, T., Cook, A. F. II, Crisp, D., Croft, S. K., Cruikshank, D., Cuzzi, J. N., Danielson, G. E., Davies, M. E., De Jong, E., Dones, L., Godfrey, D., Goguen, J., Grenier, I., Haemmerle, V. R., Hammel, H., Hansen, C. J., Helfenstein, C. P., Howell, C., Hunt, G. E., Ingersoll, A. P., Johnson, T. V., Kargel, J., Kirk, R., Kuehn, D. I., Limaye, S., Masursky, H., McEwen, A., Morrison, D., Owen, T., Owen, W., Pollack, J.

B., Porco, C. C., Rages, K., Rogers, P., Rudy, D., Sagan, C., Schwartz, J., Shoemaker, E. M., Showalter, M., Sicardy, B., Simonelli, D., Spencer, J., Sromovsky, L. A., Stoker, C., Strom, R. G., Suomi, V. E., Synnott, S. P., Terrile, R. J., Thomas, P., Thompson, W. R., Verbischer, A. and Veverka, J. (1989) Voyager 2 at Neptune: Imaging science results. *Science*, **246**, 1422–49.

27. Hammel, H. B., Beebe, R. F., deJong, E. M., Hansen, C. J., Howell, C. D., Ingersoll, A. P., Johnson, T. V., Limaye, S. S., Magalhaes, J. A., Pollack, J. B., Sromovsky, L. A., Suomi, V. E. and Swift, C. E. (1989) Neptune's wind speeds obtained by tracking clouds in Voyager images. *Science*, **245**, 1367–9.

28. Hammel, H. B., Lockwood, G. W., Mills, J. R. and Barnet, C. D. (1995) Hubble Space Telescope Imaging of Neptune's cloud structure in 1994. *Science*, **268**, 1740.

BIBLIOGRAPHY

Bergstralh, J. T., Miner, E. D. and Matthews, M. S. (eds) (1991) *Uranus*. University of Arizona Press, Tucson. (See atmosphere chapters on pp. 69–324 by Strobel *et al.*, Atreya *et al.*, Fegley *et al.*, Conrath *et al.*, Allison *et al.* and West *et al.* and the references cited in each chapter.)

Cruikshank, D. P. (ed) (1995) *Neptune and Triton*. University of Arizona Press, Tucson. (See chapters on pp. 427–682 by Bishop *et al.*, Baines *et al.*, Gautier *et al.*, and Ingersoll *et al.* and the references cited in each chapter.)

Ingersoll, A. P. (1999) Atmospheres of the giant planets. In Beatty, J. K., Petersen, C. C. and Chaikin, A. (eds) *The New Solar System* (Fourth Edition). Sky Publishing Corporation, Cambridge, MA, pp. 201–20.

10

The magnetosphere of Neptune

10.1 STRENGTH AND ORIENTATION OF THE MAGNETIC FIELD

Uranus and Neptune seem to be twins in a number of ways. They are approximately the same color. They are of very similar size, mass, temperature, and their atmospheric compositions seem almost identical, at least as far as could be determined from Earth-based observations. However, one of the strangest characteristics of Uranus was the extraordinary orientation of its magnetic field. Prior to the Uranus encounter, all known planetary magnetic fields were more or less oriented with their magnetic and rotation axes aligned. Earth's magnetic axis is within 11° of its rotation axis. Jupiter's intense magnetic field axis is less than 10° from its rotation axis. Saturn's magnetic axis has not yet been determined accurately enough to specify its misalignment with the rotation axis, but it is certainly less than 1°! Because of the strong implications of that alignment on theories of planetary magnetic field generation, a precise measurement of the tiny misalignment is a major objective of the Cassini Mission to Saturn early in the twenty-first century.

Planetary magnetic fields (the external influence of an internal magnet) are thought to be generated by internal dynamos. Exactly how those dynamos work is not fully understood, and so science's ability to predict results from dynamo theory is limited. Each of the planets (or satellites) that possesses a magnetic field is thought to generate that magnetic field as a result of the rotation of the body itself, combined with an electrically conducting liquid layer inside the body which has some amount of convective motion within it. The nature of the electrically conducting regions may vary from body to body. For example, Earth's magnetic field is thought to originate in its molten iron core. Jupiter and Saturn, as mentioned in Chapter 8, probably have regions of conducting liquid metallic hydrogen in their interiors. The conducting liquids on Uranus and Neptune may be mixtures of water, ammonia, and methane, perhaps well above the core regions of these planets.

While the above is a description of the necessary physical elements in dynamo theory, the precise methods in which these elements combine to generate the magnetic field in dynamo theory is absent. That results, in part, because there is no unanimity about the precise mechanisms, nor is there agreement about how dynamos evolve with time. We do know that Earth's magnetic field has apparently reversed several times over geological history. As a result of the strange findings of Voyager 2 during its encounter of Uranus, we also know that a planetary magnetic field does not have to be aligned with the rotation axis of the planet. While these facts do not provide a complete answer about the nature and mathematics of dynamo theory, they certainly provide observational constraints that will eventually have to be explained by a complete theory. As we continue to collect data on different magnetic fields, the answers that will lead to a satisfactory dynamo theory may well come to light. In the meantime, scientists continue, both on the theoretical and observational frontiers, to delve into this area of scientific endeavor. Progress comes slowly, but such is the nature of evolving science.

The magnetic field of Uranus was tilted by 58.6° from the rotation axis and offset from the center of the planet about one-third of the planet's radius toward the dark north polar region. Initially there was some speculation about whether the Uranus magnetic field might be in the process of reversing itself. Alternatively, perhaps the extreme tilt of the rotation axis of Uranus may have been related to the high relative tilt of its magnetic field. Perhaps the purported collision that tilted Uranus on its side might have changed the rotation axis but not the magnetic axis. No one knew quite what to expect as Voyager 2 approached Neptune. Would Neptune's magnetic field display the more expected near-alignment seen for Earth, Jupiter, and Saturn, or would it mimic the situation at Uranus?

The first of the Voyager 2 instruments to detect the presence of Neptune's magnetic field was the planetary radio astronomy investigation. That detection occurred on August 17, 2000, about 8 days before closest approach to Neptune. The detection occurred primarily in a frequency range of 700 to 850 kHz (700,000 to 850,000 cycles per second). When the team examined the data, it found evidence of similar bursts as early as July 26, 2000 (30 days before closest approach), and the bursts continued to be detectable until August 28, 2000, with one other period of bursty radio activity on September 16, 2000 (22 days after closest approach) [1].

The magnetometer team reported [2] the following timings: (1) an initial crossing of the bow shock 13.3 h before closest approach, (2) a gradual crossing of the magnetopause between 9.9 and 8.3 hours before closest approach, (3) closest approach to Neptune on August 25, 2000 at 3:55 UTC, (4) an outbound crossing of the magnetopause 28.4 h after closest approach, and (5) multiple outbound crossings of the bow shock between 64.1 and 65.1 h after closest approach. These events, along with the corresponding distances from the center of Neptune, are given in Table 10.1.

The shape of a planetary magnetic field, if one removes the distorting effects of the solar wind, is somewhat like an apple. The parts of the magnetic field corresponding to the 'stem' and 'flower' of the apple are known as the cusps of the magnetic field. These same features exist in the presence of the solar wind, except that the sunward side is compressed toward the planet and the anti-sunward side is stretched into a long magnetic tail. As measurements of the magnetic field began arriving, it rapidly became apparent that Voyager 2 had approached Neptune along

Table 10.1. Voyager 2 crossings of the Neptune bow shock and magnetopause

Boundary	Time from closest approach (h)	Range from Neptune (km)
Bow shock (in)	−13.3	864,000
Magnetopause (in)	−9.9 to −8.3	656,000 to 570,000
Closest approach	0.0	29,240
Magnetopause (out)	+28.4	1,790,000
Bow shock (out)	+64.1 to +65.1	4,990,000

one of its magnetic cusps. That meant in turn that the magnetic field of Neptune, like that of Uranus, was not aligned along the rotation axis and that the magnetosphere was larger than implied by measurements of the distance at the time of magneto-pause crossing. (The reader is reminded that the magnetopause is the boundary between the planet's magnetic field and the solar wind; the bow shock is the outer edge of the interaction region between the magnetosphere and the solar wind, in analogy with a bow wave pushed in front of a ship as it cuts through water.) Careful measurements of Neptune's magnetic field revealed that the best-fit magnetic dipole is tilted by 47.0° with respect to the planet's rotation axis and offset from the planet's center by a full 55% of its radius in a direction roughly toward the south magnetic pole. The best-fit dipole model fit is illustrated in Figure 10.1.

The actual configuration of the magnetic field is much more complex than that of a simple dipole, especially close to the planet. In describing a magnetic field, a scheme similar to that of gravitational field modeling using spherical harmonics is useful. In the case of Uranus, quadrupole magnetic harmonics best represented the field. However, the data was insufficient to permit an estimation of octupole or higher-order harmonics. The model of Connerney *et al.* [3] that best fits the Neptune data is an octupole model. Such a model requires about 80 parameters to be fit to the data, but the model thus derived fits the data almost perfectly. In this model, dubbed the O_8 model, the magnetic south pole of Neptune is near longitude 280° W and latitude 40° S. The magnetic north pole is located near longitude 60° W and latitude 50° N. Figure 10.2 is a contour map of the magnetic field intensity near the 1-bar reference level for Neptune. Also shown in the figure are the points where the magnetic poles would pierce this reference surface and the predicted areas of ultraviolet auroral activity.

10.2   IMPLICATIONS OF THE TILT AND OFFSET OF THE MAGNETIC FIELD

With no planetary analog at the time of its discovery, the highly tilted magnetic field of Uranus led to all sorts of speculation about the cause of such an unusual orienta-tion. However, when the magnetosphere of Neptune exhibited the same general characteristics, neither was thereafter considered to be an anomaly, but was

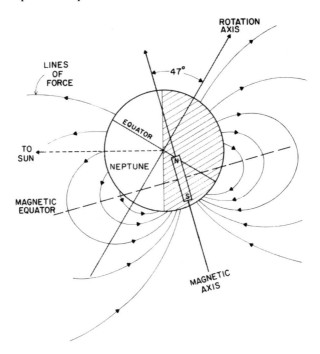

Figure 10.1. The simplest representation of the magnetic field is a magnetic dipole tilted 47.0°
from the rotation axis and offset 0.55 Neptune radii in the general direction of the magnetic
south pole. (P-35415B)

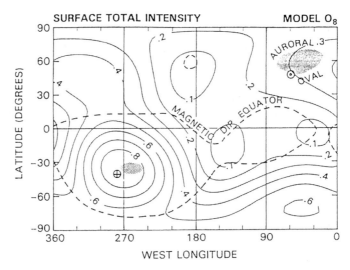

Figure 10.2. Contour map of magnetic field intensity in Gauss at the surface (1-bar pressure
level) of Neptune as predicted from the O_8 model. Also shown in the figure are the magnetic
equator, where magnetic field lines are parallel to the surface, and locations where the
magnetic poles would pierce the surface.

instead a new phenomenon in search of a logical scientific explanation. No longer did one need to invoke an explanation of special timing, such as that initially suggested for Uranus that the magnetic field just happened to be in the process of reversing polarity, because the chances of finding two planets in that circumstance were negligible. No longer did anyone attempt to formulate a theory related to the high tilt of the rotation axis of Uranus, because the axial tilt of Neptune was rather ordinary. The answer to these strange magnetic field orientations must somehow be hidden in the basic natures of these two giant planets, something that made them inherently different from Earth, Jupiter, or Saturn.

The gauntlet was quickly snatched by planet modelers, who wasted no time in arriving at an acceptable conclusion. The simple explanation was in the level within these planets where the field was generated. Earth generates its magnetic field in a molten iron layer within the core of the planet; Jupiter and Saturn each generate their magnetic fields in conducting liquid metallic hydrogen regions relatively close to the centers of these planets. The magnetic fields of Uranus and Neptune are probably generated relatively high in their interiors, near the outer boundaries of the liquid icy layer. When such fields are generated so far from the center of these planets, apparently the generating dynamos (circulating electrical currents in the conducting layers) need not be and generally are not constrained to having their centers near the planets' centers nor to having their axes closely aligned to the planetary rotation axes. The resulting magnetic fields are consequently of nearly arbitrary tilt orientations and possess offsets that can be major fractions of the total planetary radii. Such is certainly the case for the magnetic fields of Uranus and Neptune. Scientists still do not understand the details of the energy sources that drive these magnetic dynamos inside Uranus and Neptune, which is not surprising as their understanding of the corresponding dynamos within Earth, Jupiter, and Saturn is also rather limited.

One important effect of the large dipole tilt is correspondingly large changes in the magnetic field orientation as the planet rotates on its axis and as it revolves around the Sun. Figure 10.3 shows the approximate orientation with the rotation axis and magnetic axis for Neptune over the course of one Neptune revolution around the Sun. Unlike the situation with Uranus, the cusps of Neptune's magnetic field never point directly at the Sun, but reach a minimum separation of about 11.2° for a short time every eight hours near the times of Neptune's summer and winter solstices. Auroral effects will be discussed in Section 10.3.

The solar wind interaction with the magnetic field of Neptune stretches the field downwind of the planet into an extended tail. As the planet rotates, the magnetic equator moves up and down, completing a full cycle every 16.11-h rotation of the planet. The resulting tail shape will be much different from that observed for Uranus. It is likely that Neptune's magnetotail extends to great distances in the anti-sunward direction. Voyager 2's trajectory through the system was not well suited for determining the nature and extent of that magnetotail. The actual Voyager 2 trajectory plotted in magnetospheric coordinates [3] is presented in Figure 10.4, which also shows that the spacecraft passed through the cusp region of the magnetosphere on three separate occasions.

Angle Between Magnetic Axis and Sun

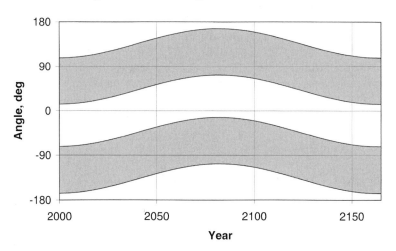

Figure 10.3. Neptune's magnetic field orientation with planet rotation and with orbital position. The magnetic axis will rotate within 11.2° of the Sun at the solstices (beginning of summer and winter) in 2000, 2082, and 2164. The gray areas show the locus of positions of the magnetic axes.

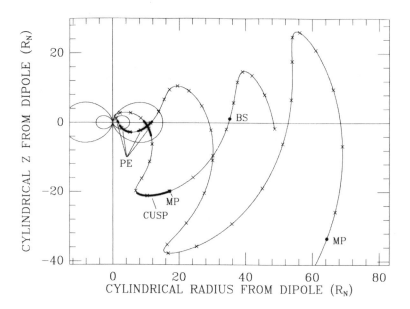

Figure 10.4. Trajectory of Voyager 2 near Neptune in magnetospheric coordinates. ρ is the cylindrical radius from the magnetic dipole axis and Z is the distance from the magnetic equator. The thickened portions of the trajectory represent those times when the spacecraft was near one of the cusps of the magnetosphere. (P-35428B)

The orbital motion of Neptune around the Sun changes the tail configuration substantially. At the equinoxes, the angle β between the Sun–Neptune line and the magnetic dipole axis varies between $43°$ and $137°$. At the northern summer solstice, the angle varies from $11°$ to $105°$. At the northern winter solstice, the angle varies from $75°$ to $169°$. At either of these solstices, the solar wind would have nearly direct access to the upper atmosphere once per rotation, and the auroral activity might be expected to be increased over its normal levels. The magnetic tail structure would also be complex. As mentioned earlier, there may be times when the tail structure might be alternating between a single plasma sheet and a bifurcated sheet (the latter when the magnetic pole is closely aligned in the Sun–Neptune direction).

10.3 AURORAE (NORTHERN OR SOUTHERN LIGHTS) ON NEPTUNE

Charged particles spiralling down Earth's magnetic field lines collide with neutral atoms in the Arctic atmosphere. If the rain of charged particles is intense enough, both visible and ultraviolet light will be emitted in quantities sufficient for remote detection. These emissions constitute an aurora, commonly called the northern (or southern) lights. During active periods on the Sun the solar wind intensity increases substantially, and the auroral displays are brighter and often extend to much lower latitudes.

Aurorae were observed by Voyager at both Jupiter and Saturn. Searches for a similar auroral ring at Uranus, designed months before the actual observations, failed to account for the unexpected large tilt of the magnetic axis and thereby missed scanning the actual regions of the magnetic poles. Nevertheless, darkside auroral emissions were observed near the north magnetic pole. The auroral region of approximately $15°$ to $20°$ in diameter is believed to be excited by electrons with energies of $\sim 10,000\,\mathrm{eV}$ [4]. Such electrons were seen in abundance by Voyager investigations, though there is evidence that the population of such electrons (or appropriate ions) along the magnetic field lines which feed into a $20°$ diameter auroral zone are insufficient to account for the observed auroral intensity [5]. This seeming inconsistency remains unresolved more than 14 years after the Voyager encounter of Uranus.

Ultraviolet emissions from the night side of Neptune were detected, but they were weak and diffuse. The first was a region extending over about $100°$ of latitude and $60°$ of longitude (an estimated 13% of Neptune) [6]. The second was more localized in both latitude and longitude and centered near the south magnetic pole. Although neither the spectral nor the spatial resolution was sufficient to positively identify this weak emission as an aurora, the data is not inconsistent with that interpretation [6]. A brighter aurora would be expected from the region above the north magnetic pole, but that region was not well observed due to the geometry of the encounter. It would have been at the very edge of the Neptune disk both during approach observations and during similar observations as Voyager retreated from the planet.

The characteristics of a classical aurora (those seen on Earth, Jupiter, and Saturn) are familiar to magnetospheric scientists. However, if one doesn't know in advance where to look for a Neptune aurora (due to the high axial tilt), the data is sparse enough to make a real study of the phenomenon difficult and perhaps impossible. Furthermore, since aurora are generally caused by electrons spiraling into the atmosphere at either end of a magnetic field line, a substantial offset of the magnetic field from the center of the planet results in stripping away most of those electrons on the side of the planet farthest removed from the center of the magnetic field. Since the Voyager encounters of Uranus and Neptune, auroral scientists have been pondering and theorizing about these effects, but it is clear that they are substantial, and without more auroral data than was provided by Voyager 2 during its encounters of Uranus and Neptune, there is little immediate hope that they will gain a full understanding of auroral effects in the presence of such unusual magnetic field configurations.

10.4   RADIO EMISSIONS FROM NEPTUNE

As mentioned earlier in this chapter, the first evidence from Voyager that Neptune has a magnetic field was in the detection of radio emissions by the planetary radio astronomy investigation on August 17, 1989, about 8 days before closest approach to Neptune [1]. These radio signals were clusters of short duration bursts at very narrow wavelengths. The bursts occurred within a frequency range extending mainly between 635 and 865 kHz (350 to 470 m wavelength). Although this was the first type of Neptune radio emission to be observed, it was only one of about five different types of emissions recorded by Voyager. The types of Neptune radio emissions are summarized in Table 10.2.

Table 10.2. Types of Neptune radio emission

Radio emission characteristics	Frequency range (kHz)	Wavelength range (km)	Comments
Narrowband, bursty (main)	550 to > 1326	< 0.02 to 0.55	First to be detected; used to measure Neptune's internal 16.11-h rotation period
Narrowband, bursty (anomalous)	450 to 520	0.58 to 0.67	Distinct from main bursty emissions and of unknown origin
Broadband, smooth (main)	~ 20 to 600	0.5 to ~ 15	Centered around times when north magnetic pole pointed toward Voyager
Broadband, smooth (high frequency)	600 to 870	0.35 to 0.5	Mainly 3 episodes near closest approach and near magnetic equator (?)
Continuum	~ 3 to > 56	< 5.4 to ~ 100	Smooth with superimposed spikes, enhanced near the magnetic equator

There is some evidence that the main narrowband bursty emissions are the result of Neptune's magnetic field merging with the interplanetary magnetic field [7]. However, there seems to be little or no correlation of burst activity with the speed or density of the solar wind. It is possible that, instead of beaming radio signals much like a rotating lighthouse beam, the bursty emissions actually turn on and off each rotation of the planet. Whether they are beamed or pulsed, careful timing of these bursty emissions yielded a rotation period for the magnetic field of Neptune (presumed to equal the rotation of the bulk of the planet's interior) of $16.108 \pm 0.006\,\text{h}$ [8].

Two other types of radio emission were detected near Neptune. The first was associated with tiny particles in the equatorial plane of the planet and is thought to be particles in an extended ring which, striking the spacecraft, are vaporized and ionized, giving off radio emissions in the process. These will be discussed further in Chapter 11. Lightning discharges in the atmosphere of Neptune are believed to generate the other type of emission. These emissions take a characteristic form known as whistlers, and are characterized by higher frequency emissions that migrate quickly with time to lower frequencies. When displayed as sounds of the same frequency, these events sound like descending-pitch whistles; hence their designation. During the Neptune encounter, 16 such events were detected [9]. No other evidence of Neptune lightning was seen by Voyager, and it is possible that lightning at Neptune is very weak, generating the observed whistler-like emissions, but not generating enough light to be seen in images of the dark side of Neptune. Kaiser *et al.* [10] suggested that the low level of lightning activity at Neptune might be related to the absence of easily polarized molecules within the atmosphere.

10.5 THE PLASMA ENVIROMENT OF NEPTUNE

A plasma is a collection of charged particles generally containing about equal numbers of positive ions and electrons. Plasmas exhibit some of the characteristics of a gas, but differ from a gas in that they are good conductors of electricity and are affected by magnetic fields. A 'cold' plasma is one whose particle energies are small enough that electromagnetic forces constrain the plasma to move with the magnetic field; a 'hot' plasma is energetic enough to escape such constraints. If the hot plasma is sufficiently dense it can distort the magnetic field through which it flows. The solar wind is an example of a hot and relatively dense plasma which distorts the Sun's magnetic field and stretches it into an enormous magnetic bubble (the heliosphere) which envelops all the known planets of the Solar System.

Unlike Uranus, Neptune possesses a large satellite, Triton, with both a tenuous atmosphere and a substantial ionosphere [11]. Triton, its atmosphere, and its ionosphere will be discussed in greater detail in Chapter 12. However, in the context of plasmas at Neptune, Triton's atmosphere and ionosphere apparently serve as the major suppliers of material to the plasma environment. Nevertheless, the trend of decreasing plasma density with increasing distance from the Sun continues, Neptune's magnetospheric plasma density being even lower than that of Uranus.

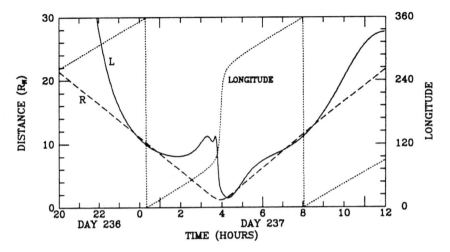

Figure 10.5. The spacecraft trajectory in L shell (solid line), radial distance from the center of the planet (dashed line), and planetary longitude (dotted line) coordinates plotted versus time.

The tenuous plasma within the magnetosphere contained both heavy and light ion species. The primary ions present within the magnetosphere are thought to be N^+ and H^+ (singly ionized nitrogen and hydrogen atoms). From a study of the relative numbers at different distances from the planet, Richardson and McNutt [12] concluded that Triton was the most likely source for both of these ion species. The maximum observed spatial density of ions was about 1 per cubic centimetre, with the heavier ions being somewhat more populous than the lighter, especially closer to the planet.

Magnetospheric scientists often discuss plasma data in terms of their characteristics with changing L shell. An L shell is a surface defined by all the magnetic field lines which cross the magnetic equator at the same distance from the center of the magnetic dipole. For a purely dipolar magnetic field unaffected by solar wind pressure, the radial distance, r, of the L shell at other magnetic latitudes, λ, is given by $r = L \cos^2 \lambda$, where L and r are generally given in units of planetary radii. Near the magnetic cusps (i.e., near the magnetic axis) L shells are very close together. The path of Voyager 2 past Neptune in terms of radial distance, L-shell distance, and planetary longitude are depicted in Figure 10.5.

For nearly aligned, non-offset magnetic fields such as those found at Earth, Jupiter, and Saturn, satellites in circular, low-inclination orbits remain at relatively constant L values. For Jupiter and Saturn, that means that satellites will tend to clear out 'hollows' in the magnetospheric plasma environment. Pioneer 11 plasma scientists inferred the presence of Saturn's G ring from such an effect, even though it was not sighted visually until Voyager 1 reached Saturn. When a magnetic field is highly tilted and offset, as in the cases of Uranus and Neptune, such particle sweeping by satellites and rings becomes extremely complex. With the very low plasma densities seen at Uranus and Neptune, no particle sweeping by satellites or rings was detectable. In fact, as mentioned earlier, Triton acts as a source for magnetospheric plasma rather than a sink.

Voyager 2 receded from Neptune along a direction well southward of the Sun–Neptune line, so no measurements of the deep magnetotail of Neptune were possible. During the Uranus encounter, the plasma detectors measured three crossings of a plasma sheet in the magnetotail. No similar crossings of a Neptune plasma sheet were detected.

10.6   ENERGETIC CHARGED PARTICLES AT NEPTUNE

Closely related to the plasma content of the Neptune magnetosphere are the charged particle measurements from the low-energy charged particle and cosmic ray investigations. The instruments associated with these investigations measured electrons with energies from 22×10^3 to 10×10^6 eV (22 keV to 10 MeV); the range for energetic ions was 28 keV to several hundred MeV. Charged particles with these high energies are much less tightly constrained by the Neptune magnetic field or by the extended solar magnetic field.

As Voyager 2 approached Uranus, upstream ions had been seen to distances of at least 40 Uranus radii (R_U), and may have extended as far upstream at $500 \, R_U$. However, no upstream ions of any sort were detected at Neptune, a unique situation among the giant planets. Even as the spacecraft crossed the bow shock, there were no dramatic indications in the data from these two investigations to signal that the event had taken place; high-energy electron and proton densities were essentially unchanged.

The situation was not much different at the crossing of the magnetopause. Only a modest increase in proton intensities and a barely detectable increase in electron intensities were seen across the transition. The lack of dramatic indications in the energetic charged particle data of either bow shock or magnetopause crossings is also unique among the giant planets encountered by Voyager. Within the magnetosheath (the region of space between the bow shock and the magnetopause) there were sporadic indications of protons and electrons streaming downwind. These were presumably particles that had leaked out of the Neptune magnetosphere and were observed at both the inbound and outbound traversals of the magnetosheath. Near the inbound magnetopause, however, particle directions were not all the same, as might be expected for conditions in the cusp region of a magnetic field.

Closer to the planet the population of high-energy electrons and protons increased to maxima of about 0.001 and 1 per cubic centimeter, respectively, just inside the orbit of Triton. Both populations sharply decreased by a factor of about 100 to 1000 at closest approach to Neptune, rising back to the maximum values near 8 Neptune radii (R_N). The explanation for the sharp dip near closest approach is particle absorption by the atmosphere of Neptune; the effectiveness of this absorption is greatly magnified by the 0.55 R_N offset of the magnetic field from the center of Neptune [13]. There is also strong evidence of energetic particle absorption by the satellite Proteus. Several other dips in population might be attributable to other satellites and to the rings; mapping of these signatures to the relevant magnetic L shells of the other satellites and rings is highly uncertain, primarily due to the uncertainty in the magnetic field structure within $3 R_N$ of the planet center.

Between the orbits of Proteus and Triton, the energetic charged particle population varies smoothly and looks remarkably symmetric between inbound and outbound legs. While such symmetry seems natural, it is certainly different than the large asymmetries noted at corresponding parts of the Uranus magnetosphere. Neptune's magnetosphere in general seems more orderly than that of Uranus.

At the L-shell distance of Triton, dramatic changes in the density and energy of ions and protons were observed. Triton has an associated torus of neutral particles which both interact with charged particles in the magnetosphere and provide additional source material for the magnetosphere. While it is clear from the data that Triton controls the outer bounds of the energetic charged particle population at Neptune, the mechanism or mechanisms by which that occurs are not understood. A Neptune orbiter may be required if these interactions are ever to be understood in detail. One unusual feature of the downstream portions of the magnetosphere was that all observed energetic charged particle motion within the supposed magnetotail was directed planetward. At Uranus, both planetward and tailward motions were seen, with tailward motions dominating beyond $55 R_U$. The significance of this characteristic of the Neptune magnetotail is uncertain, but it may indicate additional sources of charged particles deep in the Neptune magnetotail. Additionally, no magnetic disturbances were noted in the tail region of the magnetosphere; indeed, the tail seemed very quiet by comparison with the magnetotails of the other three giant planets.

10.7   WAVE–PARTICLE INTERACTIONS IN THE MAGNETOSPHERE

Charged particles interact with a magnetic field in a variety of ways. A flow of charged particles (i.e., an electrical current) creates its own magnetic field, thus distorting any pre-existing field. Any charged particle moving through a magnetic field will experience a force which tends to resist motion *across* field lines but allows relatively free motion *along* field lines. These interrelationships are described in four basic equations known as the Maxwell Equations, named after the nineteenth-century Scottish physicist and mathematician James Clerk Maxwell (1831–79).

As charged particles are accelerated or decelerated they experience an increase or decrease in energy. The increase in energy comes from the absorption of electromagnetic energy in the form of waves; the decrease in energy results in the emission of electromagnetic waves. These electromagnetic waves are often termed light waves, although they can be at any wavelength. The highest-energy waves are known as gamma rays. In order of decreasing energy, other forms are known as x-rays, ultraviolet light, visible light, infrared light, microwaves, and radio waves. The Voyager plasma wave investigation detects some of the lowest energy radio waves, whose frequencies range from 10 Hz to 56,200 Hz. Interestingly, sound waves at these frequencies span the range of sensitivity of the human ear, and one of the methods used by scientists to analyze plasma wave data is to convert the data from radio waves to sounds of the same frequency and 'listen' to the data. The 'sounds' of Jupiter, Saturn, Uranus, and Neptune were produced in this fashion, and may be familiar to many readers.

One of the more energetic interactions in a magnetic field is that which generates the ultraviolet auroral phenomena seen on Earth and other planets. Many phenomena in the magnetospheres of Solar System planets and the Sun involve low-energy interactions that result in emission or absorption of radio waves in the plasma wave frequency range. Such radio waves interact readily with charged particles and do not generally propagate over large distances. Hence, plasma wave data can be used to study nearby interactions between the charged particles and the magnetic field through which the spacecraft is passing. 'Nearby' in this context means from a few meters to a few million kilometers, depending on the frequency being monitored and the type of charged-particle interactions present.

Most of the intense radio emissions detected by the plasma wave investigation were seen in the inner magnetosphere during a 6-hour period surrounding closest approach to Neptune. The different kinds of emissions are often named for the nature of the sounds represented by the transformed radio signals (e.g., whistlers, hiss, chirp, chorus, and noise).

Although the planetary radio astronomy investigation saw evidence for a Neptune magnetosphere in data collected as early as 30 days before closest approach, the first plasma wave evidence was seen only 17 hours before closest approach. These waves, with a frequency near 562 Hz, are created when very high energy electrons escaping the Neptune bow shock and travelling along interplanetary magnetic field lines come near the spacecraft. They are an indication that the space-craft has encountered a magnetic field line that intersects the planetary bow shock.

About three hours after this initial detection, radio waves at frequencies from 10 to 178 Hz heralded the actual crossing of the bow shock. Inside the magnetosheath, radio emissions at all radio frequencies dropped to levels that were too low for detection by the plasma wave investigation, indicating a very quiet environment. Similar conditions were noted by the plasma wave instrument in the magnetosheaths of Jupiter, Saturn, and Uranus. This absence of plasma waves in the magnetosheath is not characteristic of Venus, Earth, and Mars; the reasons for this contrast between the magnetosheaths of the terrestrial and the giant planets of the Solar System is not known.

Interestingly, there was also an absence of plasma wave activity at the time of the inbound magnetopause crossing, which began three hours later. The interpreta-tion given by the plasma wave scientists was that the transition, because it occurred in the cusp region, was very gradual [14], requiring more than an hour and a half to complete. A graphical summary of the plasma waves seen in and near the Neptune magnetosphere is given in Figure 10.6.

The first plasma waves seen within the magnetosphere occurred about 3.5 hours before closest approach at a distance of about 10.2 R_N. An intense burst was seen

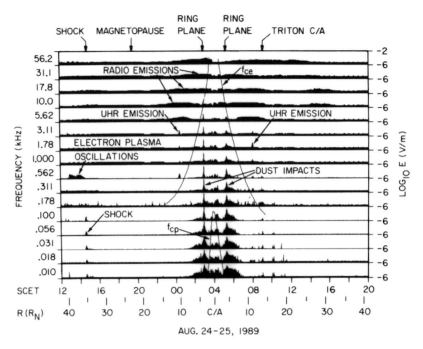

Figure 10.6. Overview of plasma waves in and near the Neptune magnetosphere from 42 R_N before to 42 R_N after closest approach. (P-35418A)

most clearly at 562 Hz (but was also visible at several higher frequencies) and lasted about 10 minutes. During this time, Voyager 2 was very near the magnetic equator, where the plasma is generally densest. Similar signals were seen at an outbound crossing of the magnetic equator at a radial distance of 11.5 R_N. Plasma waves generated by charged-particle interactions near the magnetic equator are typical of all the giant planets.

Whistlers were discussed earlier as an indication of possible lightning in the atmosphere of the planet. Whistlers can also be created by charged-particle interactions in regions of the magnetosphere where electron densities are high. Another possible region for such whistlers (other than atmospheric lightning) might be in the ionosphere of Neptune. A total of 16 whistlers were seen in the plasma wave data. Study of their characteristics has led to the conclusion that they are probably due to atmospheric lightning rather than ionospheric or other interactions. Several other types of plasma waves were observed in the inner portions of the magnetosphere, all of which were relatively weak. The plasma wave scientists have given some tentative explanations for most of the observed plasma waves, but because the electron densities in the relevant regions are poorly known, it may not be possible to provide detailed descriptions of the processes within the magnetosphere which generate these plasma waves. More observations are needed, and they can only be obtained from a properly-instrumented spacecraft in orbit around Neptune.

NOTES AND REFERENCES

1. Warwick, J. W., Evans, D. R., Peltzer, G. R., Peltzer, R. G., Romig, J. H., Sawyer, C. B., Riddle, A. C., Schweitzer, A. E., Desch, M. D., Kaiser, M. L., Farrell, W. M., Carr, T. D., de Pater, I., Staelin, D. H., Gulkis, S., Poynter, R. L., Boischot, A., Genova, F., Leblanc, Y., Lecacheux, A., Pedersen, B. M. and Zarka, P. (1989) Voyager planetary radio astronomy at Neptune. *Science*, **246**, 1498–1501.

2. Ness, N. F., Acuna, M. H. and Connerney, J. E. P. (1995) Neptune's magnetic field and field-geometric properties. In Cruikshank, D. P. (ed.) *Neptune and Triton*. University of Arizona Press, Tucson, pp. 141–68.

3. Connerney, J. E. P., Acua, M. H. and Ness, N. F. (1991) The magnetic field of Neptune. *Journal of Geophysical Research*, **91**, 19023–42.

4. Broadfoot, A. L., Herbert, F., Holberg, J. B., Hunten, D. M., Kumar, S., Sandel, B. R., Shemansky, D. E., Smith, G. R., Yelle, R. V., Strobel, D. F., Moos, H. W., Donahue, T. M., Atreya, S. K., Bertaux, J. L., Blamont, J. E., McConnell, J. C., Dessler, A. J., Linick, S. and Springer, R. (1986) Ultraviolet spectrometer observation of Uranus. *Science*, **233**, 74–9.

5. Cheng, A. F., Krimigis, S. M. and Lanzerotti, L. J. (1991) Energetic particles at Uranus. In Bergstralh, J. T., Miner, E. D., Matthews, M. S. (eds.) *Uranus*. University of Arizona Press, Tucson, pp. 831–93.

6. Sandel, B. R., Herbert, F., Dessler, A. J. and Hill, T. W. (1990) Aurora and airglow on the night side of Neptune. *Geophysical Research Letters*, **17**, 1693–6.

7. Zarka, P., Pedersen, B. M., Lecacheux, A., Kaiser, M. L., Desch, M. D., Farrell, W. M. and Kurth, W. S. (1995) Radio emissions from Neptune. In Cruikshank, D.P. (ed.) *Neptune and Triton*. University of Arizona Press, Tucson, pp. 341–87.

8. Lecacheux, A., Zarka, P., Desch, M. D. and Evans, D. R. (1993) The siderial rotation period of Neptune. *Geophysical Research Letters*, **20**, 2711–14.

9. Gurnett, D. A., Kurth, W. S., Cairns, I. H. and Granroth, L. J. (1990) Whistlers in Neptune's magnetosphere: evidence of atmospheric lightning. *Journal of Geophysical Research*, **95**, 20967–76.

10. Kaiser, M. L., Zarka, P., Desch, M. D. and Farrell, W. M. (1991) Restrictions on the characteristics of Neptunian lightning. *Journal of Geophysical Research*, **96**, 19043–7.

11. Tyler, G. L., Sweetnam, D. N., Anderson, J. D., Borutzki, S. E., Campbell, J. K., Eshleman, V. R., Gresh, D. L., Gurrola, E. M., Hinson, D. P., Kawashima, N., Kursinski, E. R., Levy, G. S., Lindal, G. F., Lyons, J. R., Marouf, E. A., Rosen, P. A., Simpson, R. A. and Wood, G. E. (1989) Voyager radio science observations of Neptune and Triton. *Science*, **246**, 1466–73.

12. Richardson, J. D. and McNutt, R. L., Jr. (1990) Low-energy plasma in Neptune's magnetosphere. *Geophysical Research Letters*, **17**, 1689–92.

13. Selesnick, R. S. and Stone, E. C. (1991) Neptune's cosmic ray cutoff. *Geophysical Research Letters*, **18**, 361–4.

14. Gurnett, D. A. and Kurth, W. S. (1995) Plasma waves and related phenomena in the magnetosphere of Neptune. In Cruikshank, D. P. (ed.) *Neptune and Triton*. University of Arizona Press, Tucson, pp. 389–423.

BIBLIOGRAPHY

Bergstralh, J. T., Miner, E. D. and Matthews, M. S. (eds) (1991) *Uranus.* University of Arizona Press, Tucson. (See magnetosphere chapters on pp. 739–958 by Ness *et al.*, Belcher *et al.*, Cheng *et al.*, Desch *et al.*, and Kurth *et al.* and the references cited in each chapter.)

Cruikshank, D. P. (ed.) (1995) *Neptune and Triton.* University of Arizona Press, Tucson. (See chapters on pp. 141–423 by Ness *et al.*, Mauk *et al.*, Schulz *et al.*, Richardson *et al.* and Gurnett and Kurth. and the references cited in each chapter.)

Van Allen, J. A. and Bagenal, F. (1999) Planetary magnetospheres and the interplanetary medium. In Beatty, J. K., Petersen, C. C. and Chaikin, A. (eds) *The New Solar System (Fourth Edition).* Sky Publishing Corporation, Cambridge, MA, pp. 39–58.

11

The rings of Neptune

11.1 DISCOVERY OF RINGS

All four of the giant planets have rings; none of the terrestrial planets (Mercury, Venus, Earth, and Mars) possesses rings. The reasons for this distinct difference must lie in the processes that took place during the formation and evolution of the planets. Voyager showed that each of the four existing ring systems has its own unique characteristics. Jupiter's ring is devoid of water ice and is probably composed of silicate dust particles with average size near 1 μm. The Jupiter ring was first discovered in Voyager imaging but has since been imaged from Hawaii's 10-m Keck telescope at an infrared wavelength of 2.27 μm. The ring particles are probably bits of dust launched by micrometeoroid impacts on nearby satellites (Metis, Adrastea, Amalthea, and Thebe). Metis and Adrastea feed the narrow main ring of Jupiter, Metis near the middle of the main ring and Adrastea near its outer edge. Exterior to the main ring are the inner and outer Gossamer rings, each slightly thicker in dimensions than its inner neighbor. The inner Gossamer ring has its outer edge at the orbit of Amalthea; the outer Gossamer ring has its outer edge at the orbit of Thebe. Interior to the main ring is a Halo ring of indistinct shape and size which is apparently much more heavily interactive with the magnetic field, indicating that its particles are probably electrically charged.

Saturn has the most extensive and magnificent ring system in the Solar System. Galileo Galilei (1564–1642) first saw the rings as cup handles in 1611. Christiaan Huygens (1629–95) later identified the rings as a detached system circling Saturn. Saturn's ring system is now divided into eight major regions, all of them more or less centered on the equatorial plane of the planet. From closest to most distant, these rings are known as D, C, B, Cassini Division, A, F, G, and E. The prominent A and B rings are composed primarily of water ice, with particle sizes ranging from dust-sized to house-sized, but mainly in the centimeter size range. Little is known about the particle sizes in the C, D, and F rings. The G and E rings are composed primarily of micrometer-sized particles of unknown composition. Saturn's main rings appear in telescopes to be broad and almost featureless, but Voyager revealed complex radial structure, even at the highest resolution, and very little azimuthal structure other than the ephemeral and enigmatic ring spokes. Some of the radial structure is explainable in terms of gravitational forces from satellites in or near the rings, but much of it is not well understood. Perhaps data from the Cassini Mission will provide the answers to some of these questions.

The Uranus rings are different in nature from either the Jupiter or the Saturn rings. Instead of the broad bright rings of Saturn or the tenuous dusty rings of Jupiter, Uranus has a series of dark, narrow rings, most likely composed of radiation-darkened methane ice particles. In order of increasing distance from the planet, these rings are designated 6, 5, 4, α, β, η, γ, δ, λ, and ϵ. The curious nomenclature arises from the circumstances of their discovery. Most were first detected as they blocked the light of the star SAO 158687 on March 10, 1977. The first set of observers found a series of five relatively distinct rings, which they named (in order of increasing distance from Uranus) α, β, γ, δ, and ϵ, the first five letters in the Greek alphabet. The second set of observers missed seeing a couple of these rings, but found another three rings closer to the planet. They named their rings 1, 2, 3, 4, 5, and 6, in order of decreasing distance from the planet. Rings 1, 2, and 3 were found to be the same as rings ϵ, γ, and β, respectively, but rings 4, 5, and 6 were previously undiscovered.

The first set of observers restudied their data, finding rings 4, 5, and 6, as well as an additional ring (η). During the Voyager 2 encounter, all nine of these rings were seen, both in images and in stellar occultation data collected by the photopolarimeter investigation. A very faint additional ring (λ) was seen in images taken of the region between the δ and ϵ rings.

The story of the discovery of Neptune's rings is chronicled in Chapter 3 of this book. Earth-based observations were not very illuminating, showing the presence of possible ring material in only about 10% of stellar occultation observations. Often, material was seen on one side of the planet and not on the other. It seemed apparent that Neptune probably had narrow rings, but that they were probably discontinuous. However, a clear understanding of their detailed nature came first as a result of data collected during the Voyager 2 encounter. We now know that there are five continuous rings: in order of increasing distance from Neptune, they are called Galle, Le Verrier, Lassell, Arago, and Adams. Galle and Lassell are relatively broad and indistinct; Le Verrier, Arago, and Adams are narrow and distinct. There may also be a partial ring sharing the orbit of the satellite Galatea, just inside the Adams ring. Curiously, the Adams ring (Figure 11.1) has five optically thicker arcs occupying about 10% of the total circumference of the ring. It is mainly these arcs that are respnsible for the confusing Earth-based stellar occultation results. From leading to trailing, the arcs have been named Courage, Liberté, Egalité 1, Egalité 2, and Fraternité. The composition and reflectivity of the Neptune rings are probably very similar to those of the rings of Uranus, but otherwise they bear little resemblance to any other known ring system.

Reasons for the differences in these planetary ring systems may be intimately related to the processes which formed them. One marvels at Nature's propensity for building rings in such a wide variety of environments. Nothing in the collected

Figure 11.1. Ring image, showing the arcs in the Adams ring. The overexposed crescent of Neptune is seen at the lower right. (P-34712)

experience of astronomers and planetologists would have permitted accurate predictions of the basic structures of a newly discovered planetary ring system on the basis of previously studied ring systems. Nor is it possible to assess the total character of any ring system from a limited dataset.

In spite of these diversities, the same laws of motion are in operation in each of the known ring systems. Each derives its major control from the gravity of the central planet, responds to secondary forces caused by the gravity of nearby satellites or other ring particles, and experiences small retarding forces which tend to cause the particles to spiral toward the planet's atmosphere. Each is constrained to orbit close to the equator of these oblate giants, appears to be much younger than its planet, and requires ongoing processes to maintain its very existence.

11.2 STRUCTURE OF THE RINGS

Large variations with azimuth are uncharacteristic of the rings of Jupiter, Saturn, and Uranus. Other than the non-circularity and smoothly varying widths of some of these rings, the enigmatic ring spokes seen within Saturn's B ring may be the only easily viewed features in these ring systems that are not symmetric around the rotation axis of the respective planet. Even the ring spokes are ephemeral in nature, generally lasting in recognizable form for at most one or two circuits of Saturn. Not so with the rings of Neptune (Figure 11.2): the arcs in the Adams ring have apparently existed in identifiable form for at least several years before the Voyager 2 encounter with Neptune. No variations in their general appearance or extent were seen during the Voyager encounter period. Their approximate

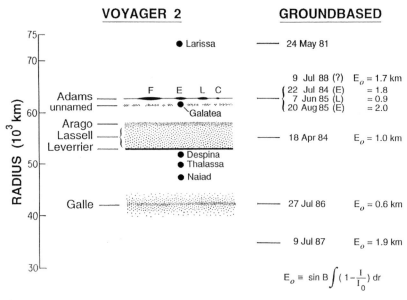

Figure 11.2. Diagram of Neptune's ring structure.

Table 11.1. Neptune ring arc dimensions (within the Adams ring)

Arc name	Approximate arc length (deg)	Separation from leading arc (deg)*	Comments
Courage	1.0 ± 0.1	0.0	Least optically thick of arcs
Liberté	4.1 ± 0.1	7.3	Well-defined arc length
Egalité	1.0 ± 0.5	19.6	The boundary between the two components of the Egalité arc is poorly defined
Egalité 2	3.0 ± 0.5	22.4	
Fraternité	9.6 ± 0.1	33.2	Longest of the arcs

* Separation in degrees measured from the centers of the arcs.

characteristics are given in Table 11.1. Dimensions of the rings and the orbital distances of nearby satellites are given in Table 11.2.

It is apparent from Table 11.2 that the rings of Neptune are not as well defined as those of Uranus. This results in large part from the much larger Earth-based star occultation dataset that exists for the Uranus rings. Three (and possibly four) of the ten well-documented ring occultation detections from Earth were of the arcs in the Adams ring. Of the six remaining, three cannot be reconciled with any known ring structure, one is near the center of the Lassell ring, and one is near the radius of the Galle ring. It is unlikely that these latter observations represent real Earth-based detections of the Lassell or Galle rings, which are probably far too optically thin to be detected from Earth.

Table 11.2. Neptune ring (and ring satellite) data from Voyager 2

Ring name (satellite)	Radial distance (km)	Width (km)	Optical depth, τ	Comments
Galle	$\sim 41,900$	~ 2000	~ 0.00008	Diffuse, indistinct edges
(Naiad)	48,227			No ring interaction?
(Thalassa)	50,075			No ring interaction?
(Despina)	52,526			May shepherd Le Verrier
Le Verrier	$53,200 \pm 20$	~ 110	0.01–0.02	Narrow, inclined $\sim 0.03°$
Lassell	53,200–57,200	4,000	-0.00015	Lower dust content
Arago	$\sim 57,200$	<100	~ 0.001	At outer edge of Lassell
Unnamed	$\sim 61,950$	<100	~ 0.001	Diffuse partial ring
(Galatea)	61,953			Probably controls arcs
Adams	$62,932 \pm 2$	15–50	0.01–0.02	Arcs have $\tau \sim 0.1$
(Larissa)	73,548			Largest of ring satellites

The tenth occultation data point, reported by Reitsema *et al.* [1], now appears to be a highly improbable occultation by the satellite Larissa more than eight years before the Voyager 2 encounter of Neptune. Their publication indicated that they thought they had observed a possible third satellite of Neptune, but later occultation measurements of the ring arcs of Neptune from Earth led most to believe that the feature seen by Reitsema *et al.* was more probably another ring arc. Only after analysis of the Voyager 2 data from the Neptune encounter was it realized that the occultation indeed *was* of a previously undiscovered satellite.

11.3 PHYSICAL PROPERTIES OF THE RINGS

Radio occultation of the Neptune rings occurred shortly after the closest approach of Voyager 2 to Neptune [2]. These occultations were done at two different wavelengths, 3.6 and 13 cm. No evidence of the rings was seen in either dataset. An upper limit of 0.01 radio optical depth was estimated. That implies that there are few if any particles in the centimeter size range or larger.

The star σ Sagittarii was observed by both the ultraviolet and photopolarimeter investigations as the Neptune rings passed between the star and the Voyager 2 spacecraft. Only the Adams ring was detected in ultraviolet light, with an optical depth of about 0.02, a total width of about 35 km, and a more opaque central core of width $\sim$ 15 km. The photopolarimeter also detected the Adams ring at a wavelength of 0.26 μm, with similar properties to those stated above for the ultraviolet mesaurements. In addition, a possible detection of the Le Verrier ring may be indicated by data near the radial position of that ring. If so, the optical depth is about 0.005 and the width near 110 km. No evidence for any of the other Neptune rings was seen in the data from these two investigations. From the radio and optical occultation data, it can be concluded that the Adams and Le Verrier rings are very likely composed primarily of material much smaller than a few centimeters in diameter. This is markedly different from the situation existing for the main rings of Saturn and Uranus, where dust-sized particles probably account for 1% or less of the total optical depth.

The imaging data leads to similar conclusions about the average size of particles in the Adams and Le Verrier rings, namely that they have a high fraction of micrometer-sized dust. In the Adams ring, the brighter arcs (Courage, Liberté, Egalité 1, Egalité 2, and Fraternité) appear to be dustier than the optically thinner portions of the ring. The Galle ring also appears to have a high dust content, whereas the Lassell ring, of comparable optical depth, appears significantly less dusty (although still more than 10 times as dusty as the main rings of Saturn or Uranus).

One characteristic common to the rings of Uranus and Neptune is their low reflectivity ($\leqslant 0.05$). It is possible that both rings are characterized by the presence of elemental carbon produced by the bombardment of methane (CH_4) ice. The rings of Neptune are apparently much less massive than those of Uranus by a factor of approximately 10,000.

Neptune has more ring satellites than Uranus, both in number and in mass. The smaller mass of the Neptune ring system is therefore not a consequence of a lesser availability of source material. The reasons for these differences are not understood, but certainly imply major differences in the evolution of these two ring systems.

11.4 INTERACTION WITH SHEPHERDING SATELLITES

The presence of the Uranus rings was well established before the arrival of Voyager 2 at Uranus; one of the primary ring objectives at Uranus was the determination of the mechanisms confining the rings to their narrow shapes. As mentioned earlier, the situation was not as clear-cut for the rings of Neptune. The primary objectives of the ring studies at Neptune were associated with verifying that the planet possesses rings and with the determination of the general nature of those rings. The paucity of ring satellites at Uranus left scientists somewhat unprepared for the unexpectedly high ratio of potential shepherding satellites to narrow rings found at Neptune.

The rings of Uranus are believed to be confined to their narrow radial widths by the gravitational action of nearby satellites. However, only one of the known satellites of Uranus (Cordelia) orbits the planet inside the outermost ring, and all of the inner nine moons of Uranus are less than 120 km in diameter. In fact, seven of the nine are less than 70 km in diameter. Their presumed composition is icy, so their masses must be at least one to three orders of magnitude (i.e., 10 to 1000 times) *smaller* than the more distant mid-sized satellites. It is therefore difficult to explain adequately the confinement mechanism for the rings of Uranus on the basis of gravitational effects of these small satellites.

Neptune's ring–satellite relationship ought to lend itself to easier explanations of the ring dynamics. Four of the eight known satellites orbit within the ring system. Naiad, Thalassa, and Despina are between the two innermost rings, Galle and Le Verrier. Galatea's orbit is just within the outermost ring, i.e., Adams. Despina and

Galatea are about 150 km in diameter, and Larissa, which orbits about 10,000 km outside the Adams ring, is close to 200 km in diameter. In spite of this 'advantage' of nearby, somewhat more massive satellites, a detailed mathematical model of ring confinement is not easy to construct.

Particles in the Adams ring circle the planet 42 times for every 43 orbits of nearby Galatea, so it is very tempting to say that Galatea's gravitational influence both confines the Adams ring radially and also confines the azimuthal spreading of the ring arcs. One attempt at such an explanation is given on pages 762–781 of Porco et al. [3]. In this explanation, the authors point out that with the 42:43 resonance, there may be a tendency for the Adams ring to develop a cyclical shape with 42 nodes, each occupying an azimuthal range of $8.571°$ ($= 360°/42$). The gravitational influence of Galatea may thus serve to restrict ring particle motion across a node. It is interesting to note that the five ring arcs would fit nicely in four consecutive regions. The details and efficiency of this confinement mechanism are not yet determined. Nor can dynamicists explain why such arcs are not seen in any other segments of the Adams ring. Furthermore, because Galatea orbits inside the orbit of the Adams ring particles, the general tendency would be to transfer energy from Galatea to the Adams ring particles. That added energy should, over geologically short time periods, cause the Adams ring to spread outward. Such outward spreading is not consistent with the observed stable arc widths over more than five years between their first detection from Earth and the Voyager 2 encounter.

It is also possible that a small satellite, perhaps no more than a few kilometers in radius and embedded either in the arc region or at a Lagrangian point [4] $60°$ ahead of or behind the arc region of the Adams ring, might be the cause of (or at least aid in) the azimuthal and radial confinement of the ring arcs. A satellite less than 10 km in diameter, particularly one of low reflectivity, could easily have gone undetected in the images shuttered by Voyager 2 and still be massive enough to make a substantial contribution to the morphology of the Adams ring.

A model similar to that invoked for Galatea on the Adams ring might help to explain the radial confinement of the Le Verrier ring by Despina, which orbits Neptune 53 times for every 52 orbits of material in the Le Verrier ring. No ring arcs have been seen in the Le Verrier ring, but that could possibly result from depletion of the source of ring material for the Le Verrier ring or from the absence of an additional small satellite in the same orbit as the Le Verrier ring. Refer again to Table 11.2 for the relative positions of the rings and the ring satellites.

Other possible ring confinement (or spreading) mechanisms might include weak electromagnetic effects, if the ring particles are small enough, and if they become sufficiently electrically charged by sunlight (photoelectric effect [5]) that their charge-to-mass ratios are relatively high. Radio data at all frequencies from 10 Hz to 1326 kHz collected during the Neptune ring plane crossings [6] indicate that a relatively thick disk of dust particles exists well outside the visible Neptune rings. It is probable that this population of dust particles has its origin in meteoritic impacts on the surfaces of satellites and larger ring particles. On the inbound leg of the encounter, at a radial distance from the center of Neptune of $3.45\,R_N$, the disk of particles had a thickness of about 538 ± 22, was shifted 146 ± 4 km north of the equatorial plane, and reached a peak rate of 443 impacts per second. The outbound leg dust disk thickness was $2{,}072 \pm 392$ km, and the dist disk was centered 948 ± 65 south of the equatorial plane; this crossing was centered at a radial distance of $4.20\,R_N$ and had a peak impact rate of 151 per second. The source of the apparent tilt of this disk with respect to Neptune's equator ($\sim 0.4°$) is possibly related to the gravitational effects of Triton, whose orbit is inclined (in the appropriate sense for the observed offsets) by more than $22°$ from Neptune's equator.

In addition to the denser disk of particles, there was a more tenuous halo of dust particles which impacted Voyager 2 at a rate of 1 to 10 particles per second during the entire time the spacecraft was within $8\,R_N$ of Neptune, including passage over the north polar region. These particles probably have much smaller masses than those in the main disk (1-μm radius versus $\sim$10-μm radius in the main disk), are probably electrically charged by the action of sunlight or by interactions with magnetospheric plasma, and are thereby lifted above and below the equatorial plane of Neptune by electromagnetic interaction with Neptune's magnetic field. This partial entrapment of tiny dust particles within the planetary magnetic field are reminiscent of the interactions in Saturn's B ring which result in radial spokes that, for short periods after their formation, co-rotate with the Saturn magnetic field.

11.5   AGE AND EVOLUTION OF THE RINGS

The large amount of dust in Neptune's main rings and in the extended disk and halo bespeak a relatively young ring system, certainly much younger than the planet and its satellites. If the upper atmosphere of Neptune were as extensive as that of Uranus, drag forces might have cleared much of the dust halo surrounding Neptune. Even without this indication of the youth of the Neptune rings, the presence of the ring arcs in the Adams ring would also indicate that the present ring system cannot have existed for more than a tiny fraction of the age of Neptune. They are very likely the result of a single and recent event, perhaps the destruction of a small satellite. The gravitational influence of Galatea alone would spread the ring arcs in an outward direction in only a few years. Detection of the ring arcs of Neptune by Earth-based stellar occultation measurements dates back to at least 1983, so the ring arcs appear to have been at least six years old at the time of the Voyager encounter. No detection of the ring arcs subsequent to the Voyager 2 encounter has been reported, as far as the authors have been able to determine, but since Neptune is moving away from the galactic plane, the number of stellar occultation opportunities continues to decrease each year. Hence, no conclusions can yet be drawn about the post-Voyager status of the ring arcs.

Atmospheric drag, while less efficient than at Uranus, will certainly deplete some of the particles from the lower rings and from the halo. Micrometeroid bombardment continues to diminish the average size of existing ring particles, in addition to replenishing the ring at the expense of nearby satellites. As the particles reach sizes of a micrometer or smaller, Poynting–Robertson drag begins to be more efficient. In this process, orbiting ring particles absorb sunlight and re-emit the light in a preferentially forward direction in the orbit [7]. This has the effect of removing energy from the particles, causing them to slowly spiral downward to lower and lower orbits until they fall into the atmosphere. As with the rings of Jupiter, Saturn, and Uranus, the Neptune ring system appears to be constantly changing in character as existing ring material is slowly removed from the system and new material is continually being added. A spacecraft returning to Neptune 100 years from now might find very little of the ring structure seen by Voyager 2, but would probably find a number of features unseen during the brief encounter of this hardy spacefarer.

NOTES AND REFERENCES

1. Reitsema, H. J., Hubbard, W. B., Lebofsky, L. A. and Tholen, D. J. (1982) Occultation by a possible third satellite of Neptune. *Science*, **215**, 289–91.
2. Tyler, G. L., Sweetnam, D. N., Anderson, J. D., Borutzki, S. E., Campbell, J. K., Eshleman, V. R., Gresh, D. L., Gurrola, E. M., Hinson, D. P., Kawashima, N., Kursinski, E. R., Levy, G. S., Lindal, G. F., Lyons, J. R., Marouf, E. A., Rosen, P. A., Simpson, R. A. and Wood, G. E. (1989) Voyager radio science observations of Neptune and Triton. *Science*, **246**, 1466–73.
3. Porco, C. C., Nicholson, P. D., Cuzzi, J. N., Lissauer, J. J. and Esposito, L. W. (1995) Neptune's ring system. In Cruikshank, D.P. (ed.) *Neptune and Triton*. University of Arizona Press, Tucson, pp. 703–804.
4. Named for Joseph Louis Lagrange (1736–1813), who first pointed out their orbital stability, Lagrangian points are orbital positions relative to a massive body (like a planet) orbiting around a more massive primary (like the Sun) where smaller 'third' bodies (like asteroids or moons) can remain in stable orbits for long periods of time. Lagrange defined five such Lagrangian points. The L3 position is centered on the massive body (as in the case of the Moon orbiting the Earth). The L1 and L2 points are located along the line between the two primary bodies, one just 'sunward' of the 'planetary' orbit and the other just anti-sunward of the planet, at distances that depend on the relative masses of the bodies involved. Although shown by Lagrange to be theoretically possible sites for stable third bodies, no objects have been found occupying the L1 or L2 points of a planet or major satellite. The L4 and L5 points are of particular interest for planetary scientists; they are stable positions in the orbit of a secondary body near which smaller tertiary bodies can remain for long periods of time. The L4 and L5 points are respectively located 60° ahead of or behind the secondary body, as in the case of the Trojan asteroids which precede or follow Jupiter in its orbit around the Sun. A small satellite orbiting approximately 60° ahead of or behind the ring arcs in Neptune's Adams ring might be partially responsible for apparent stability of the ring arcs. A slightly more extensive discussion of the restricted three-body problem and Lagrangian points can be found in Shirley, J. H. and Fairbridge, R. W. (1997) *Encyclopedia of Planetary Sciences*. Chapman & Hall, New York, pp. 382–3.
5. The photoelectric effect, first discovered by Heinrich Rudolph Hertz (1857–94), is the ejection of electrons from a material as the result of irradiation with light. Two important characteristics of the photoelectric effect are well established: (1) the number of electrons ejected is proportional to the intensity of the incident light, and (2) the electrons are emitted with a range of energies, where the maximum energy is independent of the

light intensity, but is dependent on the color of the light. The photoelectric effect is used in light meters associated with cameras and photography.

6. The best discussion of these radio emissions is given by Gurnett, D. S. and Kurth, W. S. (1995) Plasma waves and related phenomena in the magnetosphere of Neptune. In Cruikshank, D. P. (ed.) *Neptune and Triton*. University of Arizona Press, Tucson, especially pp. 399–406.

7. Poynting–Robertson drag is described in more detail in Mignard, F. (1984) Effects of radiation forces on dust particles in planetary rings. In Greenberg, R. and Brahic, A. (eds) *Planetary Rings*. University of Arizona Press, Tucson, pp. 333–66.

BIBLIOGRAPHY

Bergstralh, J. T., Miner, E. D. and Matthews, M. S. (eds) (1991) *Uranus*. University of Arizona Press, Tucson. (See rings chapters on pp. 327–465 by French *et al.* and Esposito *et al.* and the references cited in each chapter.)

Burns, J. A. (1999) Planetary rings. In Beatty, J. K., Petersen, C. C. and Chaikin, A. (eds) *The New Solar System*. Sky Publishing Corporation, Cambridge, Massachusetts, pp. 221–40.

Cruikshank, D. P. (ed.) (1995) *Neptune and Triton*. University of Arizona Press, Tucson. (See ring chapter on pp. 703–804 by Porco *et al.*, and the references cited therein.)

Greenberg, R. and Brahic, A. (eds) (1984) *Planetary Rings*. University of Arizona Press, Tucson, 784 pages (written prior to the Voyager 2 Neptune encounter).

12

The satellites of Neptune

12.1 DISCOVERY OF SIX ADDITIONAL SATELLITES

Perhaps the most enduring legacy of the Voyager mission to the giant planets is the amazing number of major discoveries credited to it. To the extent it was possible, data was made available in real time or nearly real time so that the citizenry of the world, especially those of the USA, which funded the mission, could have the thrill of seeing these new discoveries as they happened. Most of us, including those who would later pour over the data to extract the quantitative results, were fascinated by the 'other-worldly' appearance of these planets. Prior to the launch of the two Voyager spacecraft in 1977, the known natural satellites in the Solar System numbered 32: 1 at Earth, 2 at Mars, 13 at Jupiter, 9 at Saturn, 5 at Uranus, and 2 at Neptune. By the conclusion of the planetary portion of the Voyager mission in 1989, that number had grown to 61, including 3 more at Jupiter, 9 more at Saturn, 10 more at Uranus, 6 more at Neptune, and 1 at Pluto, very nearly doubling the number of known Solar System natural satellites. Of course, not all of these discoveries came directly from the Voyager mission; however, except for Pluto's Charon, those not discovered in Voyager data were in large part a result of the greater interest in the giant planets elicited by the imminent approach of Voyagers 1 and 2 to these distant worlds.

Included among the satellites discovered from Voyager data (primarily imaging data) were Jupiter's satellites Metis, Adrastea, and Thebe; Saturn's satellites Pan, Atlas, Prometheus, and Pandora; Uranus's satellites Cordelia, Ophelia, Bianca, Cressida, Desdemona, Juliet, Portia, Rosalind, Belinda, and Puck; and Neptune's satellites Naiad, Thalassa, Despina, Galatea, and Proteus. The discovery of these last five (and Larissa) is the subject of this section. Note that Saturn's Epimetheus and Neptune's Larissa were discovered from Earth-based observations, but their existence was not confirmed until they were seen in Voyager data. Prior to those confirmations, Epimetheus was thought to be the same object as Janus, and Larissa was widely suspected to be a ring arc circling Neptune.

As outlined in Chapter 1, Triton was discovered by William Lassell in 1846, very shortly after the discovery of Neptune itself, and Nereid was discovered more than a hundred years later by Gerard Peter Kuiper in 1949. Following the surprise discovery of the rings of Uranus in 1977 during what was intended to be an atmospheric occultation observation, the attention of a number of individuals turned to the possibility of rings around Neptune. Potential stellar occultations by possible ring material orbiting Neptune were identified, and the telescopic photoelectric observations of these occultations began shortly thereafter.

The first few such attempts met with little success, but in 1981 Reitsema *et al.* [1] observed a distinct occultation event by some undefined object near Neptune. The occultation data had sharply defined entry and exit times, but no corresponding event was seen on the opposite side of the planet. Reitsema *et al.* recognized that a stellar occultation by a small satellite was highly improbable, but reasoned that if it were a continuous ring, it would have been seen on both sides of the planet. Since it was not, they published the results as a possible detection of a third satellite of Neptune. Later observations also revealed one-sided occultation events. While it was possible that swarms of satellites existed, it seemed more likely to most planetary scientists that Neptune possessed partial rings or ring arcs. These later occultation events also led planetary scientists to believe that the event observed by Reitsema *et al.* was just the first of such ring arc detections, rather than a chance occultation of a new satellite of Neptune.

The first of the new satellites was spotted more than two months before Voyager 2's closest approach to Neptune, in mid-June 1989. The discovery was early enough to permit late changes to Voyager 2's observing sequence to collect a relatively close series of images of 1989N1, later known as Proteus. Although Proteus proved to be the largest of the new satellites, both in size and in orbital radius and period, it is tiny compared to Triton and orbits Neptune much closer to the planet and in less than one-fifth of Triton's orbital period. It is also much darker than either Triton or Nereid, reflecting a mere 6% of the sunlight incident on its surface. Although relatively spherical in shape, it is nonetheless irregular, possessing large-scale variations from true sphericity.

The next three satellites, Larissa (1989N2), Galatea (1989N4), and Despina (1989N3), were discovered in late July 1989, about a month before closest approach. They were each considerably smaller than Proteus and closer to the planet. As mentioned in Chapter 11, these three satellites are fortuitously placed to interact gravitationally with the Adams and Le Verrier rings of Neptune.

The last two new satellites, each with a diameter less than 100 km, were also closest to the planet and were discovered only a few days before closest approach of Voyager 2 to Neptune on August 25, 1989. They are each too small and too distant from any of the Neptune rings to significantly contribute gravitationally to their radial confinement. They may some day be primary sources of material for rings that do not presently exist.

Several other potential candidates for new satellites were identified in Voyager imaging, but the others were not found in subsequent data and were thus discarded. Such false alarms can occur as a result of noise in the imaging data or from misidentification of background objects (usually stars). Unless a sufficient number of observations was found to enable determination of a reliable set of orbital characteristics, Voyager scientists chose not to include such data in the list of newly discovered satellites.

12.2 ORBITAL AND PHYSICAL CHARACTERISTICS OF THE SATELLITES

It is interesting to note that the new satellites are neatly arranged in order of size, with the smallest (Naiad) being closest to the planet and the largest (Proteus) furthest from the planet. Unlike Triton, all of the new satellites orbit Neptune in prograde (same direction as the planet's rotation) orbits. Unlike both Triton and Nereid, the new satellites are in almost equatorial orbits, all but Naiad (whose orbital inclination is 4.7°) within 1° of the plane of Neptune s equator. Because of their closeness to the planet, all but Proteus orbit the planet in less time than the planet s rotation, three of them in less than half the Neptune rotation period of 16.1 hours. The orbital and physical characteristics of the eight known satellites of Neptune are given in Tables 12.1 and 12.2, respectively.

Some indication of the composition of the new satellites is provided by their radii and the sizes of their orbits. All six have very low reflectivity, and four of the six orbit closer to the planet than the Roche limit [2] for objects composed of water ice. It seems likely that these satellites, with diameters from $\sim 60\,\text{km}$ to $316\,\text{km}$, are composed of material with greater cohesiveness than water ice. On the other hand, the low temperatures extant at Neptune's distance from the Sun make water behave more like rock than like the water ice with which humans are familiar. Thomas et al. [3] estimate the densities of these objects to be within the limits of 0.7 to $2.0\,\text{g/cm}^3$. The masses of all the Neptune satellites except Triton are estimated in Table 12.2 on the basis of that range of densities. Triton's mass was measured directly by the gravitational influence it exerted on the Voyager spacecraft during the close encounter of that satellite. The reflectivity, expressed as a geometric albedo [4],

Table 12.1. Orbital characteristics of Neptune's satellites

Satellite number/name	Semimajor axis (km)	Inclination (deg)	Orbital eccentricity	Orbital period (h)
N8/Naiad	48,227	4.7	0.0003	7.1
N7/Thalassa	50,075	0.2	0.0002	7.5
N5/Despina	52,526	0.1	0.0001	8.0
N6/Galatea	61,953	0.1	0.0001	10.3
N3/Larissa	73,548	0.2	0.0014	13.3
N4/Proteus	117,647	0.6	0.0004	26.9
N1/Triton	354,760	156.8	0.000	141.0
N2/Nereid	5,513,400	27.6	0.753	8,643.1

Table 12.2. Physical characteristics of Neptune's satellites

Satellite number/name	Discovery (month/year)	Radius (km)	Geometric albedo	Mass $(10^{20}\,g)$	Density (g/cm^3)
N8/Naiad	Aug/1989	29 ± 6	~ 0.06	0.7–2.0	0.7–2.0
N7/Thalassa	Aug/1989	40 ± 8	~ 0.06	2–5	0.7–2.0
N5/Despina	Jul/1989	74 ± 10	0.059	12–33	0.7–2.0
N6/Galatea	Jul/1989	79 ± 12	0.063	14–40	0.7–2.0
N3/Larissa	May/1981	96 ± 7	0.056	25–73	0.7–2.0
N4/Proteus	Jun/1989	208 ± 5	0.061	260–740	0.7–2.0
N1/Triton	Oct/1846	$1,353 \pm 2$	~ 0.75	$21,398 \pm 53$	2.05
N2/Nereid	May/1949	170 ± 25	~ 0.18	140–400	0.7–2.0

is given for each of the satellites in Table 12.2. The disks of Naiad and Thalassa were not resolved in Voyager 2 imaging; their sizes are thus estimated on the basis of an assumed geometric albedo of ~ 0.06.

The names for the new satellites were selected by the Nomenclature Commission of the International Astronomical Union. The names selected are those of gods or goddesses associated with Neptune mythology or of other mythological aquatic beings. Again, this is in sharp contrast to the convention used for Uranus, which employed names from the plays of William Shakespeare or of Alexander Pope. Jupiter and Saturn also employ the mythology convention for naming of satellites, so it is the names chosen for the satellites of Uranus, not those of Neptune, which fail to follow the 'normal' pattern.

12.3   PROTEUS, LARISSA, GALATEA, DESPINA, THALASSA, AND NAIAD

Image resolution was sufficient to obtain approximate shapes for Proteus and Larissa (in addition to Triton), but for none of the other small satellites. The best three images of Proteus were centered on longitudes 65°, 94°, and 334°. These were from different enough directions to provide a good estimate of the three-dimensional appearance of Proteus. Proteus's shape can be described as a triaxial ellipsoid, with radii of $a = 218 \pm 6\,km$, $b = 207 \pm 4\,km$, and $c = 201 \pm 4\,km$. In this formulation, a is generally the axis which points toward the planet center, b is the axis along the direction of motion, and c is the axis perpendicular to the orbit plane. Although details of the spin motion of Proteus could not be determined from the data, it is assumed that Proteus, like most satellites in the Solar System (including Earth's Moon), keeps its same face toward the planet. Such a condition is known as synchronous rotation. It is customary to designate the 0° longitude and latitude for such satellites as that location which lies along a straight line connecting the center of the

Figure 12.1. Best image of Proteus. The shape is approximately spherical, reflectivity is only about 6%, and surface contrast is very low. Image scale is 1.33 km/pixel. (P-34727)

satellite to the center of the planet, with longitudes increasing in a clockwise direction as viewed from over the north pole. The Proteus longitudes cited above use that convention. The best image of Proteus obtained by Voyager 2 is reproduced in Figure 12.1. In this image, the size of the individual pixels is approximately 1.3 km. A rudimentary surface map is given in Figure 12.2 and shows a few craters and ridges. Contrast across the surface is very low, so contrast enhancement was needed to bring out even these few features.

The best image of Larissa (Figure 12.3) has an image scale of 4.1 km/pixel. Images from other perspectives were not taken at close enough ranges to resolve the disk of this small satellite, so only a two-dimensional figure can be estimated. That ellipsoidal figure has radii $a = 104 \pm 7$ km and $c = 89 \pm 7$ km, again assuming that the satellite is in locked synchronous rotation. No discernible surface features were seen in the image.

The disks of Galatea and Despina in their respective best images were each about 8 pixels wide (Figure 12.4), insufficient to say anything meaningful about their shapes or surface contrast. The corresponding best images for Thalassa and Naiad (Figure 12.5), as mentioned earlier, were at scales of about 24 and 17 km/ pixel, respectively, and no reliable disk sizes could be estimated from these images.

Phase curves (variation in brightness with phase angle) were measured for the four largest of the new satellites. They varied from 0.031 magnitudes per degree for Larissa to 0.039 magnitudes per degree for Proteus. The uncertainty in these phase functions is comparable to the differences between them, so it is safe to say that they all decrease in magnitude (increase in brightness) with the phase angle at a rate of 0.035 ± 0.004 magnitudes per degree of phase angle [5].

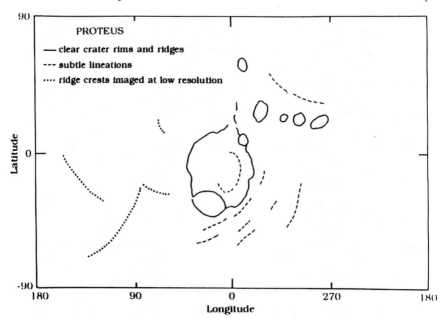

Figure 12.2. Rudimentary map of the surface features on Proteus. 0° longitude and latitude point toward Neptune, assuming locked (i.e., synchronous) rotation. The leading hemisphere in its orbit around Neptune is centered on latitude 0° and longitude 90°; the trailing hemisphere is centered on 0°, 270°.

12.4   NEREID

The most distant of Neptune's known satellites is Nereid. Its orbital period is 360.12 days, or very nearly a full Earth year. But more significantly, its orbital eccentricity is 0.753, meaning that while its mean distance from Neptune is 5,513,400 km, the difference between its greatest distance (9,665,000 km) and its smallest distance (1,362,800 km) is more than a factor of seven! Nereid's orbit is the most eccentric (out of round) of any natural satellite in the Solar System. Because of this fact and because of its great distance from Neptune, many planetary scientists argued in the past that it must be a captured asteroid. However, recent spectral detection of water ice on its surface [6] would seem to provide strong evidence against an asteroidal origin and for Nereid's having been an original satellite of Neptune. If it was, its highly eccentric orbit may be a consequence of whatever process put Triton into its present inclined retrograde orbit. More will be said about this possibility in the next section.

Because of its large orbital eccentricity, Nereid's angular motion around Neptune is highly variable. Its angular motion near its 'annual' periapsis (closest approach to the planet) is seven times as fast as at apoapsis (furthest point). It is therefore highly unlikely that Nereid's period of rotation is the same as its orbital period. Hence, one of the Voyager objectives for Nereid was to determine the

Figure 12.3. Best image of Larissa. The two-dimensional shape is approximately circular, reflectivity is only about 6%, and surface contrast is very low. Image scale is 4.1 km/pixel. (P-35027)

rotation period, either from changing surface features or from periodic variations in the integrated light from the satellite. Voyager 2 observed Nereid repeatedly over a 12-day period when the satellite was close enough to provide a reasonably strong signal (at least 50 data numbers above background) and detected no periodic variations in brightness. Only one image of Nereid was close enough to resolve the disk size, and it revealed no discernible surface markings. Voyager 2 was thus unsuccessful in determining a rotation period for this strange satellite.

Nereid is a very difficult telescopic object, but several Earth-based observers have attempted to measure Nereid's rotation period around the time of or subsequent to the Voyager encounter. Schaefer and Shaefer [7] found brightness variations of a factor of 4 with a period between 8 and 24 h. Bus and Larsen [8] found brightness variations of a factor of about 1.5, and couldn't detect a periodicity. Williams *et al.* [9] reported brightness variations of a factor of 3 and a period of 13.6 h. The

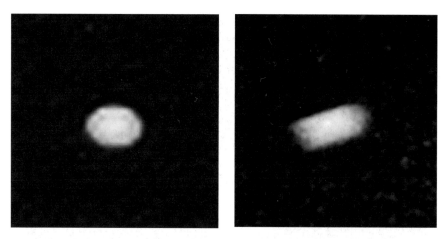

Figure 12.4. Best images of Galatea (left) and Despina (right). Image resolution is about 18 km/pixel in each of the images. Approximate sizes were estimated from these images, but no information on disk shape could be extracted.

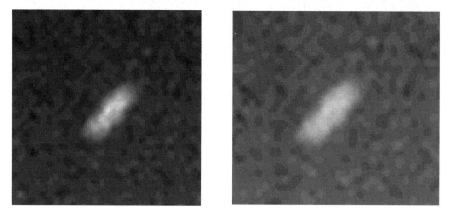

Figure 12.5. Best images of Thalassa and Naiad. The satellite disks are not fully resolved in these images, which have image scales, respectively, of 24 km/pixel and 17 km/pixel.

discrepancy between these results and the Voyager results is real. It is safe to say that as of this writing (December 2000) the rotation period of Nereid is not known, although it is almost certainly non-synchronous.

Voyager imagery did succeed in determining an approximate diameter of 340 ± 50 km and a surface reflectivity of 0.18 ± 0.02. There seem to be significant departures from spherical shape amounting to at least 20 km. The higher reflectivity and the earlier mentioned detection of water ice on its surface make Nereid a very different object from the six satellites discovered by Voyager. Nereid is not a dark carbonaceous satellite. Its surface may best be described as dirty water ice, reminiscent of Uranus's Umbriel or Oberon, although smaller than either of these satellites.

Figure 12.6. Best image of Nereid, with an imaging resolution of about 43 km/pixel. (P-34680)

In size, it is a closer match to Uranus's Miranda, which was also discovered by Kuiper. By analogy with these Uranian satellites, it is possible that Nereid is a differentiated object, with icier materials near its surface and rockier materials in its core. As the reader has probably surmised by now, much of what we presently 'know' about Nereid is little more than educated conjecture (see Figure 12.6).

12.5   TRITON

Before the discovery of Kuiper Belt objects, now known to exist in profusion, planetary scientists attempted to explain Triton's inclined retrograde orbit as a consequence of the escape of Pluto from orbit around Neptune. In such a scenario, both Triton and Pluto were thought to have been original regular satellites of Neptune. Over the years, as both observational and theoretical data have increased, it has become apparent that the theory that Pluto was once a satellite of Neptune is no longer viable. It now appears that the most likely scenario is that Pluto and Triton had their origins in the Kuiper Belt, and that Triton's unusual orbit is a consequence of its capture by Neptune rather than of its formation within the Neptune system and later orbital disruption by some undefined cataclysmic event. This section summarizes our present state of knowledge about Triton and the inferences that can be drawn from that knowledge about the present character and past history of this major Solar System satellite.

12.5.1   Voyager 2 images

Early estimates of the size of Triton were based on interpretations of the total integrated light reflected from the surface and on an erroneous mass estimate. The reasoning, based on this data, was that Triton was not likely to have a density substantially in excess of $2.0\,g/cm^3$. From this data, the diameter of Triton was estimated to be substantially larger than the actual diameter and the reflectivity of the surface substantially lower than the actual reflectivity. When Cruikshank *et al.* [10] found evidence for nitrogen gas and ice in the spectrum of Triton, planetary scientists began to suspect that the surface was brighter (and the diameter smaller) than could be reconciled with the early mass estimates. It was therefore with great anticipation that they awaited the images of Triton that were to be returned by Voyager 2.

Voyager imaging started well before the official encounter period, but on June 5, 1989, at a distance of 117 million km from Neptune, continuous science observations began. Even at that distance, the imaging narrow-angle camera, with a resolution of about 700 km/pixel, was beginning to resolve the disk of Triton. Eighty-one days later, on August 25, 1989, Voyager 2 flew within 39,800 km of the center of Triton as the climax of the Neptune encounter. Between those two dates, and for several days thereafter, hundreds of images of Triton were obtained. The best images, collected during the final approach to Triton, had an image scale of 0.4 km/pixel. One of the early images of Triton, where the disk is clearly resolved, is displayed in Figure 12.7 and a spectacular high-resolution mosaic is reproduced in Figure 12.8. Figure 12.9 is a rough geologic (tectonic) map, showing linear features on the surface.

12.5.2   Geological interpretation

Triton has what is perhaps the most diverse geological land forms found anywhere in the Solar System. There is a nearly ubiquitous irregular surface appearance, dubbed *cantaloupe terrain* because of the similarity of its appearance to that of a cantaloupe rind. This terrain is believed to be the consequence of one or more periods of deep melting and complete or nearly complete differentiation of the satellite. The source of the thermal energy that drove the melting could have been the processes associated with the capture of Triton from solar orbit, or Triton may have had enough radioactive materials in its interior to cause the necessary heating. While other considerations favor the capture theory, the geological record alone cannot differentiate between the two possibilities without major ambiguities. Either the capture process or the drag forces associated with extended Neptune gases that were

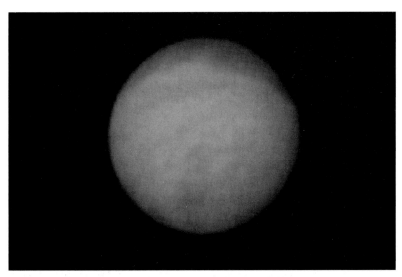

Figure 12.7. Image of Triton taken August 20, 1989, from a distance of 5.4 million km. Even at this great distance, some surface features can be seen and the approximate size and high reflectivity of the surface are already apparent. (P-34634)

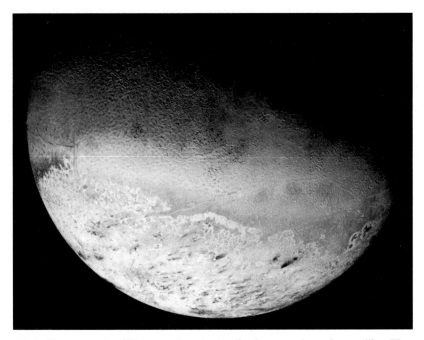

Figure 12.8. Photomosaic of Triton, taken during final approach to the satellite. The variety of features seen in this image is large, showing an extensive polar cap, dark wind streaks, cantaloupe terrain, large fractures, smooth terrain, sparse cratering, etc. (P-35317) (See color plate 7.)

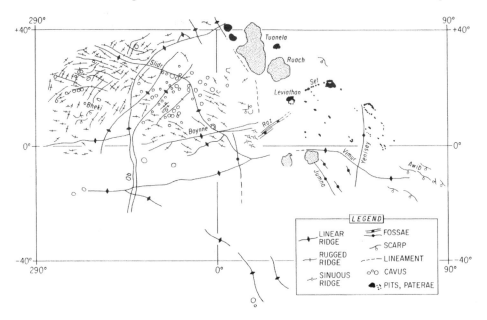

Figure 12.9. Tectonic sketch map of Triton, adapted from Croft *et al.* [11], showing the variety of types of features seen on the surface. The area south of about −10° latitude was covered with ice during the Voyager encounter, except for the regions shown as linear ridges.

present early in the planet's history would have exerted forces that tended to circularize Triton's orbit and diminish the orbit size. These forces would have cause crustal fractures, the evidence for which is apparent in the geological record. Subsequent local melting of surficial features could then have erased portions of the cantaloupe terrain, leaving behind large, almost featureless regions or filling basins with lava-like flows, probably of water ice.

Very late in its formative period, outgassing from volcanic or other processes originating in the subsurface layers gave rise to an atmosphere. Some scenarios have Triton possessing a thick atmosphere for most of its formative period. Whichever the situation, the temperature near the Triton surface cooled to the point where surface ices began to form. This in turn meant that most of the already weak sunlight was reflected away, lowering the temperature even further and overlaying the cantaloupe and altered cantaloupe terrain with reflective ice. Eventually, the temperatures cooled to the point where even the tenuous gaseous nitrogen began to freeze out on the surface, forming extensive polar ice caps. Throughout much of this late period, low-level bombardment, primarily from debris in orbit around Neptune, continued to occur, providing the light sprinkling of craters seen on the surface. Today, in addition to its unusual appearance, Triton is the coldest place encountered anywhere in the Solar System. At −235 °C, the Triton surface is a stingy 38° above absolute zero.

In addition to the nitrogen (N_2) ice detected on its surface, methane (CH_4) ice, carbon monoxide (CO) ice, carbon dioxide (CO_2) ice and water (H_2O) ice have been identified in the spectrum of Triton. That makes Triton's surface the most spectrally diverse of any icy body in the Solar System. With these large quantities of ice on the surface, the average reflectivity of Triton's surface is about 0.85, meaning that only 15% of the sunlight incident on its surface is absorbed by Triton. Sunlight is already lower by a factor of 900 ($= 30^2$) from its intensity at Earth, so the amount of heating from sunlight is minuscule. Nevertheless, some of that energy penetrates the nearly clear polar ice cap, creating a moderate greenhouse effect [12] at the bottom of the ice layer. That heating melts portions of the bottom of the nitrogen ice layer, building up pressure beneath the ice until the liquid or gaseous nitrogen finds a vent and escapes explosively into the atmosphere, carrying large amounts of entrained dust with it. What results is a plume of material, which, under the influence of prevailing winds, creates a dark downwind streak across the surface of the ice cap. The abundance of such dark streaks probably means that this process is very common in the polar region of Triton. At least four active plumes were captured in Voyager imaging; one of these is shown in Figure 12.10.

12.5.3   Physical, compositional, and photometric data

The measured diameter of Triton is $2,705.2 \pm 4.8$ km, making Triton the seventh largest satellite in the Solar System (behind the four Galilean satellites of Jupiter, Saturn's Titan, and Earth's Moon). The next largest satellite, with a diameter of 1,580 km, is Uranus's Titania; therefore the volume of Triton is more than five times that of the next largest satellite in the Solar System. Triton and the six larger

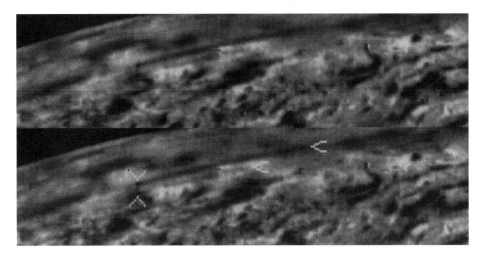

Figure 12.10. One of at least four nitrogen plumes imaged by Voyager. This one has been named the Mahilani plume. The radius of the rising column lies somewhere between 20 m and 2 km. The vertical plume rises to a height of about 8 km before being sheared by a horizontal wind that carries plume material more than 100 km downwind (to the right in the image). (P-34940)

satellites are sometimes referred to as the major satellites. Titania is the largest of the intermediate, or mid-size satellites, the smallest of which include Proteus and Nereid.

The mass of Triton was measured from the changes it caused in the radial velocity of Voyager 2 during the Triton encounter, as measured by the Doppler shift in the spacecraft radio signal. The measured mass is (2.1398 ± 0.0053) $\times 10^{22}$ kg. Comparison of the mass to the size of Triton yields a mean density of 2.0643 ± 0.0121 g/cm^3, or about twice the density of water. If most of the mass is in the form of water ice and silicate material, a density of 2.06 g/cm^3 would imply that about 65% to 72% of the total mass is in the form of silicates.

Triton's atmospheric pressure is very low, only about 0.000014 bar. It is nevertheless sufficient to support haze material and thin clouds, as well as the tiny dust particles carried aloft by the plumes. The primary source of the atmospheric gases is evaporation of the surface ices, even though the $-235\,°C$ (38 K) surface temperature keeps the evaporation rates very low. The atmospheric constituents are consequently the same nitrogen (N_2), methane (CH_4), carbon monoxide (CO), carbon dioxide (CO_2) and water (H_2O) that are present in ice form on Triton's surface. Of the five listed constituents, nitrogen is by far the most abundant, constituting well over 99% of the Triton atmosphere. The condensation–sublimation cycle of nitrogen is the dominant factor in controlling Triton's lower atmospheric and surface temperatures. Methane has an abundance of about 0.05%; carbon monoxide and carbon dioxide each have an abundance of about 0.1%. Because of the extremely low temperatures, water vapor in the atmosphere is even less than the methane abundance. In the upper atmosphere (above 8 km altitude), and perhaps throughout the atmosphere and on the surface, photochemical reactions very likely produce more complex hydrocarbons which may be responsible for some of the hazes and thin clouds observed.

As mentioned earlier, Triton's reflectivity is about 85%, but there are some areas where the reflectivity may be as high as 95%. It is probable that Triton's surface reflectivity changes with season. Triton's inclined orbit precesses around Neptune, and Triton's seasons are therefore highly variable, both in duration and in severity (see Figure 12.11). The maximum seasonal variations occur on a repetitive basis about every 688 years. The thickness and latitudinal extent of the polar ice cap(s) are expected to vary correspondingly. Earth-based measurements of the brightness of Triton over the past several decades show substantial changes in the integrated brightness of this strange satellite.

12.5.4   The question of Triton's origin

Triton is among the largest satellites in the Solar System. However, there is one stark difference between Triton and the other major satellites: Triton is the only one that has an inclined, retrograde orbit. Other satellites with retrograde orbits include the outer four satellites of Jupiter, Saturn's Phoebe, and Uranus's Caliban and Sycorax, all of which are thought to be captured asteroids. Each of these retrograde satellites

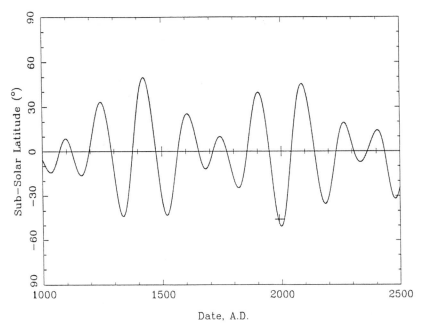

Figure 12.11. Variation of subsolar latitude on Triton with time. This variation is caused by the combination of Neptune's 164-year orbit and Triton's 688-year orbital precession period. Note that in the year 2000, the subsolar latitude is farther south than it has been for over 1,000 years.

(other than Triton) is very distant (more than 5 million km) from its respective planet, has significant ($\geqslant 0.16$) orbital eccentricity, is only a tiny fraction of Triton's size, and has (at least for those whose size is known) a very low reflectivity. Therein lies the dilemma for planetary scientists in their consideration of Triton's possible origin.

Triton's orbit is almost perfectly circular, Unlike any of the asteroids, Triton's surface is icy and highly reflective. Triton is also many times larger, and is radically different in composition and surface processes, than any known asteroid. In composition, it may be representative of Kuiper Belt objects, but again, none of the known 'KBOs' is more than a tiny fraction of Triton's size or mass. As many have noted, it appears to resemble the planet Pluto in size, density, and reflectivity.

Planetary scientists are anxious to study Pluto with an appropriately instrumented spacecraft flyby mission to see how far those similarities extend. If Pluto and Triton are near twins, it might be logical to assume a common history and origin. If they are found to be markedly different from one another, that in turn might emphasize Mother Nature's propensity for diversity.

Considering the available data, McKinnon *et al.* [13] favor capture theories over theories which have Triton formed in orbit around Neptune, but readily admit that the geological record and other evidence cannot rule out either theory. They suggest six steps that might lead to a clearer understanding of Triton's origin:

1. A continued vigorous program of Earth-based spectroscopic studies, intended to identify additional surface constituents and atmospheric chemistry.
2. Better understanding of Triton's geology, including detailed modeling of processes that shaped and continue to shape the surface.
3. More theoretic studies on the potential processes associated with capture or with formation within the Neptune system.
4. Better understanding of the accretion processes that formed and evolved Neptune, and their relationship with and effects on the satellite system, especially Triton.
5. Spacecraft exploration of Pluto, particularly observational data that might emphasize the differences between Pluto and Triton.
6. A Neptune orbital spacecraft mission, which would permit repetitive, long-term observation of processes occurring on or near Triton's surface.

NOTES AND REFERENCES

1. Reitsema, H. J., Hubbard, W. B., Lebofsky, L. A. and Tholen, D. E. (1982) Occultation by a possible third satellite of Neptune. *Science*, **215**, 289–91.
2. The Roche limit is the closest distance a fluid body can orbit a planet (or other central body) without being pulled apart by tidal forces. A solid body may survive within the Roche limit if its structural strength exceeds the tidal forces. The Roche limit, r_L, is calculated from $r_L = 2.456 R_p (\rho_p/\rho_s)^{1/3}$, where R_p is the radius of the planet, ρ_p is the density of the planet, and ρ_s is the density of the satellite.
3. Thomas, P. C., Veverka, J. and Helfenstein, P. (1995) Neptune's small satellites. In Cruikshank, D. P. (ed.) *Neptune and Triton*. University of Arizona Press, Tucson, pp. 685-99.
4. Geometric albedo is the ratio of an object's brightness at $0°$ phase (Sun and observer in the same direction from the object) to the brightness of a perfectly diffusing disk with the same position and apparent size. An object with a specular (mirror-like) surface may have a geometric albedo that exceeds 100%.
5. Phase angle, when referred to lighting conditions, is the angle between the observer and the Sun as seen from the object being viewed. An object at $0°$ phase is fully illuminated, like a full Moon. An object at $180°$ phase is not illuminated, like a new Moon.
6. Cruikshank, D. P. (1999) Triton, Pluto, and Charon. Chapter 21 in Beatty, J. K., Petersen, C. C. and Chaikin, A. (eds) *The New Solar System* (4th edn). Sky Publishing Corporation, Cambridge, Massachusetts, pp. 285–96.
7. Schaefer, M. W. and Schaefer, B. E. (1988) Large-amplitude photometric variations of Nereid. *Nature*, **333**, 436–8.
8. Bus, E. S. and Larson, S. (1989) CCD photometry of Nereid. *Bulletin of the American Astronomical Society*, **21**, 982 (abstract).
9. Williams, I. P., Jones, D. H. and Taylor, D. B. (1991) The rotation period of Nereid. *Monthly Notices of the Royal Astronomical Society*, **250**, 1–2P.
10. Cruikshank, D. P., Brown, R. H. and Clark, R. N. (1984) Nitrogen on Triton. *Icarus*, **58**, 233–305.

11. Croft, S. K., Kargel, J. S., Kirk, R. S., Moore, J. M., Schenk, P. M. and Strom, R. G. (1995) The geology of Triton. In Cruikshank, D.P. (ed.) *Neptune and Triton*. University of Arizona Press, Tucson, pp. 879–947.

12. The greenhouse effect occurs when sunlight, primarily at visible wavelengths, is transmitted through the material (atmosphere, glass, ice, etc.) overlying a partially absorbing surface and heating that surface. The resulting radiation emitted by the heated surface is at wavelengths for which the overlying material is less transparent (or even opaque), and the 'trapped' radiation further heats the surface.

13. McKinnon, W. B., Lunine, J. I. and Banfield, D. (1995) Origin and Evolution of Triton. In Cruikshank, D. P. (ed.) *Neptune and Triton*. University of Arizona Press, Tucson, pp. 807–77.

BIBLIOGRAPHY

Beatty, J. K., Petersen, C. C. and Chaiken, A. (eds) (1999) *The New Solar System* (4th edn). Sky Publishing Corporation, Cambridge, Massachusetts. (See Chapters 21, 22, and 23 on pp. 285–320.)

Bergstralh, J. T., Miner, E. D. and Matthews, M. S. (eds) (1991) *Uranus*. University of Arizona Press, Tucson. (See satellite chapters on pp. 469–735 by Pollack *et al.*, Brown *et al.*, Veverka *et al.*, Croft *et al.*, McKinnon *et al.*, and Greenberg *et al.* and the references cited in each chapter.)

Cruikshank, D. P. (ed.) (1995) *Neptune and Triton*. University of Arizona Press, Tucson. (See satellite chapters on pp. 685–99 by Thomas *et al.* and on pp. 807–1148 by McKinnon *et al.*, Croft *et al.*, Kirk *et al.*, Brown *et al.*, Yelle *et al.*, and Strobel and Summers and the references cited in each chapter.)

13

Post-Voyager observations of Neptune

13.1   EARTH-BASED TELESCOPE OBSERVATIONS

More than 11 years have expired since Voyager 2 encountered Neptune. During that period there have been major improvements in the telescopic tools available for studying the planet and its satellites. As of this writing, no new satellites of Neptune have been discovered, but it is only a matter of time until that occurs. Over the past three years, 12 new satellites of Jupiter [1], 12 new satellites of Saturn [2], and 6 new satellites of Uranus [3] have been found. Part of this plethora of new discoveries is due to improved instrumentation, but most of it is due to the ingenuity of observers in discovering new and improved methods of data reduction. Computer processing of digital data can bring out details well beyond the previous limits imposed by visual scanning of photographic plates.

Those responsible for many of the new discoveries have progressed from Jupiter to Saturn to Uranus in their quest for new objects for two primary reasons. Owing to the remoteness of Neptune from the Sun (and Earth), satellites are more dimly illuminated and angular sizes are correspondingly smaller. These two effects combine to make distant satellites of a given size much more difficult to detect than those in the nearer systems. On the other hand, because Neptune is more remote from the perturbing gravitational influence of other planets, the size of the sphere within which small satellites can remain for long periods of time in relatively stable orbits is much larger than for any other planet. Hence the area of sky that must be covered to find additional satellites of Neptune is much larger than that required for the nearer planets, and the search time is correspondingly longer. When that search gets underway, it seems likely that the retinue of Neptune satellites may double or triple.

Earth-based studies of Neptune's rings will continue to be a daunting task. Additional stellar occultations may provide more precision in the orbital period of the optically denser arcs in the Adams ring, but the detection rate of such occultation events will remain proportional to the fraction of the circumference of the Adams ring occupied by the arcs, which, at the time of the Voyager 2 encounter, was about

10%. The arc material may continue a slow azimuthal spread, but unless additional material is supplied to the arcs, the optical depth will decline accordingly, and the occultations will become more difficult to detect. Furthermore, because Neptune's rings are dim, of small radial extent, and optically thin, the prospects of imaging Neptune's rings directly from Earth are not good.

As with the other giant planets, few techniques are available for magnetospheric or internal observations of Neptune from Earth. Primary advances on those fronts will be achieved as improvements in theoretical studies are developed. Perhaps the detailed results of magnetospheric and internal studies of Jupiter and Saturn, gleaned from the orbital missons of NASA's Galileo and Cassini spacecraft, will lead to improved understanding of the corresponding processes extant at Neptune.

Direct high-resolution imaging of Neptune from Earth has experienced the most dramatic improvements since 1989. In particular, adaptive optics techniques [4] from several large telescopes now meet or exceed resolutions available from the Hubble Space Telescope (HST) in Earth orbit. For imaging of Uranus and Neptune, these techniques have reduced the primary advantage of HST to that of spectral coverage, because adaptive optics techniques only work at wavelengths where Earth's atmosphere is relatively transparent. In the near- or mid-infrared, adaptive optics techniques now routinely achieve resolutions of 0.01 or 0.02 seconds of arc. For comparison, the angular diameter of the Moon as seen from Earth is about 30 minutes of arc ($= 1,800$ seconds of arc); in other words, adaptive optics techniques can reveal detail at a resolution of $\sim 10^{-5}$ the diameter of the Moon. From Earth, Neptune has an apparent diameter of ~ 2.1 seconds of arc. Adaptive optics techniques can therefore provide resolutions of 0.5 to 1% of Neptune's diameter, comparable to some of the smaller atmospheric features noted in Voyager 2 imaging. Because Neptune imaging with adaptive optics techniques is comparable to HST imaging, the results from both will be shown in the next section.

13.2   HUBBLE SPACE TELESCOPE OBSERVATIONS

The Hubble Space Telescope (HST) was placed in orbit around the Earth in 1990, a year after the Voyager 2 encounter with Neptune. However, a small error in the shape of the primary mirror resulted in poorer-than-anticipated imaging resolution. Corrective optics were installed during the first HST servicing mission in 1993. Beginning at that time, and continuing though the present, HST has provided some of the most spectacular results since the dawn of astronomy [5]. Included among these are ever-increasing numbers of Solar System results. HST imaging was the first to clearly resolve tiny Pluto and its moon Charon, the first to provide a coarse surface map for Pluto, the first to collect convincing evidence for the existence of about 200 million inert comets in the Kuiper Belt [6], and the first to record in detail the results of Comet Shoemaker–Levy 9's collision with the planet Jupiter. Nearly Voyager-class imaging of Jupiter, Saturn, Uranus, and Neptune has also been obtained, updating the Voyager results and making it possible to estimate the long-term variability of the atmospheres of the giant planets of the Solar System.

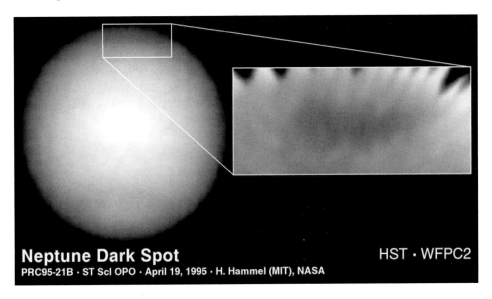

Neptune Dark Spot   HST · WFPC2
PRC95-21B · ST ScI OPO · April 19, 1995 · H. Hammel (MIT), NASA

Figure 13.1. Recent HST image of Neptune from Hammel [8].

Based on analogy with Jupiter's Great Red Spot, Neptune's Great Dark Spot was anticipated to be a long-lived resident in Neptune's atmosphere. It was with mild surprise, therefore, that Hammel *et al.* announced in the early 1990s that the Great Dark Spot had apparently disappeared from the atmosphere [7]. The later appearance of a similar feature in Neptune's northern hemisphere lends credence to the conclusion that large-scale atmospheric storms in Neptune's atmosphere can develop and disappear in times that are short in comparison to the 164-year orbital period of this blue giant. Examples of relatively recent images of Neptune from HST and from Hawaii's Infrared Telescope Facility can be seen Figures 13.1 and 13.2 respectively. Compare these with the Voyager image in Figure 9.8.

13.3 SPACECRAFT MISSIONS TO NEPTUNE

No spacecraft missions are presently funded or planned for future explorations of Neptune. If past trends continue into the future, NASA is the only space agency likely to attempt to send spacecraft to the giant planets of the outer Solar System. With that realization, it is appropriate to examine NASA's plans for future exploration of the outer planets, as of late 2000.

NASA goals and objectives are divided among five enterprises: Space Science, Earth Science, Biological and Physical Research, Human Exploration and Development of Space, and Aerospace Technology [10]. Of these, only the Space Science

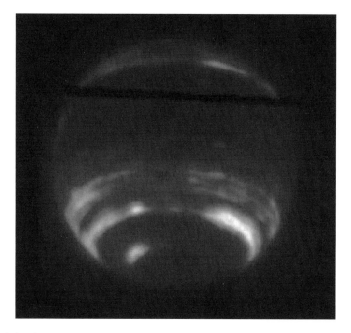

Figure 13.2. Adaptive optics image of Neptune obtained by de Pater [9] at the Infrared Telescope Facility in Hawaii.

enterprise addresses missions to the outer planets. Within the Space Science enterprise, there are four science themes: Astronomical Search for Origins, Structure and Evolution of the Universe, Sun–Earth Connection, and Solar System Exploration. It is in relation to the last of these science themes that missions to the planets are approved and funded. In its November 2000 Strategic Plan, NASA's Space Science enterprise outlines its flight program for 2003 and beyond [11]. On page 64 of that plan is the following statement:

Exploration of the outer Solar System has revealed that the outer planets and their moons are rich in organic material, that subsurface liquid water may exist in some places, and that prebiotic chemical processes occur in some of these environments. The Galileo spacecraft has returned fascinating information about the moon Europa. The Cassini–Huygens mission, now en route to Saturn, will extend this exploration through intensive investigations of the organic-rich atmosphere and surface of Saturn's giant moon, Titan.

Continuing this exploration thrust, the Outer Planets program will focus on prebiotic chemistry in likely places in the outer Solar System. Mission sequence decisions will be based on ongoing scientific discoveries and technological progress. Destinations for missions in this line include returns to Europa and Titan, reconnaissance of the Kuiper Belt, and a more comprehensive study of the Neptune system, including its moon Triton.

On the following page, under the heading, 'For Possible Implementation After 2007', the strategic plan continues:

> According to current planning, the Europa Orbiter, Pluto-Kuiper Express, Titan Explorer, and Europa Lander could be followed within the Outer Planets line by a *Neptune orbiter*. This mission is an important component of our investigation of the outer Solar System, including Neptune's moon, Triton, which may be an icy, organic-rich, captured Kuiper Belt object.

Hence there is a distinct possibility that NASA will attempt a Neptune orbiter at some unspecified epoch after 2007. The actual timing of such a mission, of course, will depend heavily on the level of NASA funding by the US Congress and on the relative priority placed by Congress, NASA, and the scientific community on such a mission.

The Pluto–Kuiper Express (PKE) mission, if it is funded and launched, may have some importance relative to a possible future Neptune orbiter mission. If Pluto proves to have many similarities to Triton (and a possible similar origin), the Pluto portion of the PKE mission could reveal much about the nature and evolution of Triton. Depending on the nature of those findings, they could either make the further investigation of Triton a high priority or they could potentially render such a mission less important.

In addition to the spacecraft mentioned as part of the Outer Planets program (including the nearly completed Galileo mission to Jupiter and the in-transit Cassini–Huygens mission to Saturn and Titan), the Voyager spacecraft, both 1 and 2, remain active as of this writing. Although they have not yet encountered the termination shock (the boundary where the solar wind suddenly slows from supersonic to subsonic speeds) or the heliopause (the outer limits of the Sun's magnetic field), the two spacecraft remain in good condition and continue to transmit data on the charged particles, plasmas, magnetic field strength and direction, and radio emissions in the outer reaches of the Solar System. While they have long since completed their planetary encounters, their work and contributions continue, and their heritage remains bright and lasting. It was a tremendous privilege to have been part of such a mission.

NOTES AND REFERENCES

1. One additional satellite (S/1999 J1) was discovered in 1999; 11 additional satellites (S/2000 J1 through S/2000 J11) were found in 2000. S/2000 J1 was later identified as a satellite first seen in 1975 and designated at that time as S/1975 J1. More detailed information, including approximate sizes and orbital elements, can be found on the internet at:

   < http://nssdc.gsfc.nasa.gov/planetary/factsheet/joviansatfact.htm > .

   As of this writing, Jupiter has 28 known satellites.

2. Twelve new satellites of Saturn (S/2000 S1 through S/2000 S12) were discovered in 2000. More detailed information, including approximate sizes and orbital elements, can be found at:

   < http://nssdc.gsfc.nasa.gov/planetary/factsheet/saturniansatfact.htm >.

   As of this writing, Saturn has 30 known satellites.
3. Six additional satellites of Uranus have been found since 1996. Caliban and Sycorex were discovered in 1997, and Stephano, Prospero, and Setebos were discovered in 1999. In addition, in 1999 Eric Karkoschka of the University of Arizona found a previously undiscovered satellite occupying basically the same orbit as Belinda in seven images obtained by Voyager 2. A compilation of information on the Uranian satellites can be found at:

   < http://nssdc.gsfc.nasa.gov/planetary/factsheet/uraniansatfact.htm >.

   As of this writing, Uranus has 21 known satellites.
4. Adaptive optics (AO) techniques make use of the fact that light coming from distant objects arrives at Earth as a flat 'wavefront'. However, as the light traverses Earth's atmosphere, small changes are introduced by the atmosphere that vary across the wavefront. Distortion caused by these non-uniform changes result in image blurring. Without explaining how those deviations from a flat wavefront are measured, AO consists of determining the deviations several times per second and correcting the shape of the primary telescope mirror to compensate for the wavefront distortions. The resulting images, if this process is done accurately, are almost as sharp as if there were no intervening atmosphere. One caveat is that the changes must be relatively wavelength independent, that is, the atmospheric distortions must be the same for all colors in the image. This assumption of wavelength independence is best in the near infrared; hence, the best AO images are generally near infrared images.
5. The best source of information about and data from the Hubble Space Telescope is the Space Telescope Science Institute in Baltimore, Maryland. Their internet address is:

   < http://www.stsci.edu/ >.

6. Cochran, A. L., Levison, H. F., Stern, S. A. and Duncan, M. J. (1995) The discovery of Halley-sized Kuiper Belt objects using the Hubble Space Telescope. *The Astrophysical Journal*, **455**, 342–6.
7. Hammel, H. B., Lockwood, G. W., Mills, J. R. and Barnet, C. D. (1995) Hubble Space Telescope Imaging of Neptune's cloud structure in 1994. *Science*, **268**, 1740.
8. Hammel, H. B. (1995) < http://oposite.stsci.edu/pubinfor/pr/1995/21.html >
9. de Pater, I. (2000)

   < http://astron.berkeley.edu/~imke/Infrared/AdaptiveOptics/ao.htm >

10. National Aeronautics and Space Administration: *Strategic Plan 2000* (2000) NASA Headquarters, Washington, DC 20546.
11. NASA Space Science: *Strategic Plan* (2000) Code S, NASA Headquarters, Washington, DC 20546.

BIBLIOGRAPHY

National Aeronautics and Space Administration: *Strategic Plan 2000* (2000) NASA Headquarters, Washington, DC 20546, 68 pp.

The Space Science Enterprise: *Strategic Plan* (2000) NASA Headquarters, Washington, DC 20546, 127 pp. This and the document listed above are available on-line at:

< http://spacescience.nasa.gov/policy_pub.htm > .

The Space Telescope Science Institute Home Page < http://www.stsci.edu/ > has up-to-date information, including images, from the Hubble Space Telescope.

14

Comparative planetology of the four giant planets

In prior chapters, we have made an effort to provide, with every major characterization of Neptune and its system, an estimate of similar conditions at the other giant planets. Because much can be learned about each of the planets from such 'comparative planetology', we have chosen to gather together in one chapter all such comparisons. These comparisons are organized below under the topics of interiors, atmospheres, rings, magnetospheres, and satellites. Most of the comparisons will be in text and tabular form, although figures will be used where they tell the story more clearly.

14.1 INTERIORS

The giant planets of the outer Solar System dwarf Earth in size and mass (see Table 14.1). By the same token, none of the giant planets has a density of more than 30% that of Earth. In fact, the density of Saturn is only 12.5% that of Earth! These major differences in size and density translate to major differences in the interior models of the giant planets relative to that of Earth. Even among the giant planets, there are two classes of interior models: those applicable to Jupiter and Saturn and those applicable to Uranus and Neptune.

Table 14.1. Size comparison between Earth ($\oplus$) and the giant planets

Planet name	Radius, equatorial (km)	$\oplus = 1$	Radius, polar (km)	$\oplus = 1$	Oblate- ness	Volume, $\oplus = 1$	Mass, $\oplus = 1$
Earth	6,378	1.000	6,357	1.000	0.003	1.0	1.0
Jupiter	71,492	11.209	66,854	10.517	0.065	1,321.3	317.8
Saturn	60,268	9.449	54,364	8.552	0.098	763.6	95.2
Uranus	25,559	4.007	24,973	3.929	0.023	63.1	14.4
Neptune	24,764	3.883	24,340	3.829	0.017	57.7	17.1

The Jupiter and Saturn interior models have much more extensive hydrogen–helium envelopes beneath their cloudy 'surfaces'. These hydrogen–helium envelopes are mostly liquid, although it must be kept in mind that the high pressures and temperatures in the interiors of these planets are such that there is no phase transition between the gaseous and liquid states; hence, there is little to distinguish liquids from gases. When the pressures become high enough, there is a transition to an electrically conducting metallic liquid hydrogen. The transition from non-metallic to metallic hydrogen probably occurs much deeper in the interior of Saturn (at a radius of about $0.5 R_S$) than it does within Jupiter (where the transition occurs only about 10% of the distance from the cloud tops to the center). It is probable that electrical currents flowing within these liquid metallic hydrogen layers deep within Saturn and Jupiter give rise to their magnetic fields.

Beneath the extensive layers of liquid and liquid metallic hydrogen, Jupiter and Saturn undoubtedly possess Earth-sized cores. Those cores may also be differentiated, with molten icy material lying above a denser molten core of heavier materials ('rock'). Because of its much lower overall density, the percentage of rocky materials in the core is probably smaller for Saturn than for Jupiter.

Pressures within Uranus and Neptune never become great enough to form metallic hydrogen. Furthermore, the higher densities of Uranus and Neptune imply that hydrogen and helium constitute a much smaller percentage of their interiors than is true for Jupiter and Saturn, perhaps only the outermost 10%. Molten icy materials may occupy as much as 80% of the radii of Uranus and Neptune, with small rocky cores occupying the innermost $\sim$ 10% of their respective radii. Because Uranus and Neptune lack a metallic hydrogen layer, and because the molten rocky core is relatively small and very possibly a poor electrical conductor, planetary scientists have speculated that the molten icy layer that occupies most of the interior of these planets may be sufficiently electrically conductive to give rise to an intrinsic magnetic field. The strange orientations of their magnetic fields could be due to electrical currents circulating relatively high in the interior, possibly not even circling the core.

Figure 14.1 shows the estimated interior structures of the four giant planets and Earth, normalized by their respective radii.

14.2 ATMOSPHERES

The atmosphere of Earth is a relatively thin outer appendage to an otherwise rocky body. Its primary constituents are nitrogen (78%) and oxygen (21%). The giant planets, by contrast, possess no surface and are in fact predominantly atmospheric in nature. Hydrogen and helium, the gases that make up the vast majority of their outer envelopes, remain gaseous at temperatures that exist in the Solar System, except at the extreme pressures found in the interiors of Jupiter and Saturn. Near the 1-bar level in each of the giant planet atmospheres, conditions exist that allow the formation of clouds. The main bank of clouds that mark the visible sizes of Jupiter

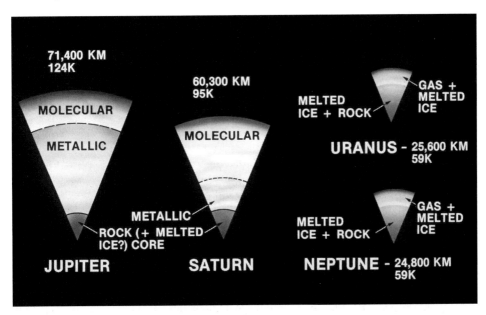

Figure 14.1. Interior models of the four giant planets and Earth.

and Saturn are thought to be composed of ammonia ice crystals; and those of Uranus and Neptune are more probably composed of ice crystals of methane (see Figure 14.2). Most of what we know about the giant planet atmospheres below those clouds is a matter of deduction on the basis of the temperature versus wavelength data from Earth, combined with theoretical calculations. Only in the case of Jupiter has the atmosphere beneath the clouds been probed by a package of instruments (the Galileo Probe) capable of measuring composition, temperature, pressure, and other characteristics of that deeper atmosphere. Those measurements partially refine, but mainly confirm, our earlier estimates of that environment. A number of atmospheric characteristics for Earth and the giant planets are compared in Table 14.2.

Each of the giant planets is characterized by winds that blow almost exclusively east and west. Near vortexes like Jupiter's Great Red Spot and Neptune's Great Dark Spot, local deviations from the purely east–west wind pattern exist, but even these are relatively limited in latitudinal extent. The latitudinal variation of these wind speeds generally displays a degree of symmetry about the equator, and the pattern of wind speeds may remain relatively unchanged for centuries. It is further-more apparent that the primary direction of these winds (parallel to the respective equators) is driven more by the rotation of the planet than by the direction of solar irradiation. Even the extreme tilt of the rotation axis of Uranus has little or no effect on the direction of the zonal winds, although more recent observations indicate that solar insolation may affect the level of turbulence.

Table 14.2. Selected characteristics of the atmospheres of Earth and the giant planets

Characteristic	Earth	Jupiter	Saturn	Uranus	Neptune
Mean solar distance (AU)	1.000	5.203	9.555	19.218	30.110
Incident flux (W/m^2)	1,367	50.50	14.97	3.70	1.51
Effective temperature (K)	288	124.4	95.0	59.1	59.3
Energy balance	1.0002	1.66	1.82	<1.1	2.69
% of H$_2$ in lower atmos.	0.00005	84	86	83	79
% of He in lower atmos.	0.00052	16	14	15	18
% of CH$_4$ in lower atmos.	0.00015	0	0	2	3
Rotation period (h)	23.9345	9.9249	10.6562	17.24	16.11

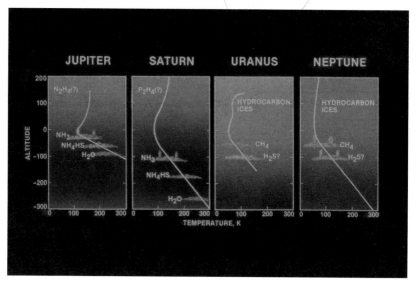

Figure 14.2. Temperatures versus altitude for the four giant planets are compared. (JPL-12274B)

14.3   RINGS

During the early 1970s, while the two Voyager spacecraft were being prepared for launch, only Saturn of all the giant planets was known to have a ring system. Early in 1977, the same year that Voyagers 1 and 2 were launched, the presence of narrow rings surrounding Uranus was disclosed during the observation of Uranus' passage in front of a distant star. Jupiter's narrow ring was discovered during the Voyager 1 encounter in 1979. With the presence of rings established for three of the four gas giant planets, there was a push during the early 1980s to determine if distant Neptune would have this characteristic. As discussed in Chapter 3, the results were disappointingly inconclusive prior to the Neptune encounter of Voyager 2 in 1989. We now know that each of the giant planets possesses a ring system and that the four systems are remarkably diverse in their characteristics. Many details on the Jupiter ring system have been added since the 1979 Voyager encounters by Galileo observations.

Jupiter's ring system is dusty and is apparently the direct result of meteoritic bombardment of the planet's three innermost satellites, Metis, Adrastea, and Amalthea. The Gossamer Belt extends planetward from each of these three satellites, increasing in thickness commensurate with their inclinations. The main ring of Jupiter is the narrowest and brightest of the three components and is about 6,400 km in width. It is likely that some of these particles become electrically charged, either through interaction with Jupiter's radiation field or through photo-ionization as incident sunlight drives electrons from surface atoms. If tiny dust particles obtain a sufficiently large charge-to-mass ratio, they can be carried to higher latitudes by Jupiter's magnetic field. This may be the process that creates the so-called 'halo' that extends northward and southward from the Gossamer Ring.

Saturn's ring system is by far the most extensive and the most massive of the giant planet ring systems. First viewing of the Saturn ring system is attributed to Galileo in 1611, but it was much later when Huygens recognized that Galileo's observation had been of a detached ring, nowhere touching the planet. Eventually, Cassini recognized that the rings were split into two main components, separated by a gap now known as the Cassini Division. The ring outside the Cassini Division became known as the A ring, the ring inside the Cassini Division became the B ring. Early in the twentieth century, a C ring was discovered interior to the B ring. Some claim to have seen a D ring interior to the C ring, but later verification of the D ring by Voyager revealed a ring far too tenuous to have been seen from Earth. Near the times of the Earth's crossing of the Saturn ring plane – an event or triplet of events that occur about every 15 years – Earth-based observers detected a very dim ring extending outward to the orbit of Saturn's satellite, Rhea, about four times farther from Saturn's center than the outer edge of the A ring. This extensive ring has become known as the E ring. Two other rings, both relatively narrow, were detected by Pioneer 11 during its 1979 encounter of Saturn. The F ring, just outside the A ring, was imaged by Pioneer 11. The G ring occupied an area further out than the F ring, but still near the inner edge of the E ring. It was detected by the absence of magnetospheric charged particles near that distance from Saturn.

Uranus's ring system, ignoring for the moment the fine dust component, consists primarily of ten relatively narrow rings. Nine of these were discovered on the basis of a single stellar occultation in 1977; the tenth was discovered by Voyager 2. From closest to farthest from the planet, the rings are designated as 6, 5, 4, α, β, η, γ, δ, λ, and ε. The ε ring is far and away the widest and most prominent (and most eccentric) of these narrow rings. It has a smoothly varying width of 20 to 95 km and is broadest at its greatest distance from Uranus and narrowest at its closest point to the planet. Rings 6, 5, and 4 possess the most highly inclined ring orbits, but still depart from the equatorial plane by less than $0.1°$.

Interior to the 6 ring is a broad, diffuse ring with a width of about 2,500 km. It is probably composed of relatively fine dust particles and has not yet been given an official name by the International Astronomical Union (IAU). While Voyager 2 was in the shadow of Uranus, the cameras looked back in the general direction of the Sun and detected a rich dusty ring system spanning most of the distance occupied by the narrow rings shown in Figure 6.53. A few of these are identified with the narrow rings, but the majority were seen solely in the single image obtained during shadow passage. The IAU has similarly chosen not to give these rings any official designation.

The discovery of, and characteristics of, the Neptune ring system are given in Chapter 11. Neptune's rings are more nearly like those of Uranus than any other ring system, but there are some very significant differences between the two systems. The Adams ring, the Le Verrier ring, and the Arago ring are narrow; the Lassell and Galle rings are broad and diffuse. The Adams ring has five distinct and discontinuous ring arcs, all contained within approximately 40° of ring longitude. These are the only portions of the Neptune ring system with sufficient optical depth to be relatively easily detected in stellar occultation measurements from Earth.

The rings are not only different in their large-scale appearance, but also in their particle sizes, compositions, and reflectivities. The Jupiter ring particles are mainly dust-sized silicate particles with moderate reflectivity. The Saturn ring particles have a mean size in the centimeters to tens of centimeters range, are predominantly water ice, and are highly reflective. The Uranus and Neptune rings are very dark, reflecting only about 5% or so of the sunlight incident upon them. Their particle size range is mainly dust-sized, and they are thought to be composed of radiation-darkened methane ice, perhaps mixed with silicate materials. The ring sizes and extents are depicted in Figure 14.3, where the dimensions of each ring system have been normal-

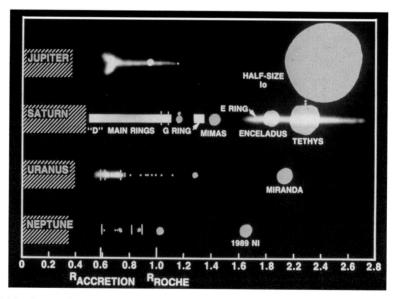

Figure 14.3. Comparison of the ring systems of the giant planets. The scale of each has been normalized to the radius of the central planet. (D-2001-0724-C2)

ized to the radius of the central planet. Table 14.3 lists some of the dimensions and orbital characteristics of the known rings.

14.4 MAGNETOSPHERES

The largest structure within the Solar System is the magnetosphere (heliosphere) of the Sun; the second largest structures are the magnetospheres of the giant planets. The Sun's magnetic field extends to more than 100 AU, i.e., more than 100 times the Earth–Sun distance. Voyagers 1 and 2 continue the search for the outer boundary of the Sun's magnetosphere, known as the heliopause, which they are expected to reach before their useful lifetimes expire. Jupiter's magnetotail (the downwind extension of Jupiter's magnetosphere) appears to extend beyond the orbit of Saturn, a distance of at least 5 AU. The lengths of the magnetotails of Saturn, Uranus, and Neptune are unknown, but extend well beyond the limits probed by Voyager 2.

Little was known about the giant planets' magnetospheres prior to the encounters of Pioneers 10 and 11 and Voyagers 1 and 2. Radio emissions at decametric wavelengths have long been detected from Earth. These emissions were known to be generated at Jupiter, and it was assumed that they were due to interactions within Jupiter's magnetosphere. At present, there exists no means of studying the magnetospheres of the giant planets from Earth. Hence, the first actual measurements of the magnetosphere of Jupiter were collected by Pioneer 10. Pioneer 11 holds that honor for Saturn; and Voyager 2 was the first to sense the magnetic fields of Uranus and Neptune.

The interiors of the giant planets are too hot to retain ferromagnetism (as in solid iron or other magnetic materials) from the time of their formation. Their magnetic fields must therefore be generated by dynamos within each of these planets. The mechanisms which drive these dynamos are poorly understood, even for Earth. However, it is believed that the necessary conditions for such dynamos include sufficiently rapid rotation of the planet, a liquid electrically conducting core or layer in the interior, and some method of inducing convection within that liquid layer. As discussed in Section 14.1, the region within Jupiter and Saturn that is believed to be responsible for the generation of their magnetic fields is a liquid metallic hydrogen layer. The corresponding layer within Uranus and Neptune is believed to be region of superheated water, methane, and ammonia.

The magnetic influence of the giant planets extends well beyond their cloud tops, as mentioned above. That influence is felt primarily in its effect on charged particles. The planetary magnetic fields deflect most of the solar wind particles, creating a wind-sock-shaped cavity within the solar wind. Exterior to the outer boundary (i.e., magnetopause) of each of those magnetospheres, a bow shock wave exists, which possesses many characteristics similar to the bow wave in front of a moving ship. Within that so-called magnetosheath, the solar wind particles are slowed to subsonic speeds. Much of the flow within the magnetosheath is turbulent, especially near its 'nose'.

Table 14.3. Orbital data for the named rings of the giant planets

Ring name	Semi-major axis (km)	Width (km)	Semi-major axis (R_X)	Orbit eccentricity	Inclination (°)	Period (h)
Jupiter:						
Halo	111,400	22,800	1.40–1.72	0.0	0.0	5.7–6.5
Main	126,000	6,400	1.72–1.81	0.0	0.0	6.5–7.1
Gossamer	178,600	98,800	1.81–3.2	0.0	0.0	7.1.–1.62
Saturn:						
D (inner)	66,000	8,500	1.09–1.24	0.0	0.0	5–6
C (inner)	74,500	17,500	1.24–1.53	0.0	0.0	6–8
B (inner)	92,000	25,500	1.53–1.95	0.0	0.0	8–11
B (outer)	117,500	–	–			
A (inner)	122,200	14,600	2.03–2.27	0.0	0.0	12–14
A (outer)	136,800	–	–			
F	140,210	30–500	2.33	0.002 6	> 0	15
G	168,000	8,000	2.72–2.85	0.0	0.0	20
E (inner)	180,000	300,000	3–8	0.0	0.0	22–108
E (outer)	480,000	–	–			
Uranus:						
6	41,837	1–3	1.637	0.001 0	0.062	6.199
5	42,235	2–3	1.652	0.001 9	0.054	6.288
4	42,571	2–3	1.666	0.001 1	0.032	6.363
α	44,718	4–13	1.750	0.000 8	0.015	6.851
β	45,661	7–12	1.786	0.000 4	0.005	7.069
η	47,176	1–2	1.846	0.000 0	0.001	7.424
γ	47,627	1–4	1.863	0.000 1	0.002	7.531
δ	48,300	3–7	1.890	0.000 0	0.001	7.691
λ	50,024	2–3	1.957	0.0	0.0	8.107
ϵ	51,149	20–95	2.001	0.007 9	0.000	8.382
Neptune:						
Galle	42,000	2,000	1.65–1.73	0.0	0.00	5.5–5.9
Le Verrier	53,200	110	2.15	0.0	0.03	8.2
Lassell	55,200	4,000	2.15–2.31	0.0	~ 0	8.2–9.1
Arago	57,200	< 100	2.31	0.0	~ 0	9.1
Adams*	62,933	15–50	2.54	0.0	0.00	10.5

* The Adams ring has five ring arcs, known, from leading to trailing, as Courage, Liberté, Egalité 1, Egalité 2, and Fraternité

Although the magnetopause resists the passage of ions from the solar wind into the magnetosphere or of magnetospheric ions out into the solar wind, it is not impervious to such ions, particularly if they have sufficiently high energy. The impending crossing of a bow shock and magnetopause by a spacecraft is often heralded by detection of both 'upstream' ions and radio emissions generated by the charged particle interactions near the magnetopause. Similarly, solar wind

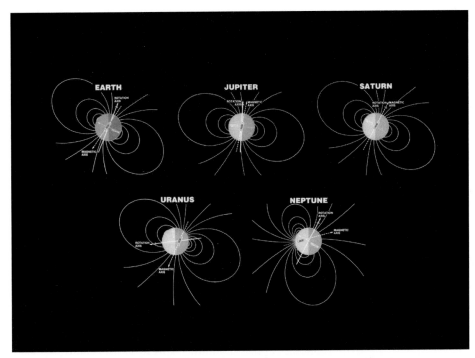

Figure 14.4. Schematic comparison of the magnetic field orientations and offsets for Earth and the giant planets. (P-35825A) (See color plate 4.)

particles that leak through the magnetopause can, if they impact molecules within the upper atmosphere, give rise to visible and ultraviolet aurorae. Because such a 'leakage' is much easier near the magnetic poles, these aurorae are generally found as ring-shaped emissions near the magnetic poles of the planet. In the cases of Earth, Jupiter, and Saturn, the magnetic poles are located relatively near the rotation poles. We often refer to such emissions on those planets, as the Aurora Borealis (north) or the Aurora Australis (south). The magnetic poles of Uranus and Neptune are remote from their respective rotational poles, so the designation of their aurorae as either northern or southern may be inappropriate. Figure 14.4 depicts schematically the relationship of the tilt of the rotation axis with respect to the orbital plane of Earth and the giant planets; it also displays the tilt and offset of the best-fit magnetic dipole from the rotation axis. For Uranus and Neptune, the tilt of the magnetic axes are high enough to permit the solar wind to flow almost unimpeded into the upper atmosphere during four different epochs in the revolutionary period of each planet. Increased ultraviolet emissions from these two planets might be expected when such conditions exist.

Electrically charged particles interact with a magnetic field in various ways, as described in Maxwell's Equations (see Section 10.7). In the deeper parts of a planet's magnetosphere, where the magnetic field strength is higher, these interactions serve

Table 14.4. Comparison of the magnetospheres of Earth and the four giant planets.

Characteristic	Earth	Jupiter	Saturn	Uranus	Neptune
Equatorial radius (km)	6,378	71,492	60,268	25,559	24,764
Tilt of equator to orbit (°)	23.45	3.12	26.73	97.86	29.56
Solar wind dens. (amu/cm^3)	8	0.3	0.1	0.02	0.008
Mag. moment(10^{12} Oe km^3)	0.08	1,560	46	4.2	2.3
Mag. moment (Earth = 1)	1	20,000	600	50	30
Surface mag. field (Oe)	0.31	4.28	0.21	0.25	0.15
Dipole tilt and sense (°)	+11	−10	0	−59	−47
Dipole offset (R_P)	0.07	0.10	0.05	0.33	0.55
Typical nose distance (R_P)	10	65	20	18	27
Max. ion density (cm^{-3})	1–4,000	>3,000	~100	3	2
Primary ion types	O$^+$, H$^+$	O$^+$, S$^+$	O$^+$,OH$^+$,H$^+$	H$^+$	N$^+$,H$^+$
Rotation period (h)	23.934	9.925	10.656	17.24	16.11

to freeze low-energy charged particles within the magnetic field, or restrict their flow to being along magnetic field lines rather than across them. Such charged particles are designated 'trapped radiation'. The Van Allen Belts, discovered by America's first artificial satellite, Explorer 1, are an example of trapped radiation within the magnetosphere of Earth. Similar 'trapped radiation' belts exist for each of the giant planets. The strongest and most compositionally diverse of the trapped ions are those of Jupiter. Some comparative information on the magnetic fields and trapped radiation for Earth and the giant planets is captured in Table 14.4.

14.5   SATELLITES

Over the recent past the number of known natural satellites in the Solar System has increased markedly. Prior to the launches of Voyagers 1 and 2, the number of known satellites was 34, including Earth's Moon, Mars's Phobos and Deimos, Jupiter's four Galilean satellites, Amalthea, and eight captured asteroids, eleven satellites of Saturn, five satellites of Uranus, and two of Neptune. As of the end of 2000, that number had grown to 91, of which 87 circle the giant planets. An in-depth discussion of the surface and bulk characteristics of all these satellites is beyond the scope of this book. Rather, we discuss and compare them as classes of objects and tabulate some of their known orbital and physical characteristics.

Seven of the Solar System's planetary satellites exceed 1,000 km in radius. These include Earth's Moon, the Galilean satellites of Jupiter (Io, Europa, Ganymede, and Callisto), Saturn's Titan, and Neptune's Triton. These are often classified as the major satellites. No two of them are alike, and they are all comparable in size (or larger than) the planets Mercury and Pluto.

Earth's Moon is the most extensively studied and best known of the satellites. It is the only one yet visited by humans and the only one for which samples have been returned to Earth. It is one of three satellites whose density is known to exceed $3\,g/cm^3$, the other two being Io and Europa. The Moon has very little water, except perhaps as ice in the permanently shaded floors of near-polar craters. It is possible that very slow evaporation of such ice deposits may give rise to an extremely tenuous near-polar hydrogen atmosphere.

Although many of these major satellites show evidence on their surfaces of past vulcanism, Io is the only one known to have presently active volcanoes. The Galileo Mission has revealed the presence of at least 100 active volcanic vents on Io, and scientists estimate that twice that number may remain as yet undiscovered. The volcanic eruptions appear to involve primarily sulfur and sulfur dioxide rather than the basaltic lavas found on Earth. Io has the highest known density among the satellites, slightly more than 3.5 times the density of water. It is possible that heat, either radiated from Jupiter during its hotter formative stages or generated by the surface flexure mechanism responsible for Io's present volcanism, has driven most of the less dense volatiles from Io.

Europa has the lowest crater density among the major satellites, ostensibly due to recent or continuing resurfacing of the water ice crust from a subsurface liquid water reservoir. That crust is criss-crossed with streaks that bespeak fracturing of the crust and lateral motion of the icy platelets, followed by deep refreezing. Europa may be the only place in the Solar System (other than Earth) to have liquid water at or near its surface. Some have speculated that a deep Europa ocean beneath the thick crust could be the best place in the Solar System to search for signs of extraterrestrial life. Although conditions on Mars may seem more Earth-like than those on Europa, liquid water boils away spontaneously at Mars's surface because of the low density of its atmosphere, and there is no evidence at the surface for large subterranean Martian oceans.

Ganymede is the largest natural satellite in the Solar System, edging out Saturn's Titan for that honor by a few tens of kilometers in radius. The Galileo spacecraft recently discovered that Ganymede possesses its own internal magnetic field, the first satellite so distinguished. That magnetic field may indicate that Ganymede has a molten metallic (or other electrically conducting material) core. The surface is heavily cratered, but large-scale albedo (i.e., dark and light) features on its surface testify to early geologic activity in this satellite's past history. However, unlike Io or Europa, little evidence exists for geologically recent activity of internal origin.

Callisto is the third largest satellite in the Solar System and is the most heavily cratered of the major satellites. Its density is lowest among the Galilean satellites of Jupiter, most likely due to icy composition of a major fraction of its interior. Although no surficial evidence exists for the presence of liquid water in the interior, Callisto's magnetic field signature has led many scientists to believe that an ocean may exist deep in the interior of this satellite. Similarly, Callisto is unique among the major satellites in its complete absence of resurfaced areas; apparently Callisto has been geologically inert for the last four billion years.

The Solar System's second largest satellite, Saturn's Titan, is the only one to possess a dense atmosphere. Titan's predominantly nitrogen atmosphere is about ten times as deep as Earth's (also predominantly nitrogen) atmosphere. At Titan's surface, the pressure is about 60% greater than Earth's sea-level atmospheric pressure. A second major constituent of Titan's atmosphere is methane (CH_4); and the very presence of methane is an indicator of a liquid methane reservoir at or near Titan's surface. This has led scientists to speculate that much of Titan's surface may be covered by a mixture of liquid ethane ($\sim 75\%$) and liquid methane ($\sim 25\%$). If there is truth to this speculation, Titan shares with Earth the distinction of being the only two bodies in the Solar System to possess both liquid and solid surface areas beneath a thick atmosphere. Considering these similarities and the presence of simple hydrocarbons in the atmosphere of Titan, one might be tempted to assume that Titan might also harbor life. However, Voyager 2 showed that temperatures near Titan's surface are only 94 K ($-179\,^\circ$C)! That extreme cold and the absence of liquid water at Titan's surface strongly mitigate against (but don't rule out) the possibility of Titan life forms.

The upper atmosphere of Titan is filled with hydrocarbon hazes that result from the action of sunlight on the atmospheric methane. Those hazes completely hid the surface of Titan from the gaze of the Voyager imaging cameras. The size of Titan's surface, the atmospheric density near the surface, and the density of Titan's interior are known almost exclusively because of radio signals transmitted through Titan's atmosphere by Voyager 1 and received at tracking stations on Earth. One of the primary goals of the Cassini Mission includes the investigation of Titan, primarily through the measurements collected by the European Space Agency's Huygens Probe carried to Titan by Cassini. The descent of Huygens through Titan's atmosphere in January 2005 will change forever our understanding of this intriguing satellite.

The most distant of the major satellites in the Solar System is Neptune's Triton. With a diameter of 2,700 km, it is the smallest of the major satellites, but certainly not the least interesting. In many ways, Triton shares more similarities with the planet Pluto than it shares with other Solar System satellites; the two bodies may indeed have similar origins in the outer Solar System. Triton's orbit around Neptune, although nearly circular, is retrograde and inclined to Neptune's equator, implying that Triton was captured by Neptune rather than being formed with the planet. It has a tenuous but easily detected nitrogen atmosphere, and active plumes of nitrogen and entrained dust were detected by Voyager 2. More details of its nature were given in Chapter 12.

Table 14.5 summarizes the orbital and physical characteristics of the major satellites of the Solar System.

A second subdivision of satellites might be those referred to as mid-size satellites. These are satellites that range from ~ 170 to ~ 790 km in radius, and there are 14 known members of this group in the Solar System. They include six Saturn satellites (Mimas, Enceladus, Tethys, Dione, Rhea, and Iapetus), five Uranus satellites (Miranda, Ariel, Umbriel, Titania, and Oberon), two Neptune satellites (Proteus and Nereid), and Pluto's Charon. They are all icy satellites approximately spherical shapes and with prograde orbits, indicating that most if not all were formed with the planet they circle. The outermost three (Proteus, Nereid, and Charon) have not been studied from close range. The size, mass, and density of

Table 14.5. Orbital and physical characteristics of the major satellites

Satellite name	Radius (km)	Density (g/cm^3)	Distance $(10^3\,km)$	Inclination (deg)	Eccentricity	Period (days)
Moon	1,738	3.34	384.4	18.3–28.6	0.0549	27.3217
Io	1,815	3.55	421.6	0.04	0.0041	1.7691
Europa	1,569	3.01	670.9	0.47	0.0101	3.5512
Ganymede	2,631	1.95	1,070	0.20	0.0015	7.1546
Callisto	2,400	1.86	1,883	0.28	0.0070	16.6890
Titan	2,575	1.88	1,222	0.33	0.0292	15.9454
Triton	1,350	2.07	436.3	157.34	0.0000	5.8768

Charon have been estimated from data taken in the 1990s when Charon's orbit was edge-on to Earth-based observers and mutual occultations were occurring. Mass estimates for Proteus and Nereid are presently derivable only by assuming a density and calculating the mass from the sizes of these satellites.

Essentially all mid-sized satellites whose surfaces have been imaged from close range show evidence for surface alteration since the time of their formation. Iapetus appears to have a surface deposit of dark material on its forward hemisphere, quite probably as a result of dark material originating from Phoebe, a smaller, more distant satellite of Saturn. Enceladus's surface appears to have melted within recent geologic times and large portions of the early cratering have been erased. Because Saturn's E ring is densest at the orbit of Enceladus, some scientists suspect that there may be active ice volcanism on the satellite's surface, and that much of the E ring particle population may be provided by such activity. Many of the other mid-size satellites have cracks (faults) across their surfaces, and in places (notably on Mimas) have some large-scale vertical and horizontal displacements along those fault lines. The densities of these objects are generally in the range of 1.0 to 2.0 g/cm^3, implying that their bulk composition may be mainly water ice. Table 14.6 summarizes their characteristics.

The small satellites of the Solar System constitute the majority of known satellites. Many, perhaps most, are captured asteroids or fragments of pre-existing satellites. There are two such satellites of Mars, 24 of Jupiter, 23 of Saturn, 16 of Uranus, and five of Neptune. Except for Phobos and Deimos of Mars (with densities around 1.8 g/cm^3), the other small satellites have unknown densities (and hence compositions) . Many of them were discovered within the year prior to the completion of this book and as they have not yet been officially named only their temporary designation is given. In Table 14.7 we tabulate their known orbital characteristics and provide data (or estimates) of their mean radii. An 'R' in the 'Period' column means that the orbit is retrograde or, in other words, the satellite has an orbital inclination greater than 90°.

Table 14.6. Orbital and physical characteristics of the mid-size satellites

Satellite name	Radius (km)	Density (g/cm³)	Distance (10³ km)	Inclination (deg)	Eccentricity	Period (days)
Mimas	198.7	1.14	185.52	1.53	0.0202	0.9424
Enceladus	249.4	1.07	238.02	0.02	0.0045	1.3702
Tethys	529.9	1.01	294.66	1.09	0.0000	1.8878
Dione	560	1.49	377.40	0.02	0.0022	2.7369
Rhea	764	1.24	527.04	0.35	0.0010	4.5175
Iapetus	718	1.03	3,561.3	7.52	0.0283	79.3302
Miranda	235.8	1.20	129.39	4.22	0.0027	1.4135
Ariel	578.9	1.66	191.02	0.31	0.0034	2.5204
Umbriel	584.7	1.40	266.30	0.36	0.0050	4.1442
Titania	788.9	1.72	435.91	0.14	0.0022	8.7059
Oberon	761.4	1.63	583.52	0.10	0.0008	13.4632
Proteus	208.9	?	117.65	0.55	0.0004	1.1223
Nereid	170	?	5,513.4	27.6	0.7512	360.136
Charon	593	1.83	19.64	0.0	0.00	6.3872

Table 14.7. Orbital characteristics and estimated sizes of the small satellites

Satellite Name	Radius (km)	Distance (10³ km)	Inclination (deg)	Eccentricity	Period (days)
Phobos	11.1	9.38	1.08	0.0151	0.3189
Deimos	6.2	23.46	1.79	0.0003	1.2624
Metis	20	127.96	~ 0	~ 0.0	0.2948
Adrastea	10	128.98	~ 0	~ 0.0	0.2983
Amalthea	86	181.40	0.36	0.0026	0.4982
Thebe	50	221.90	1.10	0.0179	0.6745
S/1975 J1	(4)	7,507	43.08*	0.242	130.02
Leda	(5)	11,165	27.46*	0.164	240.92
Himalia	(85)	11,461	27.50*	0.162	250.56
Lysithea	(12)	11,717	28.30*	0.112	259.20
Elara	(40)	11,741	26.63*	0.217	259.64
S/2000 J11	(2)	12,555	28.27*	0.248	286.95
S/2000 J3	(3)	20,216	149.68*	0.218	585.17R
S/2000 J7	(3)	20,964	148.73*	0.220	617.16R
S/2000 J5	(2)	21,132	148.69*	0.227	624.55R
Ananke	(10)	21,276	148.89*	0.244	610.45R
S/2000 J6	(2)	23,078	165.00*	0.261	718.71R
S/2000 J4	(2)	23,168	164.95*	0.270	722.98R
S/2000 J9	(2)	23,312	165.21*	0.251	729.96R
S/2000 J10	(2)	23,387	165.35*	0.238	733.70R
Carme	(15)	23,404	164.91*	0.253	702.28R
Pasiphae	(18)	23,624	151.43*	0.409	708.04R
S/2000 J2	(3)	23,745	165.21*	0.243	750.81R

S/2000 J8	(3)	23,911	152.72*	0.425	758.05R
Sinope	(14)	23,939	158.11*	0.250	724.51R
S/1999 J1	(5)	24,100	147.14*	0.283	758.76R
Pan	(10)	133.58	0.0	0.004	0.5750
Atlas	16	137.67	0.3	0.000	0.6019
Prometheus	50	139.35	0.0	0.0024	0.6130
Pandora	42	141.70	0.0	0.0044	0.6285
Epimetheus	59	151.42	0.33	0.0126	0.6942
Janus	89	151.47	0.17	0.0066	0.6945
Telesto	11	294.66	1.14	0.0009	1.8878
Calypso	10	294.66	1.47	0.0002	1.8878
Helene	16	377.40	0.20	0.0016	2.7369
Hyperion	142	1,481.1	0.64	0.1046	21.2766
S/2000 S5	(7)	11,365	46.16*	0.334	449.22
S/2000 S6	(5)	11,440	46.74*	0.322	451.48
Phoebe	110	12,944	174.75*	0.1644	548.21R
S/2000 S2	(9.5)	15,199	45.13*	0.364	686.92
S/2000 S8	(3.2)	15,645	152.85*	0.271	728.93R
S/2000 S11	(13)	16,392	33.97*	0.479	783.30
S/2000 S10	(4.3)	17,611	34.47*	0.473	871.17
S/2000 S3	(16)	18,160	45.56*	0.295	893.07
S/2000 S4	(6.5)	18,239	33.51*	0.536	925.70
S/2000 S9	(2.8)	18,709	167.50*	0.208	951.38*
S/2000 S12	(2.8)	19,470	175.81*	0.114	1,016.83R
S/2000 S7	(2.8)	20,470	175.81*	0.466	1,088.89R
S/2000 S1	(8)	23,096	173.09*	0.333	1,312.37R
Cordelia	(13)	49.77	0.08	0.0003	0.3350
Ophelia	(16)	53.79	0.10	0.0099	0.3764
Bianca	(21)	59.17	0.19	0.0009	0.4346
Cressida	(31)	61.78	0.01	0.0004	0.4636
Desdemona	(27)	62.68	0.11	0.0001	0.4736
Juliet	(42)	64.35	0.06	0.0007	0.4931
Portia	(54)	66.09	0.06	0.0001	0.5132
Rosalind	(27)	69.94	0.28	0.0001	0.5585
Belinda	(33)	75.26	0.03	0.0001	0.6235
S/1986 U10	(20)	76.4	∼ 0	0.0011	0.638
Puck	77	86.01	0.32	0.0001	0.7618
Caliban	(30)	7,165	140.89*	0.159	579.47R
Stephano	(10)	7,922	144.01*	0.228	673.56R
Sycorax	(60)	12,174	159.40*	0.523	1,283.27R
Prospero	(10)	16,567	151.75*	0.439	2,037.14R
Setebos	(10)	17,824	158.03*	0.551	2,273.34R
Naiad	(29)	48.23	4.74	0.0003	0.2944
Thalassa	(40)	50.07	0.20	0.0002	0.3115
Despina	74	52.53	0.06	0.0001	0.3347
Galatea	79	61.95	0.05	0.0001	0.4287
Larissa	96	73.55	0.20	0.0014	0.5547

*Inclination measured with respect to the ecliptic (i.e., the Earth's orbital plane).

BIBLIOGRAPHY

Beatty, J. K., Petersen, C. C. and Chaiken, A. (eds) (1999) *The New Solar System* (4th edn).
    Sky Publishing Corporation, Cambridge, Massachusetts. (See Chapters 4, 14, 15, 16, 17,
    18, 19, 20, 21, 22, 23 and the appendix entitled 'Planet, Satellite, and Small-body Char-
    acteristics' on pp. 39–58, 193–320, 387–91.)
Jet Propulsion Laboratory's Solar System Dynamics web page < http://ssd.jpl.nasa.gov/ > is a
    useful reference for the latest in planetary data.
The National Space Science Data Center web page < http://nssdc.gsfc.nasa.gov/planetary/ >
    also compiles a wide variety of planetary data and is relatively up-to-date.

Index

      6 Leaves densely soft-pubescent beneath, sweet to taste; pith
          continuous but with firm cross bands .... *Symplocos*

      6 Leaves glabrous or with a few brownish, sometimes
          glandular, hairs beneath, smooth to touch, not sweet to
          taste; pith without cross bands 7

        7 Leaves mostly >15 cm long .... *Rhododendron*

        7 Leaves mostly <15 cm long 8

          8 Leaf apex acute to acuminate; corolla actinomorphic;
            fruit a nearly round capsule; axillary bud of
            lowermost leaves hidden between the leaf petiole
            and stem .... *Kalmia*

          8 Leaf apex rounded, obtuse, or rarely acute; corolla
            zygomorphic; fruit an elongate capsule; axillary bud
            evident, not hidden .... *Rhododendron*

1 Leaves deciduous 9

    9 Crushed leaves and twigs spicy aromatic 10

      10 Plants usually with some 2- and/or 3-lobed leaves; leaves
          with two prominent lateral veins arising from near the
          leaf base .... *Sassafras*

      10 All leaves unlobed, without prominent lateral veins .... *Lindera*

    9 Crushed leaves and twigs not spicy aromatic 11

      11 Leaves with two or more prominent lateral veins arising at or
          near the leaf base 12

        12 Leaves with two prominent lateral veins, the base usually
           asymmetrical; petiole <2 cm long .... *Celtis*

        12 Leaves with >2 prominent lateral veins, the base symmetrical;
           petiole >2 cm long .... *Cercis*

      11 Leaves without prominent lateral veins arising at or near the
          leaf base 13

        13 Stipules or their scars present, sometimes minute (examine
           carefully at 10x) 14

          14 Stipules or their scars completely encircling the twig 15

            15 Subshrubs; leaves <3 mm wide .... *Polygonella*

            15 Trees; leaves >3 mm wide .... *Magnolia*

          14 Stipules or their scars not encircling the twig 16

            16 Plants with milky sap (examine broken petiole base);
             branches often with axillary spines; fruit green, warty,
             softball-sized .... *Maclura*

            16 Plants without milky sap; axillary spines absent; fruit
             otherwise 17

17　Leaves clustered at the tips of current year's growth; bud scales >1; fruit an acorn . . . . *Quercus*

17　Leaves not clustered at the tips of current year's growth; bud scale one; fruit a capsule . . . . *Salix*

13　Stipules or their scars absent 18

18　Leaves arranged in two distinct rows on opposing sides of the stem, i.e., in 1 plane 19

19　Leaves oblong-obovate, >15 cm long, with rusty-brown hairs beneath, often ill-scented when crushed . . . . *Asimina*

19　Leaves elliptic to ovate, <15 cm long, not rusty-brown pubescent beneath, not ill-scented when crushed 20

20　Leaves with simple hairs beneath, the margins distinctly ciliate and with gland-tipped teeth . . . . *Stewartia*

20　Leaves sparsely to densely stellate-pubescent beneath (10X), the margins not ciliate and without gland-tipped teeth . . . . *Styrax*

18　Leaves spirally or randomly arranged, not in distinct rows, i.e., in >1 plane 21

21　Plants with thorns, spines, or sharp-pointed spur branches 22

22　Leaves silvery beneath, densely covered with peltate scales (10X) . . . . *Elaeagnus*

22　Leaves not covered with silvery, peltate scales 23

23　Mature leaves >3 cm wide, strongly reticulate veined; sap milky (examine broken petiole base) . . . . *Bumelia*

23　Mature leaves <3 cm wide, not strongly reticulate veined; sap not milky 24

24　Spines needle-like, <1 mm wide at the base; inner bark and wood yellow; leaves glaucous beneath . . . . *Berberis*

24　Spines stout, >1 mm wide at the base; inner bark and wood not yellow; leaves not glaucous beneath . . . . *Lycium*

21　Plants without thorns, spines, or sharp-pointed spur branches 25

25　Lateral veins arching strongly upwards, becoming parallel or nearly so with the midrib; at least some

hairs forked and closely appressed to the lower leaf
surface (10x) . . . . *Cornus*

25 Lateral veins not conspicuously arching upwards **and**
the hairs not forked and closely appressed to the lower
leaf surface 26

26 Leaves silvery beneath, densely covered with
peltate scales (10x) 27

27 Low shrubs, mostly <3 m tall; fruit a 3-seeded
capsule; native, but possibly extirpated . . . . *Croton*

27 Shrubs or small trees up to 5 m tall; fruit a fleshy,
red achene; introduced . . . . *Elaeagnus*

26 Leaves not silvery, peltate-scaly beneath 28

28 Shrubs, mostly <4 m tall 29

29 Leaves with yellow glands beneath or with
prominent scale-like structures on the lower
midrib 30

30 Leaves yellow glandular beneath
(10x) . . . . *Gaylussacia*

30 Leaves with lacerate, scale-like structures
along the midrib (10x) . . . . *Menziesia*

29 Leaves without yellow glands beneath or
scales along the midrib 31

31 Leaves or buds clustered toward the tips of
twigs . . . . *Rhododendron*

31 Leaves or buds not clustered toward the
tips of twigs 32

32 Plants with at least some opposite
and/or whorled leaves . . . . *Decodon*

32 Plants with all leaves alternate 33

33 Mature leaves >8 cm
long . . . . *Pyrularia*

33 Mature leaves <8 cm long 34

34 Twigs of the current season
flexible, not easily broken;
axillary bud hidden beneath the
leaf petiole . . . . *Dirca*

34 Twigs of the current season not
flexible or leathery, easily broken;
axillary bud evident, not
hidden 35

35 Leaf blades 4x or more longer
than wide . . . . *Salix*
35 Leaf blades <4x longer than
wide 36
36 Leaves pubescent with the
hairs closely appressed to
the leaf surface (10x); fruit
a capsule . . . . *Lyonia*
36 Leaves glabrous or, if
pubescent, then the hairs
not closely appressed to the
leaf surface; fruit a
berry . . . . *Vaccinium*
28 Medium to large trees 37
37 Leaves (at least some of them) at the tip of
twigs with petioles >3.5 cm long; leaf apex
mostly rounded . . . . *Cotinus*
37 Leaves at the tip of twigs with petioles <3.5
cm long; leaf apex mostly acute or
acuminate 38
38 Leaves mostly tapering at the base;
vascular bundle scars 3; pith continuous
but with firm cross bands . . . . *Nyssa*
38 Leaves mostly rounded at the base;
vascular bundle scar 1; pith without cross
bands . . . . *Diospyros*

Key 2

1 Leaves palmately veined, i.e., with two or more **prominent** lateral veins
arising at or near the base of the leaf blade, the mid and uppermost veins
abruptly less evident 2
  2 Sap milky (examine broken petiole base) 3
    3 Leaves arranged in two distinct rows on opposing sides of the
stem, i.e., in 1 plane, glabrous or soft pubescent beneath . . . . *Morus*
    3 Leaves randomly or spirally arranged, not in distinct rows, i.e., in
>1 plane, densely velvety pubescent beneath (some leaves may be
opposite or whorled) . . . . *Broussonetia*
  2 Sap not milky 4

4   Crushed twigs and leaves spicy aromatic; at least some leaves
    lobed, but without teeth . . . . *Sassafras*

4   Crushed twigs and leaves not spicy aromatic; leaves toothed
    and/or lobed 5

    5   Axillary buds hidden within the leaf petiole; stipules or their
        scars encircling the twig . . . . *Platanus*

    5   Axillary buds evident, not hidden in the leaf petiole; stipules or
        their scars absent or, if present, not encircling the twig 6

      6   Petioles, especially along the upper surface, and stems of the
          current season stellate (10X) pubescent (some simple hairs
          may also be present) . . . . *Hibiscus*

      6   Petioles and stems of the current season glabrous, evenly
          pubescent, or with reddish gland-tipped hairs, but without
          stellate hairs 7

        7   Trees, >4 m tall 8

          8   Leaves star-shaped . . . . *Liquidambar*

          8   Leaves not lobed or, if so then, not star-shaped 9

            9   Leaf blades ± as long as wide . . . . *Tilia*

            9   Leaf blades longer than wide . . . . *Celtis*

        7   Shrubs, <4 m tall 10

          10   Petioles, stems, and lower leaf veins covered with
              reddish, gland-tipped hairs (10X) . . . . *Rubus*

          10   Petioles, stems, and lower leaf veins glabrous or
              without reddish, gland-tipped hairs 11

            11   Leaves lobed and toothed 12

              12   Axils of basal lateral veins (lower surface) with
                  stellate hairs (10X); bristles and/or spines
                  absent . . . . *Physocarpus*

              12   Axils of basal lateral veins glabrous or, if
                  pubescent, then the hairs not stellate; bristles
                  and/or spines often present . . . . *Ribes*

            11   Leaves toothed, but not lobed . . . . *Ceanothus*

1   Leaves with a single evident midrib **or** pinnately veined **or** the basal,
    lateral veins not significantly more prominent than the mid and upper
    veins, i.e., the lateral veins becoming progressively reduced and less
    evident from the leaf base to the leaf apex 13

  13   Stipules or their scars present (sometimes minute, examine carefully
      at 10X), **either** on the stem at the petiole base **or** on the petiole base
      just above its attachment to the stem 14

    14   Stipules or their scars encircling the twig or essentially so 15

15 Leaves not lobed or lobed only at the base 16

16 Leaf margin toothed, not lobed; lateral veins parallel; stipule scars almost completely encircling the twig . . . . . . *Fagus*

16 Leaf margin not toothed, sometimes 2-lobed at base; lateral veins not parallel; stipule scars completely encircling the twig . . . . *Magnolia*

15 Leaves 4–6-lobed, the apex broadly and shallowly notched . . . . *Liriodendron*

14 Stipules or their scars not encircling the twig 17

17 Leaves arranged in two distinct rows on opposing sides of the stem, i.e., in 1 plane 18

18 Mature leaves with <4 marginal teeth or lobes/cm 19

19 Lateral veins extending into each marginal tooth and beyond into a short bristle . . . . *Castanea*

19 Lateral veins not extending into each tooth or lobe and beyond into a short bristle 20

20 Leaf margin coarsely toothed 21

21 Leaf blades pustulate-punctate above (10x); leaves often lobed, the base symmetrical; sap milky (examine broken petiole base) . . . . *Morus*

21 Leaf blades not pustulate-punctate above, leaves not lobed, the base often asymmetrical; sap not milky . . . . *Tilia*

20 Leaf margin undulate or toothed only above the middle 22

22 Basal margin of leaf blade attached to one or both lowermost lateral vein(s), not to petiole; stellate hairs of lower leaf surface tending to be more dense in the axils of lateral veins (early to mid-season) . . . . *Fothergilla*

22 Basal margin of leaf blade attached to the petiole; stellate hairs of lower leaf surface evenly distributed, especially on the veins, not tufted in the axils of lateral veins (early to mid-season) . . . . *Hamamelis*

18 Mature leaves with >4 marginal teeth or lobes/cm 23

23 Leaves asymmetric at the base and/**or** the axillary bud positioned to one side of the leaf scar, i.e., not directly above the leaf scar 24

24   Leaf blade widest below the middle; lateral veins
     <10/each side of the midrib . . . . *Planera*
24   Leaf blade widest at or near the middle; lateral veins
     mostly >10/each side of the midrib . . . . *Ulmus*
23   Leaves symmetric at the base and the axillary bud
     positioned directly above the leaf scar 25
     25   Leaves with ± the majority of lateral veins forking
          before reaching the leaf margin . . . . *Ostrya*
     25   Leaves rarely with lateral veins forking (if so, mostly
          the basal ones) before reaching the leaf margin 26
          26   Leaf buds >3 mm long; twigs sometimes with the
               odor of wintergreen; bark often exfoliating;
               lenticels mostly longer than wide . . . . *Betula*
          26   Leaf buds <3 mm long, twigs not aromatic; bark
               smooth, "muscular;" lenticels circular or slightly
               longer than broad . . . . *Carpinus*
17   Leaves randomly or spirally arranged, not in distinct rows,
     i.e., in >1 plane 27
     27   Leaves clustered at the tips of current year's growth; fruit
          an acorn . . . . *Quercus*
     27   Leaves not clustered at the tips of current year's growth
          or, if so, then the fruit not an acorn 28
          28   Leaf blade >3x as long as wide 29
               29   Leaf margin divided to near the midrib; leaves
                    glandular beneath (10x); shrubs . . . . *Comptonia*
               29   Leaf margin finely toothed but not deeply divided;
                    leaves not glandular beneath; trees or shrubs 30
                    30   Bud scale 1; leaf teeth not gland-tipped . . . . . . *Salix*
                    30   Bud scales >1; leaf teeth gland-tipped
                         (10x) . . . . *Prunus*
          28   Leaf blade <3x as long as wide 31
               31   Petioles, stems, and lower leaf veins densely covered
                    with reddish gland-tipped hairs (10x) **and** the leaves
                    palmately lobed; shrubs . . . . *Rubus*
               31   Petioles, stems, and lower leaf veins without **both**
                    gland-tipped hairs and palmately lobed leaves; small
                    shrubs or trees 32
                    32   Leaf blades at their widest point ± as long as wide,
                         if lobed, then with a milky sap (examine broken
                         petiole base) 33

33 Leaf blades <3.5 cm wide . . . . *Prunus*

33 Leaf blades >3.5 cm wide 34

  34 Leaf blade widest at or near the
  base . . . . *Populus*

  34 Leaf blade widest at or near the
  middle 35

    35 Leaves velvety-pubescent beneath;
    marginal teeth >12 on each side; sap milky
    (examine broken petiole base); some
    opposite or whorled leaves also may be
    present . . . . *Broussonetia*

    35 Leaves glabrous or nearly so beneath;
    marginal teeth <12 on each side; sap not
    milky . . . . *Populus*

32 Leaf blades longer than wide **or** lobed, but without
milky sap 36

  36 Leaf margin regularly toothed, the teeth
  approximately the same size or becoming
  slightly smaller from base to apex or the leaves
  evergreen (thick and leathery) and the teeth
  sharp-pointed 37

    37 Upper leaf midrib with a row of reddish
    brown glands 38

      38 Leaf teeth conspicuously gland-tipped
      (similar to midrib glands); sharp-pointed
      spur branches absent; fruit <1.2 cm
      wide . . . . *Aronia*

      38 Leaf teeth either not gland-tipped or with
      glands reduced and less conspicuous than
      those of the midrib; sharp-pointed spur
      branches often present; fruit >1.2 cm
      wide . . . . *Malus*

    37 Upper leaf midrib without a row of reddish
    brown glandular hairs 39

      39 Plants regularly thorn bearing or with
      sharp-pointed spur branches 40

        40 Spur branches present on stems of the
        previous season(s) . . . . *Prunus*

maturity; fruit an
apple . . . . *Malus*

44 Stipules attached to the stem
below the leaf base and
leaving a scar; fruit <1 cm
wide 46

46 Inner wood of twigs
yellowish and ill-scented;
stipules deciduous,
greenish, not turning
black . . . . *Rhamnus*

46 Inner wood of twigs not
yellowish and ill-scented;
stipules usually persistent,
turning black 47

47 Leaf margins glandular-
serrate; pedicels mostly
>6 mm long; petals
separate, linear,
yellow to yellowish
green; stamens
free . . . . *Nemopanthus*

47 Leaf margins not
glandular-serrate or
rarely obscurely so;
pedicels mostly <6 mm
long (except in *I. decidua*
var. *longipes*); petals fused
at base, ovate to broadly
oblong, white to
creamy; stamens adnate
to the base of the
corolla . . . . *Ilex*

36 At least some leaves with the margin irregularly
toothed and/or lobed, the teeth/lobes of various
sizes, the margin "jagged" in appearance and the
leaves deciduous 48

48 Plants thorn bearing or with sharp-pointed
spur branches 49

58 Leaf margin distinctly ciliate **and** with gland-tipped teeth (10x); buds silky pubescent, partially hidden withing the leaf petiole; leaves arranged in two distinct rows on opposing sides of the stem, i.e., in 1 plane . . . . *Stewartia*

58 Leaf margin without cilia (or essentially so) **and** gland-tipped teeth; buds glabrous or pubescent but not hidden within the leaf petiole; leaves randomly or spirally arranged, not in distinct rows, i.e., in >1 plane 59

   59 Leaves mostly entire but occasionally with a few coarse teeth near the apex . . . . *Nyssa*

   59 Leaves with numerous teeth throughout or from the ± lower ⅓ to the apex 60

      60 Midrib of lower leaf surface distinctly pubescent with stiff hairs; the ± lower ⅓ of leaf margin entire, the remainder toothed to the apex; pith not chambered; fruit a capsule . . . . . *Oxydendrum*

      60 Midrib of lower leaf surface glabrous or with a few stellate hairs or woolly pubescent (10x); leaf margin toothed from near the base to the apex; pith chambered; fruit a winged, woody drupe . . . . *Halesia*

56 Shrubs, <4 m tall 61

  61 Leaves arranged in two distinct rows on opposing sides of the stem, i.e., in 1 plane 62

    62 Leaf margin distinctly ciliate **and** with gland-tipped teeth (10x); blade with simple hairs beneath; buds silky pubescent, partially hidden within the leaf petiole . . . . *Stewartia*

    62 Leaf margin without cilia **and** gland-tipped teeth; blade with stellate hairs beneath (10x); buds stellate pubescent, not hidden within the leaf petiole . . . . *Styrax*

  61 Leaves randomly or spirally arranged, not in distinct rows, i.e., in >1 plane 63

    63 Pith chambered, sometimes faintly . . . . *Itea*

    63 Pith not chambered 64

64 Lateral veins extending into each marginal tooth
    (especially the larger ones) or the leaves toothed
    only near the apex . . . . *Spiraea*
64 Lateral veins becoming obscure and not extending
    into each marginal tooth; leaf teeth not restricted
    to the leaf apex 65
    65 Leaves or buds clustered toward the tips of
        twigs; therefore, twigs of the previous season
        radiating in a whorled fashion . . . . . *Rhododendron*
    65 Leaves or buds not clustered toward the tips of
        twigs and twigs of the previous season not
        radiating in a whorled fashion 66
        66 Twigs and buds of the current season densely
            stellate pubescent (10x), the surface
            essentially hidden or obscured beneath the
            hairs; leaf margins serrate . . . . *Clethra*
        66 Twigs and buds of the current season
            glabrous or with simple hairs, not hidden or
            obscured; leaf margins serrulate 67
            67 Fruits (capsules) of the previous season
                persisting into the current season 68
                68 Inflorescences in axillary panicles or
                    fascicles, overwintering within the bud;
                    fruits of the current season
                    pubescent . . . . *Lyonia*
                68 Inflorescences in axillary racemes,
                    produced during the summer/fall of the
                    previous season; fruits of the current
                    season glabrous . . . . *Leucothoe*
            67 Fruits (berries) of the previous season not
                persisting into the current
                season . . . . *Vaccinium*

Key D

LEAVES SIMPLE: OPPOSITE, SUB-OPPOSITE, OR WHORLED

1 Leaves whorled 2
    2 Leaves evergreen (thick and leathery), the blade <1.5 dm
        long . . . . *Kalmia*

2 Leaves deciduous, the blade >1.5 dm long [Note: *Broussonetia* (with milky sap), *Cephalanthus*, and *Decodon* (with leaf blades <1.5 dm long), occasionally have whorled leaves and may be keyed below] . . . . *Catalpa*
1 Leaves opposite 3
　　3 Leaf margin minutely to coarsely toothed, the blade lobed or unlobed 4
　　　　4 Petioles 1 cm or less long 5
　　　　　　5 Stems green, angled, or winged; leaves semi-evergreen . . . . *Euonymus*
　　　　　　5 Stems not green, angled, or winged; leaves deciduous 6
　　　　　　　　6 Leaves coarsely toothed, <20 teeth/side 7
　　　　　　　　　　7 Leaves palmately veined . . . . *Philadelphus*
　　　　　　　　　　7 Leaves pinnately veined . . . . *Viburnum*
　　　　　　　　6 Leaves finely toothed, >20 teeth/side . . . . *Diervilla*
　　　　4 Petioles 1 cm or more long, at least on the larger leaves 8
　　　　　　8 Sap milky (examine broken petiole base); leaves densely soft, velvety pubescent beneath; leaves mostly alternate, some may also be whorled . . . . *Broussonetia*
　　　　　　8 Sap not milky; leaves glabrous or variously pubescent; all leaves opposite 9
　　　　　　　　9 Leaves palmately lobed 10
　　　　　　　　　　10 Leaves 3-lobed, densely rusty-brown pubescent beneath (10x); petioles mostly <3 cm long . . . . *Viburnum*
　　　　　　　　　　10 Leaves 3–5-lobed, not densely, rusty-brown pubescent beneath; petioles >3 cm long . . . . *Acer*
　　　　　　　　9 Leaves not palmately lobed 11
　　　　　　　　　　11 Petiole bases meeting or joined by a transverse line 12
　　　　　　　　　　　　12 Plants with at least one of the following characters: stipules absent or, if present, then attached to the leaf petiole; buds naked or valvate; lower leaf surface and/or petioles with stellate trichomes or rusty-brown hairs or scales . . . . *Viburnum*
　　　　　　　　　　　　12 Stipules absent or, if present, then attached to the petiole base; buds scaly, imbricate; lower leaf surface glabrous or with unbranched hairs 13
　　　　　　　　　　　　　　13 Leaf veins anastomosing before reaching leaf margin; petiole >3 cm long (on the larger leaves) . . . . *Hydrangea*

13   Leaf veins ending in a marginal tooth; petioles
       <3 cm long . . . . *Viburnum*
11   Petiole bases not meeting or joined by a transverse
     line   14
   14   Leaf blades <4 cm long . . . . *Forestiera*
   14   Leaf blades >4 cm long   15
     15   Leaves glabrous beneath   16
       16   Leaves coarsely and conspicuously
             toothed   . . . . *Forsythia*
       16   Leaves minutely toothed
             (serrulate) . . . . *Forestiera*
     15   Leaves pubescent beneath 17
       17   Leaf blades >1 dm wide . . . . *Paulownia*
       17   Leaf blades <1 dm wide 18
         18   Stems and lower leaf surfaces scurfy-stellate
               pubescent and densely yellow-glandular
               (10X); leaf base cuneate . . . . *Callicarpa*
         18   Stems and lower leaf surfaces with
               unbranched trichomes and without yellow
               glands; leaf base rounded . . . . *Euonymus*
3   Leaf margin not toothed, the blade unlobed 19
19   Leaves >1.5 dm long . . . . *Paulownia*
19   Leaves <1.5 dm long 20
   20   Crushed leaves strongly aromatic 21
     21   Leaf blade <2 cm wide   . . . . *Conradina*
     21   Leaf blade >2 cm wide . . . . *Calycanthus*
   20   Crushed leaves not aromatic   22
     22   Leaf blades distinctly glandular-dotted (10X) 23
       23   Stems of the current season glabrous; leaves sessile or
             subsessile; flowers yellow . . . . *Hypericum*
       23   Stems of the current season pubescent; leaves short-
             petiolate; flowers white . . . . *Ligustrum*
     22   Leaf blades not glandular-dotted 24
       24   Leaves evergreen (thick and leatherly); plants semi-
             parasitic on the branches of deciduous trees
             (mistletoe) . . . . *Phoradendron*
       24   Leaves deciduous; plants not semi-parasites or, if so, then
             root parasites 25

25   Lateral veins arching upwards, becoming parallel
     or nearly so with the midrib . . . . *Cornus*
25   Lateral veins not conspicuously arching
     upwards 26
  26   Stipules or their scars present, sometimes
       minute (examine carefully at 10x) or petiole
       bases meeting or joined by a transverse
       line 27
     27 Leaves >7 cm long 28
       28 Lower leaf surface, petioles, and twigs with
           rusty-brown scales . . . . *Viburnum*
       28 Leaf blades, petioles, and twigs glabrous or
           essentially so . . . . *Cephalanthus*
     27 Leaves <7 cm long 29
       29 Leaves glabrous beneath . . . . *Lonicera*
       29 Leaves pubescent beneath 30
         30 Leaves >4.5 cm long . . . . *Lonicera*
         30 Leaves <4.5 cm long . . . . . . *Symphoricarpos*
  26   Stipules or their scars absent **and** petiole bases not
       meeting or joined by a transverse line 31
     31 Leaves >5 cm wide . . . . *Chionanthus*
     31 Leaves <5 cm wide 32
       32 Stems of the current season
           glabrous . . . . *Nestronia*
       32 Stems of the current season pubescent
           (10x) 33
         33 Leaf blades <3 times as long as
             wide . . . . *Ligustrum*
         33 Leaf blades >3 times as long as
             wide 34
           34 Leaves arranged in two distinct rows
               on opposing sides of the stem, i.e.,
               in 1 plane, 6 cm or less long, the
               margin spiculate-scabrous
               (10x) . . . . *Buckleya*
           34 Leaves spirally or randomly arranged,
               i.e., in >1 plane, occasionally whorled
               or alternate, >6 cm long, the margin
               smooth . . . . *Decodon*

Key E

Leaves Compound and Alternate

1 Leaves ternately compound 2

  2 Plants with thorns, prickles, or bristles 3

    3 Plants with stout thorns >2 cm long; stipules or their scars
absent; petioles winged .... *Poncirus*

    3 Plants with thorns, bristles or prickles <2 cm long; stipules or their
scars present; petioles not winged .... *Rubus*

  2 Plants without thorns, prickles, or bristles 4

    4 Stipules or their scars present, sometime minute (examine carefully
at 10x); leaflets entire .... *Lespedeza*

    4 Stipules or their scars absent; leaflets variously toothed or lobed,
rarely entire 5

      5 Terminal leaflet petiolulate **(caution: poison
ivy/oak)** .... *Toxicodendron*

      5 Terminal leaflet sessile or essentially so 6

        6 Leaflets glandular-dotted above (10x), entire or crenulate;
petioles mostly >5 cm long; fruit a winged, nearly round
samara .... *Ptelea*

        6 Leaflets not glandular-dotted above, entire below the
middle, distinctly crenate-toothed above; petioles mostly
<5 cm long; fruit a red drupe .... *Rhus*

1 Leaves palmately or once- , bi- , or tripinnately compound 7

  7 Leaves palmately compound .... *Rubus*

  7 Leaves not palmately compound 8

    8 Leaves bi- or tripinnately compound 9

      9 Plants with thorns or prickles 10

        10 Leaves odd-pinnate; leaf petiole and rachis often with
prickles; fruit a black, berry-like drupe .... *Aralia*

        10 Leaves even-pinnate; leaves without prickles; fruit a
legume .... *Gleditsia*

      9 Plants without thorns or prickles 11

        11 Leaflets toothed .... *Melia*

        11 Leaflets entire 12

          12 Leaflets asymmetrical, <3 cm long .... *Albizia*

          12 Leaflets symmetrical, >3 cm long .... *Gymnocladus*

    8 Leaves once-pinnately compound 13

      13 Plants with thorns, spines, prickles, or bristles 14

14 Leaflets entire or finely crenulate 15
15 Leaflets 1.5 cm or less wide . . . . *Gleditsia*
15 Leaflets >1.5 cm wide 16
16 Crushed leaves aromatic; leaflets sessile to subsessile, glandular-dotted above (10X); fruit a reddish follicle . . . . *Zanthoxylum*
16 Crushed leaves not aromatic; leaflets petiolulate, not glandular-dotted above; fruit a legume . . . . *Robinia*
14 Leaflets distinctly toothed 17
17 Stems, leaf rachis, and pedicel densely hispid with gland-tipped hairs (10X); fruit a raspberry . . . . *Rubus*
17 Stems, leaf rachis, and pedicel not densely hispid; fruit a hip . . . . *Rosa*
13 Plants without thorns, spines, prickles, or bristles 18
18 Leaflets distinctly alternate, not in evident pairs . . . . *Cladrastis*
18 Leaflets in opposite or subopposite pairs 19
19 Leaflets entire, crenulate or with a few basal teeth 20
20 Stipules or their scars present, sometimes minute (examine carefully at 10X); leaflets usually punctate beneath (10X); fruit a punctate legume (10X) . . . . *Amorpha*
20 Stipules or their scars absent; leaflets not punctate beneath; fruit not a legume 21
21 Leaflets toothed at the base with a foul-smelling gland (10X); fruit a samara . . . . *Ailanthus*
21 Leaflets entire, without a foul-smelling gland; fruit a creamy or red drupe 22
22 Leaf rachis winged; inflorescence terminal; fruit red . . . . *Rhus*
22 Leaf rachis not winged; inflorescence axillary; fruit creamy (**caution: poison sumac**) . . . . *Toxicodendron*
19 Leaflets distinctly toothed or cleft 24
24 Low shrubs <1 m tall; wood yellow; terminal leaflet often 3-cleft . . . . *Xanthorhiza*
24 Large shrubs >1 m tall or medium to large trees; wood not yellow; terminal leaflet not 3-cleft 25

25 Leaves pubescent or glandular-pubescent beneath
(10x) 26

  26 Sap milky (examine broken petiole base); stems
with dense, velvety (to the touch) hairs; fruit a
reddish drupe .... *Rhus*

  26 Sap not milky; stems glabrous or short
pubescent, at least not velvety to the touch; fruit
a walnut, pecan, or hickory nut 27

    27 Pith chambered; median leaflets larger than
the terminal and basal; fruit a
walnut .... *Juglans*

    27 Pith not chambered; terminal leaflets larger
than the median and basal; fruit a hickory nut
or pecan .... *Carya*

25 Leaves glabrous beneath 28

  28 Sap milky (examine broken petiole base); upper
side of rachis at leaflet attachment without
glands or hairs; buds not sticky; fruit a reddish
drupe .... *Rhus*

  28 Sap not milky; upper side of rachis at leaflet
attachment bearing dark glands intermixed with
hairs; buds sticky; fruit a pome .... *Sorbus*

KEY F

LEAVES COMPOUND AND OPPOSITE

1 Leaves palmately compound .... *Aesculus*

1 Leaves trifoliolate or once-pinnate 2

  2 Leaves trifoliolate and stipules or their scars present .... *Staphylea*

  2 Leaves once-pinnate (rarely trifoliolate); stipules or their scars
absent 3

    3 Leaflets 3–5, rarely more; stems, at least the younger, green and
glaucous .... *Acer*

    3 Leaflets >5; stems not green and glaucous 4

      4 Leaflets regularly serrate; twigs with prominent lenticels;
shrubs .... *Sambucus*

      4 Leaflets entire or irregularly toothed; twigs without prominent
lenticels; trees .... *Fraxinus*

Aceraceae   *Acer* (Maple)

Small or large trees; monoecious; flowers in lateral umbels, panicles or racemes (spring–summer), fruit a pair of samaras united at base and eventually separating (spring–summer). Some species are important for lumber, maple syrup, shade, and ornamental plantings. Taxa in the *Acer saccharum* complex are not always easily differentiated.

1 Leaves compound, with 3–7 leaflets; blades of leaflets pinnately
   veined . . . . *A. negundo*
1 Leaves simple, 3–5 lobed; leaf blades palmately veined 2
   2 Buds stalked, with 2 valvate scales; flowers in terminal racemes; small
     trees, mostly in mountains 3
     3 Twigs and buds glabrous; leaves finely serrate with 3 main veins;
       bark white-striped; flowers and fruit drooping . . . . *A. pensylvanicum*
     3 Twigs and buds pubescent; leaves coarsely serrate with 5 main
       veins; bark not striped; flowers and fruit erect . . . . *A. spicatum*
   2 Buds sessile, with 4–8 scales; flowers in lateral clusters; small to large
     trees of various habitats 4
     4 Margins of leaf lobes with several-many sharp-tipped teeth; leaf
       sinuses V-shaped; fruit maturing in spring 5
       5 Terminal lobe >½ as long as the entire leaf
         blade . . . . *A. saccharinum*
       5 Terminal lobe <½ as long as the entire leaf blade 6
         6 Twigs and leaves glabrous at maturity; mature samaras
           <3 cm long . . . . *A. rubrum*
         6 Twigs and leaves pubescent at maturity; mature samaras
           >3 cm long . . . . *A. drummondii*
     4 Margins of leaf lobes entire or with a few blunt teeth; leaf sinuses
       U-shaped; fruit maturing in summer 7
       7 Leaves yellow-green beneath 8
         8 Leaf blades <8 cm long; bark white-
           gray . . . . *A. saccharum* ssp. *leucoderme*
         8 Leaf blades >9 cm long; bark brown-
           black . . . . *A. saccharum* ssp. *nigrum*
       7 Leaves green or gray-green beneath 9
         9 Leaves green and glabrous beneath except for a
           few hairs on the veins; lobes of leaves
           acute . . . . *A. saccharum* ssp. *saccharum*

9 Leaves gray-green and glaucous beneath with hairs
    on and between veins; lobes of leaves
    blunt . . . . *A. saccharum* ssp. *floridanum*

A. drummondii Hook. & H.J.Arn. *ex* Nutt., Drummond Red M.; bottomlands, sandy woods, swamps; occasional; W TN, CB, EHR. *A. rubrum* L. var. *drummondii* (Hook. & H.J. Arn. *ex* Nutt.) Sarg. Plate 21.

A. negundo L., Boxelder; riparian forests, ravines, damp successional and disturbed sites; common; statewide. Plate 22.

A. pensylvanicum L., Striped M.; rich, cool woods; common; E TN, but rare in the VR. Plate 23.

A. rubrum L., Red M.; bottomlands, swamps, riparian sites, often abundant in damp successional and disturbed sites; common; statewide. *A. rubrum* var. *trilobum* K.Koch. Plate 24.

A. saccharinum L., Silver/Water M.; riparian forests, wet to mesic woodlands and disturbed sites; frequent; statewide, with fewest records from the CU. Often planted. Plate 25.

A. saccharum Marshall ssp. *floridanum* (Chapm.) Desmarais, Southern Sugar M.; mesic bluffs and ravines; occasional; CP, WHR. *A. floridanum* (Chapman) Pax, *A. barbatum* Michx. Plate 26.

A. saccharum Marshall ssp. *leucoderme* (Small) Desmarais, Chalk M.; wooded river bluffs and ravines; infrequent; VR, U. **Special Concern.** *A. leucoderme* Small. Plate 27.

A. saccharum Marshall ssp. *nigrum* (F.Michx.) Desmarais, Black M.; mostly calcareous alluvial, riparian, and lower slope woods; occasional; statewide, most abundant in M TN. *A. nigrum* F.Michx. Plate 28.

A. saccharum Marshall ssp. *saccharum*, Sugar M.; rich slope and ravine woods, successional sites; common; statewide. Often planted. Plate 29.

A. spicatum Lam., Mountain M.; mesic woodlands and thickets; occasional; E TN. Plate 30.

Hippocastanaceae   *Aesculus* (Buckeye)

Shrubs, small or large trees; flowers yellow, creamy white or red, in terminal panicles (spring), fruit a smooth or warty-spiny, leathery capsule (summer–fall). Several species are regularly cultivated for their showy flowers; buckeyes (the large, dark brown seeds) historically have been carried as good luck charms and for medicinal reasons. An extensive hybrid swarm involving several taxa occurs in the Sequatchie Valley. Hardin, J. W. 1957. A revision of the American Hippocastanaceae. Brittonia 9:145–195.

1 Flowers pale greenish yellow; petals nearly equal in length; stamens
   long exserted; fruit spiny . . . . A. glabra
1 Flowers yellow or red; petals very unequal in length; stamens included
   or barely exserted; fruit not spiny 2
   2 Flowers red . . . . A. pavia
   2 Flowers yellow 3
      3 Large trees; pedicels stipitate-glandular; petiolules
         ±3 mm . . . . A. flava
      3 Shrubs or small trees; pedicels eglandular; petiolules
         >3 mm . . . . A. sylvatica

A. flava Sol., Yellow/Sweet B.; mesic woods, especially cove hardwood forests; frequent; EHR, E TN, with a few records from the northern WHR and the CB. A. octandra Marshall. Plate 31.

A. glabra Willd., Ohio B.; streambanks, ravines, and lower slopes, most often around limestone outcrops; common; M TN, VR. Plate 32.

A. pavia L., Red/Scarlet B.; mesic thickets and woods, streambanks; common; W TN, scattered on the WHR and across the southern tier of counties to the VR. Plate 33.

A. sylvatica Bartram, Painted B.; slopes, streambanks, ravines; frequent; CU, VR. Plate 34.

Simaroubaceae   Ailanthus (Tree-of-Heaven)

A. altissima (Mill.) Swingle*; small to medium trees; flowers greenish yellow, in large terminal panicles (late spring–early summer), fruit an elongated samara, in large clusters (summer); disturbed sites; frequent; statewide. Asia. Rapid growing but soft-wooded, quickly spreading by seeds and from root suckers; leaves foul smelling (like burned popcorn). Plate 35.

Fabaceae   Albizia (Mimosa/Silktree)

A. julibrissin Durazz.*; small to medium, bushy trees; flowers pinkish red in paniculate heads forming massive displays (spring–summer), fruit a legume (summer–

fall); often planted and spreading to ruderal sites; frequent; statewide. Asia. Rapid growing but short-lived. Plate 36.

Betulaceae   *Alnus* (Alder)

Shrubs; monoecious; flowers in catkins, staminate overwintering before opening (spring), fruit a nutlet (summer–fall). The pistillate catkin becomes a woody, cone-like structure that persists long after the nutlets are released. Hardin, J. W. 1952. The Juglandaceae and Corylaceae of Tennessee. Castanea 17:78–89; Hardin, J. W. 1971. Studies of the southeastern United States flora. 1. Betulaceae. J. Elisha Mitchell Sci. Soc. 87:39–41.

1 Pistillate flowers appearing with the leaves; nutlet winged; high-
   elevation balds . . . . *A. viridis* ssp. *crispa*
1 Pistillate flowers appearing before the leaves; nutlet wingless; over the
   state . . . . *A. serrulata*

A. serrulata (Aiton) Willd., Smooth/Hazel A.; swampy woods and thickets, stream-banks; common; statewide. Plate 37.

A. viridis (Villars) DC. ssp. *crispa* (Aiton) Turrill, Green/Mountain A.; high eleva-tion balds; Roan Mountain, Carter Co. (U). **Special Concern.** *A. crispa* (Aiton) Pursh. Plate 38.

Rosaceae   *Amelanchier* (Sarvis/Serviceberry/Shadbush)

Shrubs or small trees; flowers white to pink, mostly in racemes (early spring); fruit a reddish or purplish black pome (spring–fall). There are numerous species in east-ern North America, where hybridization, polyploidy, and apomixis are common; thus the taxonomy of the genus is not clear. Flowers and/or fruit are usually required for identification.

1 Summit of ovary and fruit tomentose . . . . *A. sanguinea*
1 Summit of ovary and fruit glabrous (or nearly so) 2
   2 Racemes erect; petals 4–12 mm long; fruiting sepals erect,
      spreading, or recurved . . . . *A. canadensis*
   2 Racemes usually drooping; petals 12–22 mm long; fruiting sepals
      tightly reflexed 3

3   Leaves at flowering <½ mature and without a reddish tinge,
    tomentose beneath; fruiting pedicel <2 cm long; fruit dryish,
    insipid . . . . *A. arborea*
3   Leaves at flowering at least ½ mature and usually with a reddish
    tinge, glabrous beneath; fruiting pedicel >2 cm long; fruit juicy,
    sweet . . . . *A. laevis*

A. arborea (F.Michx.) Fernald; Common S.; dry to mesic slopes and thickets; common; statewide, except for the western counties of W TN. *A. arborea* (F.Michx.) Fernald var. *austromontana* (Ashe) H.E.Ahles. Plate 39.

A. canadensis (L.) Medik., Canadian S.; open, upland woods; occasional; E TN. Plate 40.

A. laevis Wiegand, Smooth S.; dry to moist woods and thickets, balds, rocky openings, swamps; occasional; EHR, E TN, with a few records from W TN. *A. arborea* (F.Michx.) Fernald var. *laevis* (Wiegand) H.E.Ahles. Plate 41.

A. sanguinea (Pursh) DC., Red S.; non-calcareous substrates along riverbanks, rocky open woods, slopes; occasional; CU, VR. **Threatened.** Plate 42.

Fabaceae   *Amorpha* (False Indigo)

Shrubs; flowers dark purple, in showy, terminal racemes (spring–summer), fruit a green (becoming black) legume (summer–fall). Flowers and/or fruits normally required to separate the species.

1   Calyx and inflorescence pubescent; calyx lobes >0.8 mm long;
    legume curved . . . . *A. fruticosa*
1   Calyx and inflorescence glabrous; calyx lobes <0.8 mm long;
    legume usually straight . . . . *A. glabra*

A. fruticosa L., Tall F.I.; lake shores, old fields and thickets, especially in bottomlands; occasional; statewide. *A. nitens* F.E. Boynton, *A. tennesseensis* Shuttleworth. Plate 43.

A. glabra Desf. *ex* Poir., Mountain F.I.; old fields, mountain to bottomland woods, lake shores, streambanks, roadsides; occasional; M and E TN. Plate 44.

Vitaceae   *Ampelopsis*

Vines, climbing by tendrils; flowers white, in dichotomously branched cymes (spring–summer); fruit a berry with scant pulp (summer–fall). Related to grapes (*Vitis*), but with fewer tendrils, drier berries, white stem pith (brown-tan in grapes), and flowers in cymes (panicles in grapes).

1 Leaves compound (twice-pinnate or ternate); fruit black . . . . . . A. *arborea*
1 Leaves simple; fruit white to purple . . . . A. *cordata*

A. *arborea* (L.) Koehne, Peppervine; much-branched, bushy, sprawling, and climbing; swampy woods and river/reservoir margins; frequent; statewide, but records from M and E TN are apparently from cultivated/naturalized populations. Plate 45.

A. *cordata* Michx., Heart-Leaf P.; high climbing; mesic thickets and woods, often on alluvium; frequent; statewide, but apparently absent from much of the CP, EHR, and CU. Plate 46.

Araliaceae   *Aralia* (Devil's Walking Stick)

A. *spinosa* L.; shrubs or small trees; flowers whitish, in large panicles (early summer), fruit a purple-black drupe, in large terminal panicles (summer–fall); dry, open woods and thickets, forest borders; frequent; statewide, but not often seen in W TN. Stout prickles on the trunk, stems, and leaf stalks are characteristic. This species produces our largest compound leaf. Plate 47.

Aristolochiaceae   *Aristolochia* (Dutchman's Pipe)

High climbing, twining vines; flowers purplish brown, curved, dangling on pedicels and somewhat resembling a pipe, axillary and solitary (spring), fruit a capsule with numerous flat seeds (fall, persisting into winter).

1 Petioles and peduncles glabrous; each peduncle with a perfoliate bract
   near the base . . . . A. *macrophylla*
1 Petioles and peduncles tomentose; peduncle without a bract . . . . A. *tomentosa*

A. *macrophylla* Lam.; rocky woods and slopes; frequent; EHR, E TN, with an outlier in Giles Co. (WHR). A. *durior* Hill. Plate 48.

A. tomentosa Sims, Woolly D. P.; usually on alluvium, most often on river- and streambanks; occasional; M TN, with a few records from W TN, the CU and VR. Plate 49.

Rosaceae   *Aronia* (Chokeberry)

Shrubs; flowers white to pink-tinged, in a cyme (spring), fruit berry-like (summer–fall) and unpleasant to taste, hence the common name. The generic placement is unclear and species often are included under *Pyrus* or *Sorbus*. Hardin, J. W. 1973. The enigmatic chokeberries (*Aronia*, Rosaceae). Bull. Torrey Bot. Club 100:178–184.

1   Lower leaf surfaces, stems, rachis and pedicels pubescent; ripe fruit
    red . . . . *A. arbutifolia*
1   Lower leaf surfaces, stems, rachis and pedicels glabrous; ripe fruit
    black . . . . *A. melanocarpa*

A. arbutifolia (L.) Pers., Red C.; low woods, swamps, bogs, wet thickets; occasional; statewide, except absent from much of W TN and the CB. *Pyrus arbutifolia* (L.) L.f., *Sorbus arbutifolia* (L.) Heynh. Plate 50.

A. melanocarpa (Michx.) Elliott, Black C.; habitats similar to the first species, but sometimes found on drier sites and on bluffs and cliffs; distribution essentially the same as the preceding species. *A. prunifolia* (Marshall) Rehder, *Pyrus melanocarpa* (Michx.) Willd., *Sorbus melanocarpa* (Michx.) C.K.Schneid. Plate 51.

Annonaceae   *Asimina* (Pawpaw)

A. triloba (L.) Dunal; shrubs or small trees, often clonal; flowers purple-brown, appearing with the leaves and solitary (spring), fruit a banana-like berry, but rarely fruiting (late summer–fall); mesic woods, especially on alluvium; common; statewide. Leaves and stems with an unpleasant odor (much like fresh asphalt or diesel fuel) when crushed; ripe fruit very sweet with a custard-like pulp. Plate 52.

Berberidaceae   *Berberis* (Barberry)

Shrubs, usually with spines; flowers yellow, in umbels or racemes (spring), fruit a berry (summer).

1   Spines forking; leaves toothed; flowers in racemes . . . . *B. canadensis*

1   Spines unbranched; leaves entire; flowers solitary or in
     umbels .... *B. thunbergii*

B. canadensis Mill., Canada B.; rocky woods, usually near river- and streambanks;
occasional; E TN. **Special Concern.** Plate 53.

B. thunbergii DC.*, Japanese B.; commonly cultivated, rarely spreading to disturbed
woods and ruderal sites; infrequent; statewide, but most often seen in E TN. Plate 54.

Rhamnaceae   *Berchemia* (Rattan Vine/Supple-Jack)

B. scandens (Hill) K.Koch; high climbing, twining vines; flowers greenish white, in
panicles (spring), fruit a blackish drupe (late summer); rich woods and thickets,
usually in low ground; occasional; statewide, except U. Vines are used for making
rattan furniture. Plate 55.

Betulaceae   *Betula* (Birch)

Small or large trees, often with several stems from the base and usually with exfo-
liating bark; monoecious; staminate catkins formed in autumn and present during
winter, pistillate catkins appear with the leaves, fruit a samara (summer). Hardin,
J. W. 1952. The Juglandaceae and Corylaceae of Tennessee. Castanea 17:78–89;
Hardin, J. W. 1971. Studies of the southeastern United States flora. 1. Betulaceae.
J. Elisha Mitchell Sci. Soc. 87:39–41; Huber, F. C., J. A. DeLapp, and C. A.
Mitchell. 1977. *Betula papyrifera* var. *cordifolia* (Regel) Fern. in Tennessee. Castanea
42:324–325.

1   Bark and twigs with the noticeable odor of wintergreen when
     crushed 2
     2   Bark yellowish, scaly; margin of fruiting bracts ciliate; buds
          acute .... *B. alleghaniensis*
     2   Bark reddish, not scaly; margin of fruiting bracts glabrous; buds
          sharp-pointed .... *B. lenta*
1   Bark and twigs without the noticeable odor of wintergreen when
     crushed 3
     3   Leaf blade with 5–8 pairs of lateral veins; bark whitish to gray; a
          mountain species .... *B. cordifolia*
     3   Leaf blade with 8 or more pairs of lateral veins; bark reddish brown to
          blackish; a wide-ranging species .... *B. nigra*

B. alleghaniensis Britton, Yellow B.; streambanks, rich mesic woods; occasional; E TN. *B. lutea* F.Michx. Plate 56.

B. cordifolia Regel, Heart-Leaf/Mountain White B.; known from a north-facing slope at 4400 ft in the GSMNP, Sevier Co. (U). **Endangered.** *B. papyrifera* Marshall var. *cordifolia* (Regel) Fernald. Plate 57.

B. lenta L., Sweet B.; streambanks, rich mesic woods; occasional; E TN. Formerly the commercial source of wintergreen oil. Plate 58.

B. nigra L., River/Black B.; sloughs, and river, reservoir and stream margins; frequent; statewide. Plate 59.

Bignoniaceae   *Bignonia* (Crossvine)

B. capreolata L.; high-climbing vines with tendrils; flowers showy with corollas deep orange-reddish outside and pale orange-reddish inside, in axillary clusters (spring), fruit a flat capsule (summer–fall); rich woods and thickets, mostly in moist soils; common; statewide. *Anisostichus capreolata* (L.) Bureau. Plate 60.

Moraceae   *Broussonetia* (Paper Mulberry)

B. papyrifera (L.) Vent.*; shrubs or small trees; monoecious; staminate flowers in catkins, pistillate in globular heads (spring), fruit unknown in TN; persisting from plantings, usually in cemeteries and around old home sites, sometimes spreading from root sprouts; occasional; statewide. Asia. Plate 61.

Polygonaceae   *Brunnichia* (Ladies' Eardrops)

B. ovata (Walter) Shinners; thicket-forming vines with terminal tendrils; flowers in showy, drooping panicles (summer), fruit a winged nutlet (late summer–fall); bottomlands, swampy woods/thickets, stream/reservoir margins; common; W TN, WHR. *B. cirrhosa* Banks *ex* Gaertn. Plate 62.

Santalaceae   *Buckleya* (Pirate Bush/Sapsuck)

B. distichophylla (Nutt.) Torr.; hemiparasitic, freely branching shrubs; flowers green, in umbels (spring), fruit a drupe (summer); rich woods, streambanks, found mostly on the roots of *Tsuga canadensis* (Hemlock); infrequent; U. **Threatened.** Plate 63.

Sapotaceae   *Bumelia* (Southern Buckthorn)

B. lycioides (L.) Pers.; small trees, usually with thorny branches; flowers white, in clusters (spring), fruit a green to purple berry (summer); rich woods, thickets, fencerows, mostly in low ground; frequent; statewide except in the U, most often seen in M TN. Plate 64.

Ericaceae   *Buxella* (Box Huckleberry)

B. brachycera (Michx.) Small; evergreen shrubs, forming extensive colonies from underground stems; flowers pinkish, in axillary raceme-like panicles with deciduous bracts (spring), fruit a dark-blue, glaucous drupe (summer); wooded slopes and thickets; infrequent; northern CU. *Gaylussacia brachycera* A.Gray. Plate 65.

Verbenaceae   *Callicarpa* (American Beautyberry/French Mulberry)

C. americana L.; shrubs; flowers pinkish, in a many-flowered cyme (late spring), fruit a drupe (late summer); rich woods and thickets; frequent; counties across the southern two-thirds of the state, sometimes cultivated elsewhere. Plate 66.

Calycanthaceae   *Calycanthus* (Sweetshrub/Carolina Allspice)

Aromatic shrubs; flowers maroon, terminal and solitary (spring), fruit with numerous one-seeded achenes embedded within a leathery receptacle (fall, persisting into winter).

1   Twigs, petioles, and lower leaf surfaces
      pubescent . . . . *C. floridus* var. *floridus*
1   Twigs, petioles, and lower leaf surfaces glabrous, or with a
      few scattered hairs . . . . *C. floridus* var. *glaucus*

C. floridus L. var. *floridus*; mesic woods, thickets, and streambanks; occasional; WHR, EHR, and E TN. Plate 67.

C. floridus L. var. *glaucus* (Willd.) Torr. & A.Gray; mesic woods, thickets, and streambanks; occasional; E TN. *C. fertilis* Walter, *C. nanus* (Loisel.) Small; *C. floridus* L. var. *laevigatus* (Willd.) Torr. & A.Gray; *C. floridus* L. var. *oblongifolius* (Nutt.) Boufford & Spongberg

Menispermaceae   *Calycocarpum* (Cupseed)

C. lyonii (Pursh) A.Gray; high-climbing vines; flowers greenish white, in panicles (late spring–summer), fruit a black drupe (fall); mesic thickets and woods, most often in bottomlands; occasional; statewide except the U. The seed of each drupe is hollowed out (cup-like) on one side. Plate 68.

Bignoniaceae   *Campsis* (Trumpet Creeper/Trumpet Vine)

C. radicans (L.) Seem. *ex* Bureau; vines, climbing by aerial rootlets; flowers large and showy with tubular orange-scarlet corollas, in terminal corymbs (spring–summer), fruit an elongated woody capsule (summer, persisting into winter); low woods and thickets, fencerows, often covering tobacco barns; common; statewide. Sometimes cultivated as "humming-bird plant." Plate 69.

Betulaceae   *Carpinus* (Blue Beech/Ironwood/Hornbeam)

C. caroliniana Walter; small to medium trees; monoecious; staminate flowers in catkins, pistillate flowers clustered (spring), fruit a nutlet surrounded by leaf-like bracts (summer–fall); mesic slopes, low woods and streambanks; common; statewide. The smooth bark is quite like that of American beech but with "muscular" ridges. The more northern ssp. *virginiana* (Marshall) Furlow is sometimes separated from the more southern ssp. *caroliniana* on the basis of leaf characters; these appear to intergrade in TN. Plate 70.

Juglandaceae   *Carya* (Hickory)

Medium to large trees; monoecious; staminate flowers in elongated pendulous catkins, pistillate in terminal, few-flowered spikes (spring), fruit a nut surrounded by a dry husk (late summer–fall). Hickories are known for their tough, durable wood, which is highly desirable for lumber and historically has been used for implement parts and firewood. Numerous hybrids, including some named taxa, occur throughout the state and often make identification tenuous, especially without mature fruit, leaf, bud, and bark characteristics. Hardin, J. W. 1952. The Juglandaceae and Corylaceae of Tennessee. Castanea 17:78–89.

1 Scales of terminal buds valvate (not overlapping); fruit husks winged; kernel bitter or sweet 2
  2 Terminal buds sulfur yellow; kernel bitter . . . . *C. cordiformis*
  2 Terminal buds yellowish or reddish brown to black; kernel bitter or sweet 3
    3 Leaflets normally 11 or more; kernel sweet . . . . *C. illinoinensis*
    3 Leaflets normally 9 or fewer; kernel bitter . . . . *C. aquatica*
1 Scales of terminal buds imbricate (overlapping); fruit husks not winged; kernel sweet 4
  4 Leaflets 5 (3–7), margins with persistent tufts of hairs near the tips of the teeth; bark exfoliating in long strips or plates 5

5 Leaflets hirsute and with conspicuous peltate scales on the lower
   surfaces; twigs stout and hairy, rarely blackening on drying;
   terminal buds tan to dark brown, tomentose .... *C. ovata* var. *ovata*
5 Leaflets mostly glabrous and with a few peltate scales on the
   lower surfaces; twigs slender and glabrous, mostly blackening
   on drying; terminal buds reddish brown, mainly
   glabrous .... *C. ovata* var. *australis*
4 Leaflets 3–9, margins glabrous or ciliate but without hairs grouped
   into tufts; bark furrowed or exfoliating in long strips or plates 6
   6 Twigs stout; terminal buds >1.2 cm long; leaflets mostly 7–9,
      hirsute on lower surface with abundant hairs; bark exfoliating or
      furrowed 7
      7 Petiole and rachis lightly pubescent; leaflets acuminate; husks
         minutely hirsute; bark exfoliating .... *C. laciniosa*
      7 Petiole and rachis densely hirsute; leaflets acute; husks
         glabrous; bark ridged but not exfoliating .... *C. tomentosa*
   6 Twigs slender; terminal buds <1.2 cm long; leaflets 3–7, usually
      glabrous on lower surface except near the midrib; bark furrowed
      or smooth 8
      8 Leaf rachis and leaflet veins woolly; buds and fruits covered
         with white to yellow-golden scales .... *C. pallida*
      8 Leaf rachis and leaflet veins glabrous, occasionally with axillary
         hairs; buds and fruits without scales .... *C. glabra*

C. aquatica (F.Michx.) Nutt., Water H./Bitter Pecan; river floodplains, streambanks, bluffs, levees; common; W TN, more often seen toward the Mississippi River. Plates 71, 80a.

C. cordiformis (Wangenh.) K.Koch, Bitternut H.; floodplains, riparian forests, ravines, mesic slope forests; common; statewide. Plates 72, 80b.

C. glabra (Mill.) Sweet, Pignut/Red H.; dry, rocky sites, especially ridge and upper slope forests; common; statewide. *C. glabra* (Mill.) Sweet var. *odorata* (Marshall) Little, *C. ovalis* (Wangenh.) Sarg. Plates 73, 80c.

C. illinoinensis (Wangenh.) K.Koch, Pecan; streambanks, floodplains, ravines, lower slopes; occasional; W TN and WHR; records eastward are probably from cultivated material. Plates 74, 80d.

C. laciniosa (F.Michx.) Loudon, Big Shellbark H./Kingnut; riparian, bottomlands, ravine and lower slope forests; occasional; statewide. Plates 75, 80e.

C. *ovata* (Mill.) K.Koch var. *australis* (Ashe) Little, Carolina H.; bottomlands, slope and ravine forests; occasional; statewide, but with few occurrences in W TN and the U. C. *carolinae-septentrionalis* (Ashe) Engl. & Graebn. Plates 76, 80f.

C. *ovata* (Mill.) K.Koch var. *ovata*, Shagbark H.; bottomlands, ravine, slope and ridge forests; common; statewide. C. *ovata* (Mill.) K.Koch var. *pubescens* Sarg. Plates 77, 80g.

C. *pallida* (Ashe) Engl. & Graebn., Sand H.; well-drained sandy or rocky soils on bluffs, ridges and slopes; common; E TN, with scattered occurrences westward. Plates 78, 80h.

C. *tomentosa* (Poir.) Nutt., Mockernut H.; well-drained slope and ridge forests, often in xeric sites; common; statewide. C. *alba* (L.) K.Koch. Plates 79, 80i.

Fagaceae   *Castanea* (Chestnut)

Shrubs or trees; monoecious; staminate flowers in catkins, pistillate flowers one-several near the base of the catkins (spring); fruit a nut, up to three within a spiny involucre or bur (summer–fall).

1   Lower leaf surfaces glabrous; each bur with more than
    one nut . . . . C. *dentata*
1   Lower leaf surfaces tomentose; each bur with one nut . . . . C. *pumila*

C. *dentata* (Marshall) Borkh., American C.; formerly large trees; mesic to dry slope and ridge forests; frequent (formerly); statewide. **Special Concern.** Once valuable for lumber and nuts, but now mostly persisting as stump sprouts due to the devastating chestnut blight. The sprouts only occasionally fruit. Plate 81.

C. *pumila* (L.) Mill., Allegheny Chinquapin; shrub or small trees; dry slope and ridge forests; common; EHR, E TN. Plate 82.

Bignoniaceae   *Catalpa* (Catalpa/Cigar Tree)

Medium to large trees; flowers white with purple-yellow spots, in terminal panicles (spring), fruit an elongated, cigar-like capsule (summer–fall). Historically and currently cultivated, thus the natural range of both species is uncertain. Regularly defoliated by "catalpa-worms" (catalpa sphinx, the larval form of an American hawk moth).

1 Leaf apex long-tapering; corolla >4 cm broad, the lower lobe notched;
   seed wings mostly rounded at the tips, the tuft of hairs separate or
   slightly fused at the base .... *C. speciosa*
1 Leaf apex short acuminate; corolla 2–4 cm broad, the lower lobe entire;
   seed wings pointed at the tips, the tuft of hairs mostly fused at
   the base .... *C. bignonioides*

C. bignonioides Walter*, Southern C.; cultivated and escaping to vacant lots and other ruderal areas; infrequent; statewide. Apparently native well to the south of TN. Plate 83.

C. speciosa (Warder *ex* Barney) Engelm., Northern C.; mesic woods and thickets; frequent; W TN, where it is apparently native; planted, persisting, and naturalized statewide. Plate 84.

Rhamnaceae   *Ceanothus* (New Jersey Tea)

C. americanus L.; low shrubs; flowers whitish, in terminal panicles or corymbs (early summer), fruit a drupe (summer); dry, open woods, thickets, roadsides; frequent; statewide, fewest records from W TN. Plate 85.

Celastraceae   *Celastrus* (Bittersweet)

Twining, sometimes sprawling vines; flowers greenish, racemose (spring); fruit a globular, leathery, orange-yellow capsule splitting to expose the seeds which have fleshy, scarlet coverings (summer–fall). Both species are often cultivated for their brilliant fruit/seeds.

1 Inflorescences terminal; leaves elliptic to obovate;
   native .... *C. scandens*
1 Inflorescences axillary; leaves roundish to obovate;
   introduced .... *C. orbiculatus*

C. orbiculatus Thunb.*, Oriental B.; planted, persisting, naturalized; occasional; statewide. Asia. Plate 86.

C. scandens L., American B.; rocky slopes, streambanks, thickets; frequent; state-wide, but uncommon in M TN. Plate 87.

Ulmaceae   *Celtis* (Hackberry/Sugarberry)

Trees (sometimes shrubby) with grayish and often conspicuously warty bark; monoecious; flowers greenish, axillary, pistillate and staminate solitary or clustered (spring), fruit a variously colored drupe (late summer–fall). Identification is not always easy since the group is taxonomically complex, morphologically variable, and apparently hybridization is common.

1 Leaf margins mostly entire or with a few scattered large teeth; blades elliptic to ovate-lanceolate . . . . *C. laevigata*
1 Leaf margins conspicuously to sparingly but at least partially serrate; blades ovate 2
   2 Leaf margins serrate toward apex only . . . . *C. tenuifolia*
   2 Leaf margins serrate to well below the middle . . . . *C. occidentalis*

C. laevigata Willd., Southern H./S.; various habitats including floodplains and rocky woods but also in fencerows and old fields; common; statewide. Plate 88.

C. occidentalis L., Northern H.; rich woods, bottomlands, rocky slopes, especially in calcareous regions; frequent; statewide, but more scattered than the previous species. Plate 89.

C. tenuifolia Nutt., Dwarf H.; occasional; statewide, but rare in W TN. *C. georgiana* Small. Plate 90.

Rubiaceae   *Cephalanthus* (Buttonbush)

C. occidentalis L.; much-branched shrubs, often thicket forming; flowers white, in spherical peduncled heads (late spring–summer), fruit an indehiscent nutlet (fruiting heads summer–winter); margins of swamps and reservoirs (often forming impenetrable thickets along the outer edge of reservoir fluctuation zones), around ponds, streambanks and other wetlands; common; statewide. Plate 91.

Fabaceae   *Cercis* (Redbud)

C. canadensis L.; small trees; flowers pinkish, appearing before the leaves in umbel-like clusters (early spring), fruit a flat legume (summer–winter); mesic woodlands, thickets and successional fields, more often on limestone; common; statewide. Regularly planted. Plate 92.

Ericaceae   *Chimaphila* (Spotted Wintergreen/Pipsissewa)

C. maculata (L.) Pursh; low, nearly herbaceous, evergreen subshrubs; flowers white, fragrant, single or more often in groups up to 5 on upright, naked peduncles (summer), fruit a capsule (summer–fall); a variety of habitats including dry hardwood or pine-hardwood forests, ravines, streambanks, upland wet woods; common; statewide, except infrequent in W TN. Leaves are alternate but crowded and often appear to be whorled. Plate 93.

Oleaceae   *Chionanthus* (Fringe-Tree/Old Man's Beard)

C. virginicus L.; tall shrubs or small trees; flowers whitish in open panicles (spring), fruit a bluish purple drupe (summer); mesic, often bluffy woods, thickets, streambanks; frequent; E TN, with a few occurrences in M TN. Sometimes planted. Plate 94.

Fabaceae   *Cladrastis* (Yellow Wood)

C. kentukea (Dum.Cours.) Rudd; small to medium trees; flowers white, in showy panicles (spring), fruit a linear legume (summer); rich, mesic, usually calcareous woods; occasional; statewide, but apparently absent from many counties in the CB, VR, and U. Sometimes planted. *C. lutea* (F.Michx.) K.Koch. Plate 95.

Ranunculaceae   *Clematis* (Clematis/Virgin's Bower)

Vines, sometimes woody at base only, usually clambering over other plants; flowers white, in panicled heads (summer–fall); fruit a flattened achene with an elongated, persistent, and plumose style, in heads (summer–fall). The genus includes several herbaceous taxa and large-flowered cultivars are regularly seen in plantings. Essig, F. B. 1990. The *Clematis virginiana* (Ranunculaceae) complex in the southeastern United States. Sida 14:49–68.

1 Leaflet margins entire, sometimes lobed . . . . *C. terniflora*
1 Leaflet margins toothed 2
   2 Leaves 3-foliolate; fruits >35/head . . . . *C. virginiana*
   2 Leaves 5-foliolate to biternate; fruits <35/head . . . . *C. catesbyana*

C. catesbyana Pursh, Catesby's V.B.; open, disturbed sites, usually on limestone; common; M TN and on the CU, most often seen in the CB. Plate 96.

C. terniflora DC.*, Sweet Autumn C.; cultivated and naturalized in fencerows and other disturbed sites; occasional; M and E TN. Asia. *C. dioscoreifolia* H.Lév. & Vaniot, *C. maximowicziana* Franch. & Sav., *C. paniculata* Thunb. Plate 97.

C. virginiana L.; moist, disturbed, open sites, especially fencerows, roadsides, and along streams; common; statewide. Plate 98.

Clethraceae   *Clethra* (White Alder/Pepper-Bush)

Shrubs, rarely small trees; flowers white, in racemes (summer); fruit a hairy capsule (summer–fall).

1 Leaves acuminate, oblong-elliptic, 8–20 cm long . . . . *C. acuminata*
1 Leaves acute, oblong-obovate, <10 cm long . . . . *C. alnifolia*

C. acuminata Michx., Mountain W. A.; rich mountain woods; frequent; upper CU and U. Plate 99.

C. alnifolia L., Coastal W. A.; wet woods and thickets; Coffee Co. (EHR). **Threatened.** Plate 100.

Menispermaceae   *Cocculus* (Coralbeads/Snailseed)

C. carolinus (L.) DC.; sprawling and climbing (twining) vines; flowers greenish, in racemes (summer), fruit a bright red drupe, in showy clusters (late summer–fall); mesic woods and thickets, fencerows; frequent; statewide, with fewest records from W TN and the CU. The "seed" of each drupe is ridged and coiled, resembling a snail. Plate 101.

Myricaceae   *Comptonia* (Sweet Fern)

C. peregrina (L.) J.M.Coult.; low, fragrant shrubs; monoecious or dioecious; staminate flowers in flexuous catkins, pistillate in globular catkins (spring), fruit a hard nutlet surrounded by long persistent bracts (spring–summer); gravel and boulder bars along and in the Big South Fork of the Cumberland River, Scott Co. (CU). **Endangered.** Plate 102.

Lamiaceae   *Conradina* (Cumberland Rosemary)

C. verticillata Jennison; low, branching shrubs with lower branches often rooting; flowers in upper leaf axils, corollas lavender, spotted within (spring), fruit a nutlet (spring–summer); sandy and gravelly creek/riverbanks and bars; infrequent; EHR, CU. **Threatened federally and in TN.** Plate 103.

Cornaceae   *Cornus* (Dogwood)

Shrubs or small trees; flowers white to cream-colored, in cymes or heads (spring), fruit a drupe, often colorful (summer–fall). Interrelationships, often resulting from hybridization, between and within non-bracted species groups are yet unclear and the taxonomy/nomenclature remain confusing. Wilson, J. S. 1965. Variation of three taxonomic complexes of the genus *Cornus* in eastern United States. Trans. Kansas Acad. Sci. 64:747–817.

1   Leaves or leaf scars alternate; fruit blue-black . . . . *C. alternifolia*
1   Leaves or leaf scars opposite; fruit red, blue, or white 2
   2   Upper leaf surface very rough to touch (like fine sandpaper); fruit
    white . . . . *C. drummondii*
   2   Upper leaf surface smooth to touch; fruit red or bluish 3
      3   Pith of second-year stems white; leaves lanceolate or ovate-
      lanceolate . . . . *C. foemina*
      3   Pith of second-year stems tan or brown; leaves mostly ovate 4
         4   Inflorescence subtended by 4 showy bracts; fruit red; hairs on
         lower leaf surface white; young stems greenish brown; plants
         of mesic to dry sites . . . . *C. florida*
         4   Inflorescence lacking showy bracts; fruit bluish; hairs on
         lower leaf surface brownish; young stems reddish; plants of
         wet sites . . . . *C. amomum*

C. alternifolia L.f., Alternate-Leaf D; dry woods, rocky slopes; frequent; EHR, E TN, rare on the WHR and in the CB. Plate 104.

C. amomum Mill., Silky D.; swamps, streambanks, reservoir shorelines, wet woods and fields; common; statewide, but with few records from W TN. Plate 105.

C. drummondii C.A.Mey., Rough-Leaf D.; swamps, streambanks, reservoir shorelines, wet woods and fields; occasional; W and M TN, with a few scattered occurrences eastward to Roane Co. (VR). Plate 106.

C. florida L., Flowering D.; rich woods, fencerows, successional fields; common; statewide. Forma *rubra* (Weston) Palmer & Steyerm. is a common cultivar with pink-red bracts. Plate 107.

C. foemina Mill., Stiff D.; swamps, streambanks, reservoir shorelines, wet woods and fields; common; statewide except in the U. *C. stricta* Lam. Plate 108.

Betulaceae   *Corylus* (Hazelnut)

Shrubs; monoecious; staminate catkins develop in autumn and flowers open in spring, clustered pistillate flowers develop in spring; the fruit is an edible nut enclosed in a tight-fitting husk that consists of two bracts (late summer–fall). Commercial filberts and hazelnuts are members of this genus and are cultivated in various parts of the world, including the United States. Hardin, J. W. 1952. The Juglandaceae and Corylaceae of Tennessee. Castanea 17:78–89; Hardin, J. W. 1971. Studies of the southeastern United States flora. 1. Betulaceae. J. Elisha Mitchell Sci. Soc. 87:39–41.

1   Twigs and petioles bristly with glandular hairs (10x); fruiting bracts downy, separate and not beaked . . . . *C. americana*
1   Twigs and petioles mostly glabrous or at most with a few glandular hairs; fruiting bracts bristly, connate, and with a beak 2–5x the fruit length . . . . *C. cornuta*

C. americana Walter, American H.; thickets, old fields, woodlands, often along gully banks; common; statewide, with fewest records from the CB and U. Plate 109.

C. cornuta Marshall, Beaked H.; rich woods, thickets, forest borders; occasional; E TN. Plate 110.

Anacardiaceae   *Cotinus* (Smoke-Tree)

C. obovatus Raf.; shrubs or small trees with odoriferous yellow wood; flowers whitish yellow, in loose panicles (spring), fruit a drupe (summer); calcareous rocky woods and bluffs; infrequent; CU. **Special Concern.** The smoky appearance is due to the hairy, filamentous fruit stalks. Plate 111.

Rosaceae   *Crataegus* (Hawthorn/Haw/Thornapple)
Contributed by Ron L. Lance

Large shrubs or small trees; flowers white, in terminal corymbose clusters or sometimes only 1–3 flowered (spring–early summer), fruit a green, yellow, or red pome (summer–fall). A taxonomically difficult genus with many poorly understood taxa previously described, additionally complicated by polyploidy, aneuploidy, hybridization, and apomixis. A simplistic approach to the burdensome taxonomy is to accept a broad definition of one or two "major" species within each of the 13 series known to occur in Tennessee, including as synonyms all seemingly "minor" species that are most closely related. Phipps, J. B., P. G. Robertson, P. G. Smith, and J. Rohrer. 1990. A checklist of the subfamily *Maloideae* (*Rosaceae*). Can. J. Bot. 68:2209–2269; Lance, R. L. 1997. The hawthorns of the southeastern United States. Printed by the author, Fletcher, N.C.

In the key, all references to leaves infer those of short-shoots (floral leaves), unless otherwise noted. Leaves on other shoots such as elongating terminal twigs, sprouts, and juvenile plants typically exhibit wider variation and will not necessarily conform to the descriptions used. The criterion of "elevated calyces" refers to the trait of mature fruit appearing to have a short neck which bears the calyx lobes. Additional field collections and observations are critical to a better understanding of *Crataegus* taxonomy. Collections ideally should include flowers and fruits from the same plant.

1   Leaf veins extending to sinuses as well as to points of lobes 2
   2   Leaves very thin, deeply incised-lobed; fruit ellipsoid . . . . . *C. marshallii*
   2   Leaves firm, shallowly lobed or maple-like; fruit globose 3
      3   Leaves spatulate, <2 cm wide, base cuneate . . . . *C. spathulata*
      3   Leaves ovate, 2–5 cm wide, base not cuneate . . . . . . *C. phaenopyrum*
1   Leaf veins extending only to points of lobes or to larger teeth 4
   4   Leaves mostly ±3 cm long . . . . *C. uniflora*
   4   Leaves >3 cm long 5
      5   Petioles and lowermost margin of blades distinctly glandular,
         especially early in the season 6
         6   Leaves conspicuously lobed on terminal shoots; calyx elevated
            on fruit . . . . *C. intricata*
         6   Leaves shallowly lobed or merely toothed; calyx sessile on
            fruit 7
            7   Fruit 2–3-seeded; seeds channeled or hollowed on inner
               face; flowering May-June, after leaves are essentially full
               size . . . . *C. calpodendron*
            7   Fruit 3–5-seeded; seeds flat on inner face; flowering
               April–May, with expanding leaves 8
               8   Flower pedicels with spreading or pilose pubescence and
                   conspicuous stalked glands; calyx lobes deeply glandular-
                   serrate, persist on fruit; mature leaves pubescent above

and below; veins not conspicuously impressed on upper
surface . . . . *C. harbisonii*
  8 Flower pedicels appressed-pubescent to glabrate; all
     parts sparingly glandular; calyx lobes remotely toothed,
     usually deciduous from mature fruit; mature leaves
     pubescent below only, becoming glabrate; veins
     impressed on upper surface . . . . *C. collina*
 5 Petioles and lowermost margin of blades eglandular, or with
    only a few remote glands in spring 9
   9 Leaf bases cuneate or wedge-shaped 10
     10 Leaf veins impressed above, prominent
        beneath . . . . *C. punctata*
     10 Leaf veins not as above, often obscure 11
        11 Leaves shallowly lobed, widest at or below middle;
           fruit soft mature . . . . *C. viridis*
        11 Leaves rarely lobed, widest above the middle; fruit mealy,
           hard or dry when mature . . . . *C. crus-galli*
   9 Leaf bases rounded, flattened, or cordate, occasionally abruptly
      narrowed 12
     12 Petioles and lower leaf surfaces closely and softly
        pubescent . . . . *C. mollis*
     12 Petioles and lower leaf surfaces sparsely pubescent
        to glabrous (but young leaves may be scabrate
        above) 13
        13 Mature fruit succulent; calyx sessile . . . . . . *C. macrosperma*
        13 Mature fruit hard, dry or mealy; calyx
           elevated . . . . *C. pruinosa*

C. calpodendron (Ehrend.) Medik., Pear H.; calcareous slopes, rocky woods, stream-sides; infrequent; M and E TN. *C. tomentosa* L., *C. chapmanii* (Beadle) Ashe. Plate 112.

C. collina Chapm., Hill H.; moist lowlands to hills and rocky slopes; infrequent; M and E TN except the EHR and the U. Plate 113.

C. crus-galli L., Cockspur H.; thin woodlands, fields, thickets; frequent; statewide. *C. algens* Beadle, *C. arborea* Beadle, *C. effulgens* Sarg., *C. fecunda* Sarg., *C. mohrii* Beadle. This is essentially a glabrous species, but in cases where pedicels and leaves are pubescent, referral to a second species in this series, *C. berberifolia* Torr. & A.Gray, may be warranted; including *C. engelmanii* Sarg., *C. sinistra* Beadle, *C. tetrica* Beadle, *C. torva* Beadle. Plate 114.

C. harbisonii Beadle, Harbison's H.; limestone hills and thickets, calcareous or basic soils; very rare, Davidson Co. (CB); also collected in 1948 in Obion (MV), Shelby, and Weakley Counties (CP) and Lawrence Co. (southern WHR), but no extant specimens have been located and these populations are presumed extirpated. **Endangered.** *C. ashei* Beadle. Plate 115.

C. intricata Lange, Entangled H.; dry woods, hills, fields, and rocky slopes; occasional in M and E TN. A highly variable and poorly understood series with at least 15 described taxa possibly represented in the state, here considered in synonomy. *C. biltmoreana* Beadle, *C. boyntonii* Beadle, *C. rubella* Beadle, *C. sargentii* Beadle, *C. straminea* Beadle. Plate 116.

C. macrosperma Ashe, Fanleaf H.; mountain uplands and moist woods; occasional in acid to basic soils, infrequent to rare in calcareous sites; statewide, but rare in the CP. This species has been synonomized under *C. flabellata* (Bosch.) K.Koch in some texts. *C. roanensis* Ashe, *C. basilica* Beadle. Plate 117.

C. marshallii Eggl., Parsley Hawthorn.; dry to moist woodlands; infrequent; primarily on the CP, southern WHR and in the VR. Plate 118.

C. mollis (Torr. & A.Gray) Scheele, Downy H.; bottomlands or basic to calcareous uplands; infrequent, WHR, CB. *C. gravida* Beadle, *C. cibaria* Beadle. Plate 119.

C. phaenopyrum (L.f.) Medik., Washington H.; fields, openings, thin woodlands; occasional; CP, WHR, CB, CU. Commonly planted and apparently spreading in vicinity of cultivation. *C. youngii* Sarg. Plate 120.

C. pruinosa (H.L.Wendl.) K.Koch, Frosted H.; low to upland woods, occasional; statewide; Normally a glabrous species, but pubescent in var. *virella* (Ashe) Kruschke. *C. gattingeri* Beadle. Plate 121.

C. punctata Jacq., Dotted H.; moist uplands; frequent; higher elevations of the U, also known from the CB, EHR, and CU. *C. amnicola* Beadle, *C. verruculosa* Sarg. Plate 122.

C. spathulata Michx., Little-Hip H.; low woods and pine-hardwood forests; infrequent; WHR, CU, VR. Plate 123.

C. uniflora Münchh., One-Flower H.; xeric, rocky and sandy soils; occasional; WHR, EHR, VR, U. Plate 124.

C. viridis L., Green H.; moist woodlands; frequent in lowlands in W TN, infrequent M and E TN. Cultivated fairly commonly. *C. ingens* Beadle, *C. interior* Beadle, *C. lanceolata* Sarg. Plate 125.

Euphorbiaceae   *Croton* (Alabama Croton)

C. alabamensis E.A.Sm. *ex* Chapm.; broad-crowned shrubs with overwintering leaves and noticeably silvery-scaly new shoots; monoecious; flowers pale green, in terminal racemes (spring), fruit a capsule (summer); calcareous bluffs, wooded ravines and slopes; reported from Coffee Co. (EHR). **Endangered—Possibly Extirpated.** Otherwise known from a few sites in middle Alabama. Plate 126.

Lythraceae   *Decodon* (Swamp Loosestrife)

D. verticillatus (L.) Elliott; shrubs, often woody at base only; flowers pink to purple, clustered in upper leaf axils (late summer), fruit a capsule (fall); swamps and other shallow waters; occasional; MV, CB, EHR, VR. Plate 127.

Saxifragaceae   *Decumaria* (Climbing Hydrangea)

D. barbara L.; vines, sprawling or more usually high-climbing by aerial roots; flowers white and fragrant, in terminal cymes (spring), fruit a capsule (summer); rich woodlands, especially in wet or swampy sites; occasional; W TN, concentrated toward the TN River, and from the southern tier of counties eastward to Polk and then north to Blount. Plate 128.

Caprifoliaceae   *Diervilla* (Bush Honeysuckle)

Shrubs; flowers pale yellow, becoming darker (often reddish) with age, in terminal or axillary cymes (spring), fruit a capsule (summer). All eastern North American species of Bush-Honeysuckle occur in TN but are generally rare. Hardin, J. W. 1968. *Diervilla* (Caprifoliaceae) of the southeastern U.S. Castanea 33:31–36.

1   Petioles >5 mm; leaves ciliate on the margins; twigs round in cross
    section . . . . *D. lonicera*
1   Petioles <5 mm (or absent); leaves eciliate on the margins; twigs
    quadrangular in cross section. 2
    2   Branchlets, leaves, pedicels, and calyx densely
        pubescent . . . . *D. rivularis*
    2   Branchlets, leaves, pedicels, and calyx essentially
        glabrous . . . . *D. sessilifolia*

D. lonicera Mill.; dry woods and thickets, creek sides, rocky slopes; infrequent; WHR, CU, and U. **Threatened.** Plate 129.

D. rivularis Gatt.; damp, rocky woods, streambanks; infrequent; E TN. **Threatened.**
D. sessilifolia Buckley var. *rivularis* (Gatt.) H.E.Ahles.

D. sessilifolia Buckley; rocky slopes and streambanks, mostly >3000 ft; infrequent;
E TN. Plate 130.

Ebenaceae   *Diospyros* (Persimmon)

D. virginiana L.; small to medium trees with heavy and hard wood; polygamous,
often monoecious; flowers pale yellow, staminate flowers often clustered, pistillate solitary (spring), fruit a plum-like berry (fall); old fields, fencerows, dry to
mesic woods, thickets; common; statewide. Fruits are edible when fully ripe, usually after frost. *D. virginiana* var. *pubescens* (Pursh) Dippel. Plate 131.

Thymelaeaceae   *Dirca* (Leatherwood)

D. palustris L.; shrubs; flowers yellow, in lateral clusters (spring), fruit a greenish
drupe (summer); rich woods, often in ravines and along streams; frequent; M and
E TN. The wood is brittle but the bark is very tough and flexible. Plate 132.

Elaeagnaceae   *Elaeagnus* (Autumn/Russian Olive)

Evergreen or deciduous shrubs or small trees; flowers white-yellow, axillary and
solitary or in umbellate clusters, fruit a fleshy achene (see key for seasons); planted
for ornament and wildlife food.

1   Leaves evergreen; lower leaf surfaces with silvery scales interspersed
    with reddish brown scales; fall flowering, fruiting the following
    spring . . . . *E. pungens*
1   Leaves deciduous (often late falling); lower leaf surfaces mostly with
    silvery scales but with a few scattered brown scales; spring flowering,
    fruiting the following summer–fall . . . . *E. umbellata*

E. pungens Thunb.*, Autumn O.; planted, persisting, rarely escaping; occasional;
statewide. Asia. Plate 133.

E. umbellata Thunb.*; Russian O.; planted, naturalized, old fields, roadsides, other
disturbed sites; occasional; statewide. Seen more often than the preceding species.
Asia. Plate 134.

Ericaceae   *Epigaea* (Trailing Arbutus)

E. repens L.; prostrate, trailing, evergreen subshrubs; flowers fragrant, white to pink, in axillary and terminal clusters (early spring), fruit a thin-walled, hairy capsule (spring–summer); rocky, xeric, pine or mixed pine-hardwood forests; frequent; E TN, and a single historic record from the WHR. Plate 135.

Celastraceae   *Euonymus*

Shrubs or small trees; flowers greenish purple, in axillary cymes (spring), fruit a lobed capsule that opens to expose the showy, normally red, aril-covered seeds (late summer–fall).

1 Flowers 4-parted; fruit smooth 2
    2 Young stems with 2–4 corky wings; petioles <5 mm long; leaves
       turning bright red in autumn; introduced . . . . *E. alatus*
    2 Young stems without wings; petioles >5 mm long; leaves turning
       yellow in autumn; native . . . . *E. atropurpureus*
1 Flowers 5-parted; fruit warty 3
    3 Leaves acuminate-pointed, sessile; upright
       shrubs . . . . *E. americanus*
    3 Leaves obtuse, petiolate; procumbent shrubs . . . . *E. obovatus*

E. alatus (Thunb.) Siebold*, Burning Bush; widely cultivated and occasionally escaping into fencerows, disturbed woods, and vacant lots; statewide. Asia. Plate 136.

E. americanus L., Hearts-a-Busting/Strawberry Bush; rich woods, ravines, streambanks; common; statewide. Often scarce due to browsing by deer. Plate 137.

E. atropurpureus Jacq., Wahoo/Burning Bush; rich woods and thickets; occasional; statewide, but most common in M TN. Plate 138.

E. obovatus Nutt., Running Strawberry Bush; rich, damp woods and thickets; infrequent; VR, U, with outliers on the WHR and EHR. **Special Concern.** Plate 139.

Fagaceae   *Fagus* (American Beech)

F. grandifolia Ehrh.; large, often hollow trees; monoecious; staminate flowers in a pendulous, rounded head, pistillate flowers paired on a short peduncle (spring), fruit a triangular nut, usually two in a prickly involucre (summer–winter); rich woods; common; statewide. *F. grandifolia* Ehrh. var. *caroliniana* (Loudon) Fernald & Rehder. Plate 140.

Oleaceae   *Forestiera* (Swamp Privet)

Shrubs; dioecious or polygamous; flowers greenish yellow, in fascicles or panicles (spring); fruit a drupe (summer).

1 Leaf tip acuminate; leaves 4–8 cm long; plants of wet
   sites .... *F. acuminata*
1 Leaf tip rounded to acute; leaves 1–4 cm long; plants of dry
   sites .... *F. ligustrina*

F. acuminata (Michx.) Poir., Eastern S.P.; swamps, low woods; infrequent; more common in W TN, but known from a few populations in M TN and on the CU. Plate 141.

F. ligustrina (Michx.) Poir., Upland S.P.; dry glades, woods, and bluffs; infrequent; M TN, CU, but most common in the CB. Plate 142.

Oleaceae   *Forsythia* (Forsythia/Golden Bells)

Forsythia viridissima Lindl.*; shrubs, usually with multiple stems; flowers showy, yellow, axillary and solitary or in clusters (early spring), fruit a capsule but rarely fruiting; widely cultivated and persisting around old home sites, spreading by root sprouts into nearby fencerows and thickets; common; statewide. Asia. Plate 143.

Hamamelidaceae   *Fothergilla* (Witch Alder)

F. major (Sims) Lodd.; low shrubs; flowers whitish, in terminal, densely flowered spikes (early spring), fruit an ovoid capsule (spring); dry woods and thickets; infrequent; E TN. **Threatened.** Plate 144.

Oleaceae   *Fraxinus* (Ash)

Medium to large trees; polygamous; flowers in crowded fascicles, panicles, or racemes on the preceding year's wood (spring), fruit a flattened samara (summer–fall). American Ash is an especially valuable lumber tree and historically has been the wood used to make baseball bats. Hardin, J. W. 1974. Studies of the southeastern United States flora. IV. Oleaceae. Sida 5:274–285.

1 Twigs quadrangular (nearly square) in cross-section ..... *F. quadrangulata*
1 Twigs round in cross-section 2

2 Leaflets pale beneath; wing of samara terminal or decurrent only
along the upper ⅓ . . . . F. americana
2 Leaflets green beneath; wing of samara decurrent to near the middle
of fruit body or beyond 3
   3 Petiolule of lower leaflets winged nearly to base; samara wing <7
   mm wide; fruit body <2 mm wide . . . . F. pennsylvanica
   3 Petiolule of lower leaflets wingless; samara wing >7 mm wide; fruit
   body >2 mm wide . . . . F. profunda

F. americana L., American/White A.; rich, mesic woods, ranging from bottomlands
to slopes; common; statewide. *F. americana* L. var. *biltmoreana* (Beadle) J.W.Wright
ex Fernald. Plate 145.

F. pennsylvanica Marshall, Green A.; bottomlands, streambanks, swampy woods,
sometimes in dense stands in successional bottomlands; common; statewide, but
infrequent on the CU and in the U. *F. pennsylvanica* Marshall var. *subintegerrima*
(Vahl) Fernald. Plate 146.

F. profunda (Bush) Bush, Pumpkin A.; swamps, inundated bottomland woods; occasional; W TN. *F. tomentosa* F.Michx. Plate 147.

F. quadrangulata Michx., Blue A.; dry to mesic woods, especially on limestone outcrops and bluffs; common; M and E TN, except in the U. Plate 148.

Ericaceae   *Gaultheria* (Teaberry/Mountain Tea/Wintergreen)

G. procumbens L.; low, barely woody, aromatic, evergreen subshrubs with creeping
and upright stems bearing crowded leaves near the apex; flowers axillary, white
(summer), fruit a capsule enclosed by a fleshy red calyx and appearing berry-like
(summer–fall); xeric to mesic woods; occasional; EHR, E TN. Plate 149.

Ericaceae   *Gaylussacia* (Huckleberry)

Shrubs, usually clonal from underground stems; flowers pale, sometimes tinged
with pink, purple, or red, in lateral, bracted racemes (spring), fruit an edible drupe,
but not juicy or tasty (summer–fall). Luteyn et al. (1996) report *G. frondosa* (L.)
Torr. and A.Gray (Dangleberry) from TN without specific locality.

1 Bracts of racemes foliaceous, persisting as fruit
matures . . . . *G. dumosa*
1 Bracts of racemes small, deciduous before fruit matures 2

2   Leaves with glands on both surfaces . . . . *G. baccata*
2   Leaves with glands on lower surface only . . . . *G. ursina*

G. baccata (Wangenh.) K.Koch, Black H.; dry woods and thickets; frequent; EHR, E TN, also Stewart Co. (WHR). Plate 150.

G. dumosa (Andrz.) Torr. & A.Gray, Dwarf H.; sandy, boggy to xeric sites; infrequent; EHR, CU. **Threatened.** Plate 151.

G. ursina (M.A.Curtis) Torr. & A.Gray *ex* A.Gray, Bear/Mountain H.; wooded slopes; occasional; U. Plate 152.

Loganiaceae   *Gelsemium* (Yellow Jessamine)

G. sempervirens (L.) A.St.-Hil.; twining or trailing vines; flowers fragrant, yellow, in short axillary clusters (early spring), fruit a capsule (spring); mesic woods and thickets; infrequent; CU, VR. **Special Concern.** Plate 153.

Fabaceae   *Gleditsia* (Locust)

Medium trees; polygamous; flowers yellowish green, in spikes or racemes (spring), fruit a legume (summer). The stout thorns are a good field character.

1   Leaf rachis, petiolule, and midrib pubescent on the lower leaflet surface; fruits 4-many seeded, often twisted, >5 cm long . . . . *G. triacanthos*
1   Leaf rachis, petiolule, and leaflet midrib glabrous or essentially so on the lower leaflet surface; fruit 1(–3) seeded, not twisted, 1–5 cm long . . . . *G. aquatica*

G. aquatica Marshall, Water L.; swamps and low woods; infrequent; extreme W TN. Plate 154.

G. triacanthos L., Honey L.; margins of woods, old fields; frequent; statewide, but naturalized in U. Plate 155.

Fabaceae   *Gymnocladus* (Kentucky Coffee-Tree)

G. dioica (L.) K.Koch; large trees; flowers whitish in terminal racemes (spring), fruit a large, woody legume (summer–winter); slope forests, especially around limestone outcrops, cedar glades, swampy woods, floodplains; occasional; statewide, except the EHR, CU, and U; commonly planted as an ornamental (seeds were used as a coffee substitute), and this may account for some records. Plate 156.

Styracaceae   *Halesia* (Silverbell)

H. tetraptera Ellis; small to large trees; flowers pinkish white in lateral clusters (spring), fruit a winged and woody drupe (summer–fall); mesic woods, especially mountain coves; occasional in counties adjacent to the lower TN River, infrequent on the CU, frequent in the VR and U. *H. carolina* L. Plate 157.

Hamamelidaceae   *Hamamelis* (Witch Hazel)

H. virginiana L.; shrubs or rarely small trees; flowers yellow, in axillary clusters (fall–winter), fruit a capsule (winter–spring); dry to moist woods and thickets; frequent; M and E TN. The bark is the source of the astringent witch hazel. Plate 158.

Araliaceae   *Hedera* (English Ivy)

H. helix L.*; evergreen, climbing vines with abundant aerial roots; flowers whitish-green in terminal umbels (summer), fruit a berry (summer–fall); commonly cultivated, persisting, and spreading, especially in urban woodlots and around dumps with yard debris; statewide. Rarely flowering, so most spreading is vegetative. Eurasian, with various cultivars. Plate 159.

Malvaceae   *Hibiscus* (Rose of Sharon)

H. syriacus L.*; much-branched shrubs; flowers terminal and large, white to rose-purple, solitary in leaf axils (summer), fruit a capsule (summer–fall); commonly planted and persisting, especially around old farmsteads and cemeteries, sometimes spreading into nearby fencerows and thickets; statewide. Asia. Plate 160.

Saxifragaceae   *Hydrangea* (Hydrangea)

Shrubs, often with multiple stems; flowers white to purplish, in panicles or compound cymes, the marginal flowers usually sterile and consisting of a showy calyx (spring–summer), fruit a capsule (summer–fall). Differentiation of taxa in the *H. arborescens* complex is based on characters that are not always distinct and identification is sometimes tentative. Several species-cultivars, including both native and introduced (mostly Asian) taxa, appear occasionally in cultivation. Pilatowski, R. E. 1982. A taxonomic study of the *Hydrangea arborescens* complex. Castanea 47:84–98.

1   Leaf blades deeply lobed; sterile flowers purplish, scattered throughout the inflorescence (a panicle) . . . . *H. quercifolia*

1   Leaf blades toothed but not lobed; sterile flowers (when present)
    whitish, arranged around the edges of the inflorescence
    (corymbiform) 2
    2   Lower leaf surface green (glabrous or puberulent); sterile flowers, if
        present, <1 cm in diameter . . . . H. arborescens
    2   Lower leaf surface gray or white (variously pubescent); sterile
        flowers, if present >1 cm in diameter 3
        3   Lower leaf surface gray, the hairs not masking the
            epidermis . . . . H. cinerea
        3   Lower leaf surface white, the hairs felty and masking the
            epidermis . . . . H. radiata

H. arborescens L., Wild H.; rich woods and bluffs; common; statewide, but not often
seen in W TN. Plate 161.

H. cinerea Small, Ashy H.; shady ledges, bluffs, rich woods; occasional; statewide,
but most abundant on the WHR and in eastern W TN. *H. arborescens* L. var. *deamii*
St.John; *H. arborescens* L. ssp. *discolor* (Seringe) E.M.McClint. Plate 162.

H. quercifolia Bartr., Oak-Leaf H.; riverbanks, bluffs, dry to mesic woods; occasional; essentially statewide in southernmost counties from Shelby to Marion and
along the lower TN River. Commonly planted. Plate 163.

H. radiata Walter, Snowy H./Silverleaf; rich woods and bluffs; infrequent; U.
H. arborescens L. ssp. *radiata* (Walter) E.M.McClint. Plate 164.

Clusiaceae   *Hypericum* (St. John's Wort)

Low, compact or erect shrubs; flowers yellow, variously cymose (summer), fruit a
capsule (fall). The small, punctate secretory glands on the leaves and flowers are
reliable field characters for the genus. Gillespie, J. P. 1959. The Hypericaceae of
Tennessee. Castanea 24:24–32; Adams, P. 1973. Clusiaceae of the southeastern
United States. J. Elisha Mitchell Sci. Soc. 89:62–71.

1   Petals and sepals 4; calyx with unequal pairs, the outer broad, the inner
    narrow or obsolete 2
    2   Styles and carpels 3; upper leaves clasping or cordate, >1 cm
        wide . . . . H. crux-andreae
    2   Styles and carpels 2; leaves narrowed at the base, <1 cm wide 3
        3   Plants erect, branching above ground level, often 1–1.5 m tall;
            leaves widest near the middle . . . . H. hypericoides

3   Plants forming low, compact mats with decumbent stems,
rarely >2 dm tall; leaves widest above the middle . . . . . *H. stragulum*
1 Petals and sepals 5; calyx nearly equal 4
  4   Leaf blades with a distinct hyaline margin, the base slightly
auriculate and membranous, leaves without a line of articulation
with the stem . . . . *H. nudiflorum*
  4   Leaf blades without a hyaline margin, the base neither auriculate or
membranous, leaves with a distinct line of articulation with the
stem 5
    5   Capsules mostly 5-celled; styles mostly 5 . . . . *H. lobocarpum*
    5   Capsules mostly 3- (rarely 2- or 4-) celled; styles 3, rarely
4 or 5 6
      6   Flowers showy, 2.5–4 cm wide, terminal and solitary or
in 3-flowered cymules; capsules 6–9 mm wide . . . . . *H. frondosum*
      6   Flowers smaller, <2 cm wide, in 3-many flowered terminal
and/or subterminal cymules; capsules 1.5–6 mm wide 7
        7   Mature capsule slender-conic, <6 mm long and 3 mm wide;
seeds reddish brown . . . . *H. densiflorum*
        7   Mature capsule lance-ovoid, >7 mm long and 3.5 mm wide;
seeds dark brown or black . . . . *H. prolificum*

H. crux-andreae (L.) Crantz, St. Peter's Wort; sandy, usually damp soil; occasional; M and E TN except the CB, frequent on the CU. *Ascyrum stans* Michx., *H. stans* (Michx.) W.P.Adams & N.Robson. Plate 165.

H. densiflorum Pursh; open, wet areas, streambanks, balds; occasional; M and E TN except the CB, most common in the VR. Plate 166.

H. frondosum Michx.; bluffs, glades, calcareous outcrops; frequent; statewide. Plate 167.

H. hypericoides (L.) Crantz, St. Andrew's Cross; dry woods; common; statewide. *Ascyrum hypericoides* L. Plate 168.

H. lobocarpum Gatt.; bottomlands, wet, open areas; infrequent; CP, EHR. Carpel number is variable in this taxon and it may be best recognized as a variety or synonymous with *H. densiflorum*. *H. densiflorum* Pursh var. *lobocarpum* (Gatt.) Svenson.

H. nudiflorum Michx.; streambanks and damp, open woods; infrequent; CU. Plate 169.

H. prolificum L.; damp or dry slopes, roadsides, open woods; frequent; statewide, but scarce in W TN. Plate 170.

H. stragulum W.P.Adams & N.Robson; dry woods, barrens, roadsides, slopes; common; statewide. *Ascyrum hypericoides* L. var. *multicaule* (Michx. *ex* Willd.) Fernald, *H. hypericoides* L. var. *multicaule* (Michx. *ex* Willd.) Fosberg. Plate 171.

Aquifoliaceae   *Ilex* (Holly)

Contributed by Ross C. Clark

Evergreen or deciduous shrubs or small trees; polygamous, mostly dioecious; flowers axillary, solitary or in fascicles or cymes (late spring); fruit a red or rarely yellowish drupe (fall), the seeds surrounded by a hard, bony endocarp. Various cultivars are seen in horticultural settings.

1 Leaves evergreen . . . . *I. opaca*
1 Leaves deciduous 2
 2 Plants in flower 3
  3 At least some corolla lobes ciliate 4
   4 Leaves abruptly acuminate; longest petioles ±1 cm
     long . . . . *I. ambigua* var. *ambigua*
   4 Leaves attenuate-acuminate; longest petioles >1 cm
     long . . . . *I. ambigua* var. *montana*
  3 Corolla lobes not ciliate 5
   5 Sepals 4 6
    6 Leaves crenate or crenate-serrulate, apex various; pistillate
      pedicel ±3x as long as the fruit . . . . . . *Ilex decidua* var. *decidua*
    6 Leaves definitely serrate **and** with gradually acuminate
      apex; pistillate pedicel >3x as long as the
      fruit . . . . *I. decidua* var. *longipes*
   5 Sepals in most flowers 5 or more . . . . *I. verticillata*
 2 Plants with fruit 7
  7 Endocarps (the seed-like structures in the fruit) with obvious
    longitudinal ribs 8
   8 Sepals 4; pedicels not bracteolate 9
    9 Leaves crenate or crenate-serrulate, apex various; pedicel
      ±3x as long as the fruit . . . . *I. decidua* var. *decidua*
    9 Leaves definitely serrate **and** with gradually acuminate tips;
      pedicel >3x as long as the fruit . . . . *I. decidua* var. *longipes*
   8 Sepals variable in number (often >4); pedicels with
     bracteoles 10

10   Leaves abruptly acuminate; longest petioles ±1 cm
long ....*I. ambigua* var. *ambigua*
10   Leaves attenuate-acuminate; longest petioles >1 cm
long .... *I. ambigua* var. *montana*
7   Endocarps smooth .... *I. verticillata*

I. ambigua (Michx.) Torr. var. *ambigua*, Carolina H.; mesic woods; infrequent; WHR, CU, VR, U. *I. beadlei* Ashe. Plate 172.

I. ambigua (Michx.) Torr. var. *montana* (Torr. & A.Gray) H.E.Ahles, Mountain H.; mesic woods; common; EHR, E TN. *I. monticola* A.Gray, *I. montana* Torr. & A.Gray, *I. ambigua* (Michx.) Torr. var. *monticola* (Wood) Wunderlin & Poppleton. Plate 173.

I. decidua Walter var. *decidua*, Possum Haw; mesic woods and thickets, most often in low ground; occasional W and M TN, infrequent CU, VR. Plate 174.

I. decidua Walter var. *longipes* (Chapm. *ex* Trel.) H.E.Ahles, Buckbush; upland woods and thickets; infrequent; mostly in southern counties of W and M TN (except CB) and the CU. *I. longipes* Chapm. *ex* Trel. Plate 175.

I. opaca Aiton, American H.; mesic woods; common; statewide, except infrequent in M TN; often planted. Plate 176.

I. verticillata (L.) A.Gray, Common Winterberry; mesic woods and streambanks; infrequent; statewide, except CB. Plate 177.

Saxifragaceae   *Itea* (Virginia Willow/Sweet Spire)

I. virginica L.; slender shrubs; flowers white and numerous in terminal, often arching racemes (spring), fruit a small woody capsule (summer); swamps, wet woodlands, along stream and reservoir margins; infrequent; statewide. Plate 178.

Juglandaceae   *Juglans* (Walnut)

Medium to large trees; monoecious; staminate flowers in elongate catkins, pistillate flowers in short spikes (spring), fruit a nut surrounded by a somewhat fleshy indehiscent husk, therefore, drupaceous (fall). Walnuts are highly valued for timber, furniture, dyes, and as food for human beings and other animals.

1   Pith of older twigs dark brown; leaf scars with a mustache-like cluster of
hairs across the top; fruit glandular-hairy, ellipsoid; bark grayish .... *J. cinerea*

1   Pith of older twigs light brown; leaf scars lacking hairs across the top; fruit glabrous, nearly round; bark dark brown .... *J. nigra*

J. cinerea L., Butternut/White W.; rich woods, mesic sites, and riverbanks; once frequent statewide but becoming rare due to infestation by a canker-producing fungus. **Threatened.** Plate 179.

J. nigra L., Black W.; mixed hardwoods, slopes, and disturbed sites; frequent; statewide, but not common in W TN. Plate 180.

Ericaceae   *Kalmia* (Laurel)

Evergreen shrubs or small trees; flowers white to pinkish, in terminal corymbs (early summer), fruit a capsule (fall).

1   Leaves opposite or whorled, pubescent beneath, <2 cm wide; flowers axillary .... *K. angustifolia* var. *carolina*
1   Leaves alternate, rarely whorled, glabrous beneath, >2 cm wide; flowers terminal .... *K. latifolia*

K. angustifolia L. var. *carolina* (Small) Fernald, Sheep L./Lambkill; bogs and wet thickets; known only from Shady Valley, Johnson Co. (U). **Endangered—Possibly Extirpated.** *Kalmia carolina* Small. Plate 181.

K. latifolia L., Mountain L.; thin woods and margins of roadsides and waterways; M and E TN, more common eastward. Plate 182.

Ericaceae   *Leiophyllum* (Sand Myrtle)

L. buxifolium (P.J.Berg.) Elliott; evergreen, low, compact shrubs; flowers whitish pink, in terminal clusters (late spring), fruit a capsule (fall); high elevation shrub/heath balds; infrequent; U, mostly in the GSMNP. Plate 183.

Fabaceae   *Lespedeza*

L. bicolor Turcz.*, Bush Clover; shrubs; flowers purplish, in a paniculate cluster of racemes (summer), fruit a 1-seeded legume (fall); ruderal sites; frequent; statewide. E Asia. Usually planted for wildlife food and erosion control and commonly spreading from seeds. Plate 184.

Ericaceae   *Leucothoe*

Evergreen or deciduous shrubs; flowers white, in axillary racemes (spring–early summer), fruit a capsule (summer–fall). The inflorescences overwinter and develop during the summer–fall on shoots of the previous season.

1 Leaves evergreen . . . . *L. fontanesiana*
1 Leaves deciduous 2
    2 Capsules lobed, seeds winged; racemes usually recurved; petioles of
        some leaves >3 mm long . . . . *L. recurva*
    2 Capsule not lobed, seeds unwinged; racemes usually straight; petioles
        <3 mm long . . . . *L. racemosa*

L. fontanesiana (Steud.) Sleumer, Dog Hobble; often forming thickets along streams and moist slopes; frequent; E TN. *L. axillaris* D.Don var. *editorum* Fernald & B.G.Schub., *L. editorum* Fernald & B.G.Schub. Plate 185.

L. racemosa (L.) A.Gray, Fetterbush; wet woods, gravel bars and streambanks; infrequent; WHR, CU, VR. **Threatened.** Plate 186.

L. recurva (Buckley) A.Gray, Fetterbush; woodlands and streambanks, mostly at high elevations of upper E TN; infrequent; U. Plate 187.

Oleaceae   *Ligustrum* (Privet)

Evergreen or semi-evergreen shrubs or small trees; flowers white, in panicles (early summer), fruit a drupe (fall–winter). Hardin, J. W. 1974. Studies of the southeastern United States flora. IV. Oleaceae. Sida 5:274–285.

1 Twigs densely pubescent; leaves pubescent on the lower midrib,
    especially when young . . . . *L. sinense*
1 Twigs minutely pubescent; leaves glabrous . . . . *L. vulgare*

L. sinense Lour.*, Chinese P.; widely planted and escaping to ruderal sites and disturbed woodlands; frequent; statewide. China. A noxious weed. Plate 188.

L. vulgare L.*, Common/European P.; ruderal sites and open woods; infrequent; statewide. Europe.

Lauraceae   *Lindera* (Spicebush)

L. benzoin (L.) Blume; shrubs or small trees; dioecious; flowers yellowish, in axillary clusters (early spring, usually before the leaves), fruit a bright red drupe (early fall); low woods and streambanks; common; statewide. The stems and leaves are strongly "spicy" aromatic. Plate 189.

Caprifoliaceae   *Linnaea* (Twinflower)

L. borealis L.; evergreen, with trailing stems and short erect branches; flowers pink to white, 2–4 (rarely 6) on an upright peduncle (late summer), fruit dry, indehiscent (fall); known only from a single 1892 collection by Albert Ruth from Sevier Co. ("in mountain woods"); U. **Endangered—Possibly Extirpated.** Plate 190.

Hamamelidaceae   *Liquidambar* (Sweet Gum)

L. styraciflua L.; large trees; monoecious; staminate heads in terminal racemes, pistillate flowers in globose heads (spring); fruit fused into a woody head of spiny capsules (summer–winter); moist or wet woodlands and fields; common; statewide, absent at high elevations. The common name is derived from the chewy, somewhat sweet "gum" that exudes from damaged bark. Plate 191.

Magnoliaceae   *Liriodendron* (Tulip Poplar/Yellow Poplar)

L. tulipifera L.; large tree; flowers tulip-shaped, petals greenish yellow with orange blotches at base, solitary (spring), fruit a "cone-like" aggregate of samaras (fall); low rich woods; common; statewide. The state tree of Tennessee prized for lumber and widely cultivated. The young twigs are aromatic but with a bitter taste. Plate 192.

Caprifoliaceae   *Lonicera* (Honeysuckle)

Deciduous or evergreen vines or shrubs; flowers various shades of yellow, orange, red, or white, in cymes or few-several flowered axillary clusters (spring–late summer), fruit a berry (summer–fall).

1 Shrubs 2
   2 Leaves acuminate at the tip, pubescent beneath, especially on the
      main veins . . . . *L. maackii*
   2 Leaves rounded or with a minute mucro at the tip, glabrous
      beneath . . . . *L. canadensis*

1 Vines 3
   3 Inflorescences axillary; leaves not connate; stems usually pubescent;
      fruits in fused pairs . . . . *L. japonica*
   3 Inflorescences terminal; leaves connate below the inflorescences;
      stems glabrous; fruits not fused 4
      4 Corolla scarlet, >3 cm long, only slightly
         zygomorphic . . . . *L. sempervirens*
      4 Corolla yellow, orange, or purplish, 1.5–3 cm long, strongly
         zygomorphic 5
         5 Uppermost connate leaves glaucous above; corolla pale
            yellow . . . . *L. reticulata*
         5 Uppermost connate leaves greenish above; corolla yellow to
            orange or purple tinged 6
            6 Corolla pale yellow to purplish, gibbous just above the base,
               the tube <1.5 cm long, gradually tapering to the
               base . . . . *L. dioica*
            6 Corolla yellow, not gibbous, the tube >1.5 cm long, slender,
               abruptly contracted below the corolla lobes . . . . *L. flava*

L. canadensis Bartram, Fly H.; damp woods at high elevations; infrequent; U. **Special Concern.** Plate 193.

L. dioica L., Wild H.; open woods and riverbanks; infrequent; EHR, VR, U. **Special Concern.** Plate 194.

L. flava Sims, Yellow H.; reported from Franklin (EHR or CU) and Hamilton (CU or VR) Counties by Duncan (1967). **Special Concern.** Plate 195.

L. japonica Thunb.*, Japanese H.; ruderal sites, especially along fencerows; common; statewide. East Asia. Plate 196.

L. maackii (Rupr.) Maxim.*, Amur H.; open, disturbed woodlands and thickets, especially on neutral soils; infrequent, but rapidly spreading and a troublesome weed in the CB and VR. Asia. Plate 197.

L. reticulata Raf., Grape H.; Reported from Davidson Co. by Duncan (1967); not seen in TN in over 100 years. **Endangered—Possibly Extirpated.** *L. prolifera* (G.Kirchn.) Rehder.

L. sempervirens L., Coral/Trumpet H.; margins of woods and thickets, disturbed sites; common; statewide. Widely planted and escaping. Plate 198.

Solanaceae   *Lycium* (Matrimony Vine)

L. barbarum L.*; small, weakly spiny shrubs, sometimes vine-like; flowers purple, in axillary clusters (late spring–fall); fruit a reddish berry (late spring–fall); ruderal sites, especially along railways; infrequent; M and E TN. Eurasia. *L. halimifolium* Mill. Plate 199.

Ericaceae   *Lyonia* (Maleberry)

L. ligustrina (L.) DC.; shrubs; flowers white, axillary, in fascicles or racemes (early summer), fruit a capsule (fall); low moist woods and streambanks; infrequent; EHR and E TN with a single disjunct population in Carroll Co. (CP). Plate 200.

Moraceae   *Maclura* (Osage Orange/Horse Apple/Bodock)

M. pomifera (Raf.) C.K.Schneid.*; Medium trees; monoecious; flowers in globose heads (spring), fruit a softball-size, green, warty multiple of achenes with a milky latex (fall); open areas, fencerows, and woodland borders; infrequent; statewide, but more common in CB. The wood is heavy, durable, and used for fenceposts and for bows by Native Americans. The fruit is reputed to be an insect repellent (especially cockroaches), the milky latex gums, like rubber cement, the mouthparts of insects that eat it. Native west of TN, now widespread throughout the central and eastern U.S. Plate 201.

Magnoliaceae   *Magnolia* (Magnolia)

Evergreen or deciduous, medium trees; flowers white, terminal and solitary (spring–summer); fruit a "cone-like" aggregate of follicles (fall), the conspicuous orange to red seeds often suspended by slender threads from the open follicles. The wood is of minimal commercial value, but most species are planted for their large, showy, often fragrant flowers.

1 Leaves evergreen, or at least leathery 2
  2 Leaves leathery, tardily deciduous, whitish beneath, green
    above . . . . *M. virginiana*
  2 Leaves evergreen, glabrous to reddish brown pubescent beneath,
    dark green above . . . . *M. grandiflora*
1 Leaves deciduous 3
  3 Leaves auriculate or deeply cordate at the base 4
    4 Leaves whitish glaucous beneath; twigs and buds
      pubescent . . . . *M. macrophylla*

4 Leaves green or only slightly glaucous beneath; twigs and buds
   glabrous . . . . M. fraseri
3 Leaves cuneate to rounded at the base 5
   5 Leaves clustered at the tips of twigs; twigs and buds
      pubescent . . . . M. acuminata
   5 Leaves scattered, not in distinct clusters at the tips of twigs; twigs
      and buds glabrous . . . . M. tripetala

M. acuminata (L.) L., Cucumber M.; coves and rich woods; frequent; statewide,
except in the CB, rare in W TN and on the WHR. Plate 202.

M. fraseri Walter, Fraser M.; coves and rich woods; infrequent; U, and Scott Co.
(CU) in dry, pine-oak woodlands. Plate 203.

M. grandiflora L.*, Evergreen M.; occasionally planted statewide but rarely repro-
ducing from seeds. Southeastern U.S. Coastal Plain. Plate 204.

M. macrophylla Michx., Big-Leaf M.; ravines, rich woods, and streambanks; occas-
sional; M and E TN, most common on the CU. This species produces our largest
simple leaf. Plate 205.

M. tripetala (L.) L., Umbrella M.; coves and rich woods; infrequent; WHR, EHR, E
TN. Plate 206.

M. virginiana L., Sweetbay M.; open, wet woods and thickets; infrequent; CP, U,
with all records from the southern border of the state. **Threatened.** Plate 207.

Rosaceae	Malus (Apple)

Shrubs or small trees, often with sharp-pointed spur branches; flowers pink to
white, in umbel-like clusters (spring), fruit a fleshy pome, commonly known as an
apple (summer). Hybridization occurs between native taxa and there are many
cultivars of hybrid origin.

1 Leaves convolute in bud, woolly pubescent beneath at maturity;
   anthers yellow; calyx pubescent externally; spur branches
   absent . . . . M. pumila
1 Leaves folded in bud; glabrous at maturity; anthers pinkish; calyx
   glabrous externally; spur branches often present 2
   2 Leaves not lobed at base, the margin crenate or serrate; terminal bud
      1–3 mm long . . . . M. angustifolia

2   At least some leaves lobed near base, the margin serrate to doubly
    serrate; terminal bud 3–5 mm long . . . . *M. coronaria*

M. angustifolia (Aiton) Michx., Southern/Wild Crab A.; thin woods, roadsides and ruderal sites; common; statewide, but less frequent in the CB. *Pyrus angustifolia* Aiton. Plate 208.

M. coronaria (L.) Mill., Sweet/Wild Crab A.; thin woods and ruderal sites; infrequent; WHR, CU, U. *Pyrus coronaria* L. Plate 209.

M. pumila Mill.*, Common Apple; cultivated, persistent and escaping; frequent; statewide. Eurasian. *Pyrus malus* L. Plate 210.

Meliaceae   *Melia* (China Berry)

M. azedarach L.*; small to medium trees; flowers purple to white, in large axillary panicles (spring), fruit a yellow, globose drupe (summer); persistent and occasionally spreading from cultivation; infrequent; mostly CP. Asia. Plate 211.

Menispermaceae   *Menispermum* (Moonseed)

M. canadense L.; twining vines; flowers small, greenish white, in elongate panicles (summer), fruit a dark drupe (fall); moist woods and stream margins; frequent; statewide. The fruit is poisonous. Plate 212.

Ericaceae   *Menziesia* (Minnie-Bush)

M. pilosa (Michx.) Juss.; shrubs; flowers greenish yellow to white, roseate at the tip, in terminal corymbs (summer), fruit a capsule; woodlands and shrub balds at high elevations; U. **Special Concern.** The fused, lacerate, scalelike structures on the lower midrib are excellent field characters. Plate 213.

Rubiaceae   *Mitchella* (Partridge Berry)

M. repens L.; evergreen, scarcely woody, low creepers; flowers white, paired, terminal (late spring), fruit red, from two fused berries (summer); woodlands and streambanks; common; statewide, except rare in the CB. Plate 214.

Moraceae   *Morus* (Mulberry)

Medium to large trees; monoecious or dioecious; both staminate and pistillate flowers in spikes (early spring), fruit a multiple of drupelets (early summer). The fruits are eaten by birds, humans, and other animals.

1 Leaves smooth above, glabrous below except for a few scattered hairs
   along the midrib . . . . M. alba
1 Leaves scabrous above, pubescent below . . . . M. rubra

M. alba L.*, White M.; persistent after plantings and in ruderal sites; infrequent;
statewide. China. This species was introduced during colonial times in an unsuccessful attempt to establish a silkworm industry. Plate 215.

M. rubra L., Red M.; moist woods and streambanks; common; statewide. Plate 216.

Aquifoliaceae   Nemopanthus (Appalachian Mt. Holly)

N. collinus (Alexander) R.C.Clark; shrubs or small trees; dioecious; flowers yellow
to yellowish green, 1-several/node (spring), fruit a red or yellow drupe (summer);
high-elevation woods; known only from GSMNP (U). **Special Concern.** Ilex collina Alexander. Clark, R. C. 1974. Ilex collina, a second species of Nemopanthus in the
southern Appalachians. J. Arnold Arb. 55:435–440. Plate 217.

Santalaceae   Nestronia (Conjurer's Nut)

N. umbellula Raf.; clonal, hemiparasitic shrubs; mostly dioecious; staminate flowers
greenish, in axillary umbels, pistillate flowers solitary in leaf axils (spring), fruit a
globose drupe (summer); open woods, often with pine; infrequent; EHR, U.
Endangered. Horn, D. D., & R. Kral. 1984. Nestronia umbellula Raf. (Santalaceae),
a new state record for Tennessee. Castanea 49:69–73. Plate 218.

Rosaceae   Neviusia (Alabama Snow Wreath)

N. alabamensis A.Gray; clonal shrubs; flowers white, cymose, stamens showy
(spring), fruit a drupe-like achene (summer); limestone slopes; infrequent; CB,
EHR. **Threatened.** Long, A. A. 1989. Disjunct populations of the rare shrub
Neviusia alabamensis Gray (Rosaceae). Castanea 54:29–39. Plate 219.

Nyssaceae   Nyssa (Gum, Tupelo)

Medium to large trees; dioecious or polygamous; flowers greenish white, in heads,
racemes, or solitary (spring), fruit a blue-black or brownish drupe (summer).
Burckhalter, R. E. 1992. The genus Nyssa (Cornaceae) in North America: A revision. Sida 15:323–342.

1 Fruits in clusters of 3–5; plants of lowland to upland
   habitats . . . . N. sylvatica

1   Fruits 1 or 2/peduncle; plants of wet, usually inundated habitats 2
   2   Leaves lanceolate, 5–14 cm long; fruit oblong, blue-
     black . . . . *N. biflora*
   2   Leaves ovate to obovate, 10–25 cm long; fruit dull yellow to
     brown . . . . *N. aquatica*

N. aquatica L., Water T.; swamps, riverbanks, and floodplains; infrequent; more common in W TN, but also known from Stewart (WHR), Coffee (EHR), and Marion (CU) Counties. Plate 220.

N. biflora Walter, Swamp T./G.; swamps, riverbanks, and wet woods; more common in W TN but a few scattered populations on the W and EHR. Scarcely distinct from the following species. *N. sylvatica* Marshall var. *biflora* (Walter) Sarg. Plate 221.

N. sylvatica Marshall, Black G./T.; mostly on well-drained upland sites; common; statewide. Plate 222.

Cactaceae   *Opuntia* (Cactus/Prickly Pear)

O. humifusa (Raf.) Raf.; low, suffrutescent succulents with evergreen stems; flowers yellow (summer), fruit a berry (summer–fall); open, rocky, often thin, basic soils; common in the CB but infrequent elsewhere over the state. Occasionally culti-vated. *O. compressa* (Salisb.) Macbr. Plate 223.

Betulaceae   *Ostrya* (Hophornbeam/Ironwood)

O. virginiana (Mill.) K.Koch; small to medium trees; monoecious; both staminate and pistillate flowers in catkins (spring), the pistillate shorter than the staminate, fruit a nut, in clusters resembling hops (summer); low woods and dry slopes; com-mon; statewide. Plate 224.

Ericaceae   *Oxydendrum* (Sourwood)

O. arboreum (L.) DC.; medium trees; flowers white, paniculate-racemose (early summer), fruit a capsule (summer–fall), the flowers pendulous but the fruits erect; dry woods and disturbed sites; frequent; statewide but more common in M and E TN. Highly prized for sourwood honey. Plate 225.

Vitaceae   *Parthenocissus* (Virginia Creeper)

P. quinquefolia (L.) Planch.; vines, climbing by adhesive discs at the end of branched tendrils; flowers yellow-green, in axillary or terminal clusters (early summer), fruit a thin-fleshed berry (fall); woodlands and disturbed sites; common; statewide. Sometimes confused with poison ivy but distinguished by the presence of tendrils and palmately (mostly with 5 leaflets) compound leaves. Plate 226.

Scrophulariaceae   *Paulownia* (Empress/Princess Tree)

P. tomentosa (Thunb.) Steud.*; medium trees; flowers fragrant and showy, purple, appearing before the leaves in large terminal panicles (late spring), fruit a capsule (fall); widely planted for its showy flowers but frequently spreading to roadsides and waste places. The wood is highly prized in Asia. China. Plate 227.

Celastraceae   *Paxistima* (Mountain Lover)

P. canbyi A.Gray; evergreen shrubs; flowers greenish, solitary to few in upper leaf axils (late spring), fruit a capsule (summer); calcareous bluffs along the Holston River in Hawkins Co. (VR). **Endangered.** Plate 228.

Saxifragaceae   *Philadelphus* (Mock Orange)

Erect or arching shrubs; flowers white, in determinate racemes, cymules, or solitary (spring); fruit a capsule (summer). A highly variable complex in need of further study. Hu, Shiu-Ying. 1954–1956. A monograph of the genus *Philadelphus*. J. Arnold Arbor. 35:275–333; 36:52–109, 325–368; 37:15–83.

1 Axillary bud exposed . . . . *P. hirsutus*
1 Axillary bud concealed 2
   2 Flowers solitary or in 3's . . . . *P. inodorus*
   2 Flowers >4, in determinate racemes 3
      3 Leaves and hypanthium densely
        pubescent . . . . *P. pubescens* var. *pubescens*
      3 Leaves and hypanthium glabrous or essentially
        so . . . . *P. pubescens* var. *intectus*

P. hirsutus Nutt., Cumberland M. O.; creek and river bluffs, moist slopes; frequent; M and E TN. *P. sharpianus* S.Y.Hu. Plate 229.

P. inodorus L.; Appalachian M. O.; creek and river bluffs, moist slopes; frequent; statewide. Plate 230.

P. pubescens Loisel. var. *intectus* (Beadle) A.H.Moore; calcareous creek and riverbanks; infrequent; M TN, VR. *P. intectus* Beadle.

P. pubescens Loisel. var. *pubescens*, Ozark M. O.; creek and riverbanks; frequent; M TN, VR, U. Plate 231.

Viscaceae   *Phoradendron* (Mistletoe)

P. leucarpum (Raf.) Reveal & M.C. Johnston; hemiparasitic shrubs on flowering trees; dioecious; flowers greenish, in spike-like clusters (fall), fruit a white berry with sticky pulp (fall–winter); frequent; statewide. This is the common mistletoe used for Christmas decorations. Most parts are toxic if consumed. *P. serotinum* (Raf.) M.C. Johnston. Plate 232.

Rosaceae   *Physocarpus* (Ninebark)

P. opulifolius (L.) Maxim.; shrubs with shredding bark; flowers white or rarely pinkish, in corymbs (spring), fruit an inflated follicle (summer); streambanks and moist cliffs; infrequent; M and E TN. Occasionally cultivated. Plate 233.

Ericaceae   *Pieris* (Mountain Fetterbush)

P. floribunda (Pursh *ex* Sims) Benth. & Hook.; evergreen shrubs; flowers white, paniculate-racemose (early summer), fruit a capsule (fall); high elevation balds and thickets; known only from the GSMNP (U). **Threatened.** Plate 234.

Ulmaceae   *Planera* (Water Elm/Planer Tree)

P. aquatica (Walter) J.F.Gmel.; shrubs or small trees; monoecious; staminate flowers clustered in axils of bud scales, pistillate flowers solitary in axils of new leaves (spring), fruit a burlike nutlet (fall); swamps and river bottoms; infrequent; W TN, WHR. Plate 235.

Platanaceae   *Platanus* (Sycamore)

P. occidentalis L.; medium to large trees; monoecious; both staminate and pistillate flowers in pendulous, globose heads, the staminate heads smaller (spring), fruit a multiple of achenes (summer–winter); wet areas along lakes and streams, also bottomlands and upland wet sites; common; statewide. Plate 236.

Polygonaceae   *Polygonella* (Jointweed)

P. americana (Fisch. & C.A.Mey.) Small; subshrubs; flowers white, in paniculate racemes (summer), fruit an achene (fall); sandy/gravel bars along streams and small rivers; Morgan Co. (CU). **Endangered.** Plate 237.

Rutaceae   *Poncirus* (Trifoliate Orange)

P. trifoliata (L.) Raf.*; shrubs or small trees; flowers white, showy, in axillary fascicles (spring), fruit a berry, resembling a small, fuzzy orange, sour to taste and
scarcely edible (fall); persistent around home sites and occasionally spreading;
infrequent; statewide. With its dense, stout thorns, this species has been planted
as a security fence. China. Plate 238.

Salicaceae   *Populus* (Poplar, Aspen, Cottonwood)

Medium to large trees; dioecious; flowers in drooping catkins appearing before
the leaves (spring), fruit a capsule, seeds with a silky down (spring–summer). The
wood is soft, light, and not particularly desirable for lumber.

1 Leaves white-woolly beneath and sometimes lobed . . . . *P. alba*
1 Leaves glabrous or pubescent beneath (but not white-woolly) and not
   lobed 2
    2 Petiole adjacent to the leaf blade round or essentially so 3
        3 Larger leaves >15 cm long, the lower surface pubescent along the
           midrib; buds not sticky; native species . . . . *P. heterophylla*
        3 Larger leaves <15 cm long, the lower surface glabrous along
           the midrib (except when young); buds sticky; introduced
           species . . . . *P. x jackii*
    2 Petiole adjacent to the leaf blade distinctly flattened 4
        4 Leaf teeth >14 on each side of the blade . . . . *P. deltoides*
        4 Leaf teeth <14 on each side of the blade . . . . *P. grandidentata*

P. alba L.*, White/Silver-Leaf P.; planted, persisting and spreading, home sites,
ruderal areas; frequent; statewide. Rarely fruiting but developing dense thickets
from root sprouts. Europe. Plate 239.

P. deltoides Bartram *ex* Marshall, Cottonwood; riparian forests, bottomlands, ravines;
frequent; statewide, except in the U. Plate 240.

P. grandidentata Michx., Big-Tooth A.; dry to mesic woods, barren edges; occasional; M and E TN, except in the VR. **Special Concern.** Plate 241.

P. heterophylla L., Swamp C.; bottomlands, upland and lowland swamps; occasional;
W TN, Pennyroyal Plain of M TN, and on the EHR (White Co.). Plate 242.

P. x jackii Sarg.*, Balm-of-Gilead; planted, persisting, spreading from root sprouts
at home sites, ruderal areas; infrequent; M and E TN, rarely collected. Apparently
sterile and thought to be of hybrid origin. *P. x gileadensis* Rouleau. Plate 243.

Rosaceae   *Potentilla* (Mountain White Potentilla)

P. tridentata (Sol.) Aiton; subshrubs; flowers white, in cymes (summer), fruit a head of achenes (fall); open areas at high elevations in Carter and Unicoi Counties (U). **Special Concern.** Plate 244.

Rosaceae   *Prunus* (Cherry/Peach/Plum)

Shrubs, small and large trees; flowers white to pink, solitary or in racemes, corymbs, or fascicles (spring–early summer), fruit a fleshy drupe (summer–fall). Several species/cultivars are grown for food and as ornamentals.

1   Mature plants <1 m tall . . . . *P. pumila*
1   Mature plants >1 m tall 2
  2   Leaf blades <1.5x as long as wide . . . . *P. mahaleb*
  2   Leaf blades >1.5x as long as wide 3
    3   Flowers/fruits more than 15/raceme 4
      4   Large trees; leaves shiny above; inner bark aromatic;
          widespread . . . . *P. serotina*
      4   Shrubs or small trees; leaves dull above; inner bark not
          aromatic; mostly above 3000 ft . . . .*P. virginiana*
    3   Flowers/fruits solitary, in corymbs, umbels, fascicles, or few-
        flowered (<15) racemes 5
      5   Flowers/fruits solitary, sessile or nearly so; fruit
          velvety . . . . *P. persica*
      5   Flowers/fruits in corymbs, umbels, or fascicles, pedicellate; fruit
          glabrous or nearly so 6
        6   Teeth of leaf blades with a distinct gland or callous
            thickening, especially when young (10x) 7
          7   Mature leaf blades mostly <2 cm wide; usually shrubby
              and clonal . . . . *P. angustifolia*
          7   Mature leaf blades mostly >2 cm wide; small trees, rarely,
              if ever, clonal 8
            8   Mature leaf blades >4x as long as wide, restricted to
                the Unakas . . . . *P. pensylvanica*
            8   Mature leaf blades <4x as long as wide; widespread
                taxa 9
              9   Glands on leaf margins adjacent to the sinus of
                  teeth; flowers opening before or at the beginning
                  of leaf expansion . . . . *P. munsoniana*

         9 Glands on leaf margins terminating teeth; flowers opening well into or after leaf expansion .... *P. hortulana*
     6 Teeth of leaf blades without a distinct gland or callous thickening 10
        10 Leaf blades densely soft pubescent beneath .... *P. mexicana*
        10 Leaf blades glabrous beneath except for a few hairs on the midrib and/or for tufts in vein axils 11
            11 Leaves doubly serrate, abruptly acuminate; fruits yellow to red .... *P. americana*
            11 Leaves simply serrate to crenate, acute or gradually acuminate; fruits dark red to nearly black .... *P. umbellata*

The following species are often collected in flower before the leaves expand; this supplementary key is adapted from Smith (1994):

  1 Petals ±7.5–11 mm long; inner surface of calyx lobes densely pubescent; leaves with sharp, glandless teeth 2
    2 Calyx tube glabrous to sparsely pubescent 3
      3 Lateral margins of calyx lobes with at least some glandular teeth .... *P. americana*
      3 Lateral margins of calyx lobes lacking glands .... *P. umbellata*
    2 Calyx tube moderately to densely pubescent .... *P. mexicana*
  1 Petals ±4–7 mm long; inner surface of calyx lobes sparsely to moderately pubescent (at least the lower half); leaves with blunt, gland-tipped teeth (the glands may be deciduous); 4
    4 Lateral margins of calyx lobes not glandular-toothed (apex of calyx lobes may have 1-few glandular teeth); pedicels of flowers ±2–6 mm long .... *P. angustifolia*
    4 Lateral margins of calyx lobes glandular-toothed; pedicels of flowers ±5–15 mm long 5
      5 Flowers opening before the leaves emerge .... *P. munsoniana*
      5 Flowers opening after the leaves emerge (<½ expanded) .... *P. hortulana*

P. americana Marshall, American/Wild Pl.; woodland borders, thickets, fencerows; common; statewide. Plate 245.

P. angustifolia Marshall, Chickasaw Pl.; woodland borders, thickets, barrens, successional fields; common; statewide, but rare in the U. Plate 246.

P. hortulana L.H.Bailey, Wild Goose Pl.; woodland borders and thickets, especially in moist soils; occasional; M and E TN. Plates 247, 250a.

P. mahaleb L.*, Perfumed C.; cultivated and spreading to roadsides, fencerows, thickets, and trash dumps; occasional; statewide. Eurasia. Plate 248.

P. mexicana S.Watson, Mexican/Big-Tree Pl.; moist to dry woods and thickets; occasional; W and M TN, CU. *P. americana* Marshall var. *lanata* Sudw. Plate 249.

P. munsoniana W.Wight & Hedrick, Wild Goose Pl.; thickets and fencerows; occasional; statewide, but most common in M TN. Plate 250b.

P. pensylvanica L.f., Pin/Fire C.; dry woods and openings, especially after fire or other disturbance; frequent; U. Plate 251.

P. persica (L.) Batsch*, Common Pe.; cultivated and spreading to roadsides, fencerows, thickets, and trash dumps; common; statewide. Asia. Plate 252.

P. pumila L., Sand C.; thickets and fields; infrequent; Coffee and Franklin Counties (EHR). **Threatened.** Plate 253.

P. serotina Ehrend., Wild Black/Black C.; dry or mesic woods, fencerows, thickets, and successional fields; common; statewide. Plate 254.

P. umbellata Elliott, Hog Pl.; open woods and thickets; infrequent; M TN, CU. Plate 255.

P. virginiana L., Choke C.; thickets and woodland borders; infrequent; U. **Special Concern.** Plate 256.

Rutaceae   *Ptelea* (Wafer Ash/Hop Tree)

P. trifoliata L.; aromatic shrubs; polygamous; flowers greenish white to yellowish, in terminal corymbiform cymes (spring), fruit a flat, nearly round samara (summer–fall); streambanks and moist, rocky slopes; infrequent; M and E TN. Plate 257.

Fabaceae   *Pueraria* (Kudzu)

P. montana (Lour.) Merrill var. *lobata* (Willd.) Maesen & S.M.Almeida*; scarcely woody vines; flowers purplish, in axillary, paniculate racemes (late summer), fruit a legume (fall); common; statewide. Introduced and originally cultivated for erosion control and forage; now a troublesome weed, especially in the southeastern U.S. Asia. *P. lobata* (Willd.) Ohwi. Plate 258.

Santalaceae   *Pyrularia* (Buffalo Nut)

P. pubera Michx.; hemiparasitic shrubs; dioecious; flowers green in terminal racemes, the staminate racemes longer than the pistillate (spring), fruit a green, pyriform drupe (summer); mesic woods; infrequent; CU, U. The fruit is poisonous. Plate 259.

Rosaceae   *Pyrus* (Common Pear)

P. communis L.*; small to medium trees; flowers white, in corymbs (spring), fruit a pyriform pome (summer); persistent around former home sites and orchards, occasionally escaping into disturbed sites; infrequent; statewide. Eurasia. Plate 260.

Fagaceae   *Quercus* (Oak)

Medium to large trees; monoecious; staminate flowers in catkins, pistillate flowers scattered singly or in clusters (spring), fruit a nut (acorn), each partially or completely surrounded by a floral cup or involucre (fall). Our oaks are within two major groups: the white oaks and the red/black oaks (see key for characters). Hybridization is common between many species (within a group), and leaves often are highly variable in shape and size. Mature leaves and fruits, and often twigs and buds, are required for identification. This is our largest genus of trees and most species are highly valuable for lumber and mast.

1   Leaf margins toothed, scalloped or lobed but without marginal bristles; acorns maturing in one year; inner surface of nut shell glabrous (white oaks) 2
  2   Leaf margins with coarse teeth or scalloped but not distinctly lobed, sinuses <⅓ of the distance from margin to midrib 3
    3   Lower leaf surfaces hairy between veins, velvety to touch 4
      4   Peduncle of acorn 2–10 cm long, longer than the leaf petiole . . . . *Q. bicolor*
      4   Peduncle of acorn <2 cm long, shorter than the leaf petiole . . . . *Q. michauxii*
    3   Lower leaf surfaces glabrous or with a few hairs between veins, but not velvety 5
      5   Leaf teeth rounded or leaf margins scalloped; veins on lower leaf surfaces with tufts of spreading hairs . . . . *Q. montana*
      5   Leaf teeth acute; veins on lower leaf surfaces without tufts of hairs (hairs, if present, appressed and not tufted) 6

6   Each leaf with ≥10 secondary veins on each side; trees;
usually in calcareous or limestone sites ...... *Q. muhlenbergii*

6   Each leaf with ≤9 secondary veins on each side; shrubs/small
trees; mostly on sandy sites .... *Q. prinoides*

2   Leaf margins shallowly or deeply lobed, some-all sinuses >⅓ of the
distance from margin to midrib 7

7   Acorn ≥2.5 cm long, cup with a conspicuous fringe on the
margin .... *Q. macrocarpa*

7   Acorn <2.5 cm long, or if ≥2.5 cm, cup without a conspicuous
fringe on the margin 8

8   Leaves glabrous and glaucous beneath .... *Q. alba*

8   Leaves pubescent beneath 9

9   Twigs glabrous; acorn >½ to completely covered by the
cup .... *Q. lyrata*

9   Twigs pubescent; acorn <½ covered by the cup .... *Q. stellata*

1   Leaf margins with awned teeth (if lobed) or with a bristle at the apex (if
unlobed); acorns requiring two years to reach maturity; inner surface of
nut shell tomentose (red/black oaks) 10

10   Leaves entire (unlobed and/or untoothed) 11

11   Leaves obtrullate, or widest above the middle .... *Q. nigra*

11   Leaves widest near the middle and tapering gradually toward
each end 12

12   Leaves <2.5 cm wide at middle, glabrous beneath .... *Q. phellos*

12   Leaves >2.5 cm wide at middle, pubescent
beneath .... *Q. imbricaria*

10   Leaves lobed, toothed, or both 13

13   Leaves broadest near the tip, lobes and teeth indistinct 14

14   Leaf blades >10 cm long, about as broad as long, rusty
tomentose beneath .... *Q. marilandica*

14   Leaf blades <10 cm long, about 2x as long as wide, not rusty
tomentose beneath .... *Q. nigra*

13   Leaves broadest near the middle, lobes and/or teeth
conspicuous 15

15   Lower leaf surface uniformly pubescent 16

16   Leaves rusty-brown pubescent beneath; lobes not curved,
broader toward the tip than at the base .... *Q. velutina*

16   Leaves whitish yellow pubescent beneath; lobes usually
curved, broader toward the base than at the tip 17

17 Leaf lobes mostly 3–5, the central lobe usually much longer than the lateral lobes; base of leaf blade rounded or U-shaped . . . . *Q. falcata*

17 Leaf lobes mostly 7–11, the central lobe rarely longer than the lateral lobes; base of leaf blade cuneate to truncate . . . . *Q. pagoda*

15 Lower leaf surface mostly glabrous, sometimes with tufts of hairs in vein axils 18

18 Terminal buds pubescent throughout . . . . *Q. velutina*

18 Terminal buds glabrous or pubescent on the upper ½ only 19

19 Terminal buds pubescent on upper ½ only 20

20 Leaf blade sinuses extending <½ of the distance to the midrib; lobes acute or oblong; nut without concentric grooves at the apex . . . . *Q. rubra*

20 Leaf blade sinuses extending >½ of the distance to the midrib; lobes becoming broader toward the tip; nut with concentric grooves at the apex . . . . . . *Q. coccinea*

19 Terminal buds glabrous (occasionally with a few hairs at the apex) 21

21 Twigs and/or terminal buds yellowish, gray, or gray-brown 22

22 Acorn cup saucer- or cup-shaped, covering <⅓ of the nut; inner surface of cup glabrous . . . . *Q. shumardii*

22 Acorn cup goblet-shaped, covering >⅓ of the nut; inner surface of cup pubescent . . . . *Q. texana*

21 Twigs and/or terminal buds brown, red, or dark reddish brown 23

23 Lower leaf surface glabrous or sometimes with inconspicuous tufts in vein axils that are visible only with 10X magnification . . . . *Q. rubra*

23 Lower leaf surface with conspicuous tufts in vein axils that are clearly visible to the naked eye 24

24 Lower branches of the tree strongly drooping; acorn cup saucer-shaped, covering <⅓ of the nut; nut <15 mm long . . . . *Q. palustris*

24 Lower branches of the tree spreading and not drooping; acorn cup goblet-shaped, covering >⅓ of the nut; nut >15 mm long . . . . . *Q. texana*

Q. alba L., White O.; dry to mesic woods; common; statewide. Probably the most frequently encountered tree in the state. Plate 261a.

Q. bicolor Willd., Swamp White O.; bottomlands, streambanks, riverine and upland swamps; occasional; known from fewer than 20 counties representing all regions except the U, extremely local W of the TN River. Plate 261b.

Q. coccinea Münchh., Scarlet O.; upland forests; common; statewide, but not often seen in W TN and the CB. Plate 262a.

Q. falcata Michx., Southern Red O.; dry woods, successional fields, fencerows; common; statewide. *Q. falcata* Michx. var. *triloba* (Michx.) Nutt. Plate 262b.

Q. imbricaria Michx., Shingle O.; mesic to dry slopes but more often in bottomlands and on streambanks; frequent; statewide, but infrequent in W TN and the VR. Plate 263a.

Q. lyrata Walter, Overcup O.; bottomlands, streambanks, swamp forests; occasional; statewide, except in the U. Plate 263b.

Q. macrocarpa Michx., Mossycup/Bur O.; calcareous sites, mostly in lowlands and on footslopes; occasional; W TN, WHR, CB. Plate 264a.

Q. marilandica Münchh., Blackjack O.; xeric slopes and ridges, successional fields, fencerows; frequent; statewide, except absent from much of W TN and the CB. Plate 264b.

Q. michauxii Nutt., Swamp Chestnut/Basket/Cow O.; bottomlands, streambanks, swamps; frequent; statewide. Plate 265a.

Q. montana Willd., Chestnut/Mountain O.; dry ridges and upper slopes, often in xeric sites; common; M and E TN, with a few sites in W TN. *Q. prinus* L. Plate 265b.

Q. muhlenbergii Engelm., Chinquapin/Yellow Chestnut O.; bottomlands, ravine, and slope forests; common; statewide. Plate 266a.

Q. nigra L., Water O.; bottomlands, streambanks, swampy forests; common; W TN, then scattered eastward across the state in southern counties. Plate 266b.

Q. pagoda Raf., Cherrybark O.; streambanks, bottomlands, and other wet woods; common; W TN and lower Cumberland River Valley (WHR), with a few scattered stations on the CU and in the VR. *Q. falcata* Michx. var. *pagodifolia* Elliott. Plate 267a.

Q. palustris Münchh., Pin O.; bottomlands, swamps, including upland wet sites; W and M TN, CU. Commonly planted. Plate 267b.

Q. phellos L., Willow O.; bottomlands, streambanks, swamps, upland wet woods, especially on the Pennyroyal Plain; frequent; statewide, except the U. Often seen in lawn plantings. Plate 268a.

Q. prinoides Willd., Dwarf Chestnut O.; bluffs, sandy woods; infrequent; CB, VR. Plate 268b.

Q. rubra L., Northern Red O.; mesic slopes and other well-drained sites; common; statewide. Plate 268c.

Q. shumardii Buckley, Shumard Red O.; bottomlands, streambanks, lower slopes, sometimes on wet upland sites; frequent; statewide, except the U. *Q. shumardii* var. *schneckii* (Britt.) Sarg. Plate 269a.

Q. stellata Wangenh., Post O; dry sites, especially ridges and upper slopes, old fields; frequent; statewide, but not often seen in W TN. Plate 269b.

Q. texana Buckley, Texas Red/Nuttall's O.; bottomlands and streambanks, usually on alluvium; common; MV and extreme western CP. *Q. nuttallii* E.J.Palmer. Plate 270a.

Q. velutina Lam., Black O; dry sites, especially ridges and upper slopes; common; statewide. Plate 270b.

Rhamnaceae   *Rhamnus* (Buckthorn)

Shrubs or small trees; flowers greenish white, solitary or in umbellate, axillary clusters (spring–early summer); fruit a black drupe (fall).

1 Low shrubs <1 m tall; petals absent . . . . *R. alnifolia*
1 Shrubs or small trees >1 m tall; petals present 2
    2 Leaf tip acute to subacuminate, veins conspicuous beneath; sepals, petals, and stamens 5 . . . . *R. caroliniana*
    2 Leaf tip short acuminate, veins inconspicuous beneath; sepals, petals, and stamens 4 . . . . *R. lanceolata*

R. alnifolia L'Hér., Alder B.; damp calcareous woods, with *Thuja*; Campbell (now probably extirpated) and Claiborne Counties (VR). **Endangered.** Plate 271.

R. caroliniana Walter, Carolina B.; margins of open woods and disturbed sites; common; statewide, but infrequent in W TN. Plate 272.

R. lanceolata Pursh, Lance-Leaf B.; open, calcareous woodlands; infrequent and restricted to the CB except for an outlier in Claiborne Co. (VR). Plate 273.

Ericaceae   *Rhododendron* (Rhododendron, Azalea)

Evergreen or deciduous shrubs or small trees; flowers white, pink, purple, or various shades of yellow and red (spring–early summer), fruit a capsule (summer–fall). Perhaps our most popular and florally attractive genus of woody plants. The deciduous azaleas are often morphologically variable and identification is complicated by hybridization and polyploidy. Kron, K. A. 1993. A revision of *Rhododendron* section *Pentanthera*. Edinb. J. Bot. 50:249–364.

1 Leaves evergreen 2
    2 Lower leaf surface lepidote . . . . *R. minus*
    2 Lower leaf surface not lepidote, with glandular or eglandular hairs
       that become matted and obscure 3
        3 Leaf apex acuminate to acute, leaf blade length-width ratio
           ±2.6–5.5; calyx lobes 2–6 mm long; pedicels and ovary with
           gland-tipped hairs . . . . *R. maximum*
        3 Leaf apex rounded, obtuse, or acute, leaf blade length-width ratio
           ±1.5–3; calyx lobes 0.5–1.7 mm long; pedicels and ovary with
           multicellular eglandular hairs . . . . *R. catawbiense*
1 Leaves deciduous 4
    4 Flowers appearing before or with the leaves 5
        5 Upper corolla lobe with a contrasting, often darker, blotch at
           the base 6
            6 Flowers yellow, orange, or red; flower bud scales with
               glandular margins . . . . *R. calendulaceum*
            6 Flowers white, with a yellow blotch on the upper corolla lobe;
               flower bud scales with ciliate margins . . . . *R. alabamense*
        5 Upper corolla lobe without a contrasting or darker blotch
           at the base 7
            7 Flower bud scales glabrous; leaves glabrous or sparsely
               pubescent . . . . *R. periclymenoides*
            7 Flower bud scales densely covered with unicellular hairs; leaves
               moderately to densely pubescent . . . . *R. canescens*
    4 Flowers appearing after the leaves have expanded 8
        8 Stems glabrous . . . . *R. arborescens*
        8 Stems pubescent 9
            9 Flowers yellow, red, or orange . . . . *R. cumberlandense*
            9 Flowers white . . . . *R. viscosum*

R. alabamense Rehder, Alabama A.; open woods, dry slopes; infrequent; M TN, CU. Plate 274.

R. arborescens (Pursh) Torr., Sweet/Smooth A.; streambanks, swamps, balds; frequent; EHR, E TN. Plate 275.

R. calendulaceum (Michx.) Torr., Flame A.; woodlands, balds; infrequent; VR, U. Highly variable and late-blooming individuals are easily confused with the closely related *R. cumberlandense.* Plate 276.

R. canescens (Michx.) Sweet, Southern Pinxterbloom A.; moist or dry woods; frequent; statewide. Plate 277.

R. catawbiense Michx., Mountain Rosebay, Purple Laurel; exposed, rocky ridges and balds; infrequent; E TN. Plate 278.

R. cumberlandense E.L.Braun, Cumberland A.; mixed hardwoods, ridges; infrequent; EHR, E TN. Plate 279.

R. maximum L., Rosebay, Great Laurel; streambanks and rich woods; frequent; EHR, E TN, and a single disjunct population along the Tennessee River in Benton Co. (WHR). Plate 280.

R. minus Michx., exposed ridges, cliffs, balds; infrequent; VR, U. *R. carolinianum* Rehder. Plate 281.

R. periclymenoides (Michx.) Shinners, Wild A., Pinxterbloom; dry or moist woods; frequent; statewide, but more common in E TN. *R. nudiflorum* (L.) Torr. Plate 282.

R. viscosum (L.) Torr., Clammy A.; shrub balds; known only from the GSMNP (U). Plate 283.

Anacardiaceae   *Rhus* (Sumac)

Shrubs or small trees; polygamous; flowers greenish yellow, in racemes or panicles (spring– summer), fruit a red drupe (fall).

    1 Leaflets 3, fragrant; flowers in racemes . . . . *R. aromatica*
    1 Leaflets >3, not fragrant; flowers in panicles 2
        2 Leaf rachis winged; leaflets entire or nearly so . . . . *R. copallinum*
        2 Leaf rachis not winged; leaflets toothed 3
            3 Twigs velvety-pubescent . . . . *R. typhina*
            3 Twigs glabrous . . . . *R. glabra*

R. aromatica Aiton, Fragrant S.; open woods and glades, usually on neutral soils; frequent; statewide except MV and U, most common in the CB. Plate 284.

R. copallinum L., Winged S.; thin woods and open areas; common; statewide. Plate 285.

R. glabra L., Smooth S.; woodland margins and open areas; common; statewide. Plate 286.

R. typhina L., Staghorn S.; woodland margins, thickets, open areas; infrequent; M and E TN. Plate 287.

Saxifragaceae   *Ribes* (Gooseberry/Currant)

Shrubs, sometimes spiny or prickly; flowers yellowish, in racemes or in small clusters (spring–early summer), fruit a red or bluish black berry, sometimes with prickles or bristles (summer–fall).

1 Spines or prickles absent at nodes 2
    2 Calyx tube 2 cm or more long; fruits black, without bristles; basal leaf
       lobes entire or nearly so . . . . *R. odoratum*
    2 Calyx tube <2 cm long; fruit dark red, with glandular bristles; basal
       leaf lobes coarsely toothed . . . . *R. glandulosum*
1 Spines or prickles present at the nodes, rarely absent 3
    3 Ovary/fruit bristly . . . . *R. cynosbati*
    3 Ovary/fruit not bristly 4
       4 Leaves glandular-spotted, pubescent beneath . . . . *R. curvatum*
       4 Leaves not glandular-spotted, glabrous or nearly so
          beneath . . . . *R. rotundifolium*

R. curvatum Small; open, often moist, rocky woodlands; infrequent; CU. Plate 288.

R. cynosbati L., Dogberry; moist woods and heath balds; infrequent; E TN. Plate 289.

R. glandulosum Grauer, Skunk C.; moist, high elevation woods; infrequent; U. Plate 290.

R. odoratum H.L.Wendl., Buffalo C.; known only from bluffs along the Cumberland and Red Rivers in Montgomery Co. (WHR). **Threatened.** Plate 291.

R. rotundifolium Michx., Appalachian G.; rich, high-elevation woods; infrequent; U. Plate 292.

Fabaceae   *Robinia* (Locust)

Shrubs to large trees; flowers white, pink or nearly purple, in racemes (early summer), fruit a legume (fall). These species are often cultivated, and black locust, resistant to decay, is used for fence posts. Numerous named and perhaps some unnamed, mostly sterile and clonal hybrids and introgressants occur in TN, especially in the U. They appear after disturbance and are often short-lived; to attempt to place them in a key format would be futile. See Isely (1998) for a detailed discussion of the "varieties" of the *R. hispida* complex.

1   Flowers pink to nearly purple; young stems, pedicels and calyces
     variously glandular-hispid; shrubs . . . . *R. hispida*
1   Flowers white; young stems, pedicels, and calyces not glandular or
     hispid; medium to large trees . . . . *R. pseudoacacia*

R. hispida L., Bristly L.; thin woods and ruderal sites; frequent; statewide. Commonly cultivated. Plate 293.

R. pseudoacacia L., Black L.; mesic slopes, roadsides, and ruderal sites; common; statewide, but naturalized west of the EHR. Plate 294.

Rosaceae   *Rosa* (Rose)

Shrubs, stems usually prickly and arching, upright, somewhat climbing, or trailing; flowers white to red, solitary or in corymbs or panicles (spring–summer), the individual fruit is a bony achene, several embedded in a fleshy hypanthium at maturity, the entire structure called a hip (summer–fall). Taxonomy is confusing because of variability within species and weak barriers to hybridization. As a result, several hundred taxa have been described from the northern hemisphere. Many cultivars of native and introduced taxa, including the complex garden roses, persist from plantings. Only native taxa and introduced taxa known to escape cultivation are included. Flowers/fruit are often required for identification.

1   Styles united into a column, protruding from the
     hypanthium 2
     2   Leaflets usually 3; petals pink; native . . . . *R. setigera*
     2   Leaflets usually ±5; petals white; introduced . . . . *R. multiflora*
1   Styles distinct, only the stigmas protruding from the hypanthium 3
     3   Sepals unlike in size and shape, the outer pinnatifid;
          introduced . . . . *R. eglanteria*
     3   Sepals alike, none pinnatified; native 4
          4   Leaflets finely toothed, teeth <0.5 mm high; mostly in wet
               habitats . . . . *R. palustris*

    4 Leaflets coarsely toothed, teeth >0.5 mm high; mostly in dry
       habitats 5
       5 Nodal prickles stout, often recurved and flattened toward the
          base; internodal prickles commonly lacking . . . . . . *R. virginiana*
       5 Nodal prickles slender, straight, round at the base; internodal
          prickles usually abundant . . . . *R. carolina*

R. carolina L., Carolina/Pasture R.; open woods, fields, and thickets; common; statewide. Plate 295.

R. eglanteria L.*, Sweetbrier R.; naturalized in ruderal sites; infrequent; M and E TN. Europe. Plate 296.

R. multiflora Thunb. *ex* Murray*, Multiflora R.; naturalized in disturbed sites and fields; frequent; statewide. Originally planted for wildlife food and cover, now an invasive weed. Asia. Plate 297.

R. palustris Marshall, Swamp R.; swamps, wet woods, fields, and thickets; occasional; statewide. Plate 298.

R. setigera Michx., Prairie/Climbing R.; open woods, clearings, thickets, roadsides; occasional; statewide, but infrequent in W TN and on the CU. Plate 299.

R. virginiana Mill., Virginia R.; damp to dry thickets and fields; infrequent; WHR, CB. Plate 300.

Rosaceae   *Rubus* (Raspberry/Dewberry/Blackberry)

Shrubs or subshrubs, with erect, arching, or trailing stems, usually with prickles or bristles, or both; flowers white, pink, or purplish, solitary or in cymes, racemes, panicles, or corymbs (late spring–summer), fruit an aggregate of drupelets (summer). A taxonomically complex genus in TN; apomixis, polyploidy, and hybridization, especially in disturbed sites, is frequent.

    1 Leaves simple, palmately lobed; petals purplish; receptacle nearly
       flat . . . . *R. odoratus*
    1 Leaves pinnately or palmately compound; petals white or pink;
       receptacle conical 2
       2 Petals equal to or shorter than the sepals; fruits separating from the
          receptacle at maturity (raspberries) 3
          3 Pedicels without gland-tipped bristles; fruit black or yellow; stems
             prickly and usually glaucous . . . . *R. occidentalis*

3   Pedicles with slender, gland-tippled bristles; fruit red; stems
bristly, often with a few prickles and usually not glaucous 4
   4   Stems, pedicels, and sepals densely glandular-bristly pubescent;
fruits shiny and sticky, not pubescent . . . . *R. phoenicolasius*
   4   Stems, pedicels, and sepals only slightly glandular-pubescent;
fruits grayish pubescent . . . . *R. idaeus* subsp. *strigosus*
2   Petals longer than the sepals; fruits retained on the deciduous or
persistent receptacle 5
   5   Leaves white or grayish tomentose beneath
(blackberries) . . . . *R. bifrons*
   5   Leaves not white or grayish tomentose beneath 6
      6   Stems trailing (dewberries) 7
         7   Petals ±1 cm long; stems with bristles; stem prickles, if
present, thin and with narrow bases . . . . *R. hispidus*
         7   Petals ±1.5 cm long; bristles present or absent; stem
prickles stiff, broad based, and recurved 8
           8   Flowers mostly solitary; bristles usually
present . . . . *R. trivialis*
           8   Flowers mostly 2 or more/branch, occasionally solitary;
bristles absent . . . . *R. flagellaris*
      6   Stems erect or arching (blackberries) 9
         9   Pedicels with gland-tipped bristles . . . . *R. allegheniensis*
         9   Pedicels without gland-tipped bristles 10
           10   Leaves soft-pubescent beneath . . . . *R. argutus*
           10   Leaves glabrous or with a few scattered hairs 11
              11   Stems of the current season unarmed or with a few
slender prickles . . . . *R. canadensis*
              11   Stems of the current season with stiff
prickles . . . . *R. betulifolius*

R. allegheniensis Porter *ex* L.H.Bailey, Allegheny B.; thickets and disturbed sites; ocasional; WHR and E TN. Plate 301.

R. argutus Link, Southern B.; disturbed sites; common; statewide. *R. pensilvanicus* Poir. Plate 302.

R. betulifolius Small, Highbush B.; disturbed sites; infrequent; M and E TN. Perhaps not distinct from *R. argutus*.

R. bifrons Vest *ex* Tratt.*; persistent in disturbed sites after cultivation; occasional; statewide. Europe. Plate 303.

R. canadensis L., Smooth B.; high-elevation clearings; infrequent; U. Plate 304.

R. flagellaris Willd., Northern D.; disturbed sites; common; statewide. *R. enslenii* Tratt. Plate 305.

R. hispidus L., Swamp D.; disturbed, often damp clearings; frequent; CB, EHR, E TN. Plate 306.

R. idaeus L. ssp. *strigosus* (Michx.) Focke; high-elevation clearings; known only from the GSMNP (U). *R. idaeus* L. ssp. *sachalinensis* (H.Lev.) Focke. Plate 307.

R. occidentalis L., Black R.; disturbed sites; common; statewide. Plate 308.

R. odoratus L., Flowering R.; clearings adjacent to rich woods; occasional; E TN, most common in the U. Plate 309.

R. phoenicolasius Maxim.*, Wineberry; persistent and escaping from cultivation; occasional; M and E TN. E Asia. Plate 310.

R. trivialis Michx., Coastal Plain D.; disturbed sites; infrequent; W TN, WHR, CU, U. Plate 311.

Salicaceae   *Salix* (Willow)

Small shrubs to medium trees; dioecious; flowers in catkins (spring–early summer), fruit a capsule (summer). An extract from the bark of willow was one of the precursors of aspirin. Argus, G. W. 1986. The genus *Salix* (Salicaceae) in the southeastern United States. Syst. Bot. Monog. 9:1–170.

1 Leaves green or pale beneath 2
  2 Bud apex sharp pointed; bud scale margin free and
    overlapping . . . . *S. nigra*
  2 Bud apex blunt; bud scale margin fused 3
    3 Leaves lanceolate or elliptic-lanceolate, the margin
      serrate . . . . *S. eriocephala*
    3 Leaves linear, the margin distinctly glandular-
      denticulate . . . . *S. exigua*
1 Leaves glaucous beneath 4
  4 Bud apex sharp pointed; bud scale margin free and
    overlapping . . . . *S. caroliniana*
  4 Bud apex blunt; bud scale margin fused 5
    5 Leaf margin serrate or serrulate 6
      6 Native shrubs 7

    7 Leaves densely sericeous beneath; stipules reduced to small
glands or absent . . . . *S. sericea*

    7 Leaves glabrous or essentially so beneath; stipules or their
scars evident . . . . *S. eriocephala*

  6 Introduced trees 8

    8 Leaves narrowly lanceolate, the margins serrulate, blades
sericeous beneath . . . . *S. alba*

    8 Leaves very narrowly lanceolate, the margins spinulose-
serrulate, blades nearly glabrous beneath . . . . . . *S. babylonica*

5 Leaf margin entire or crenate 9

  9 Stipules or their scars present; leaf blade 5–9 cm long, petiole
3–7 mm long . . . . *S. humilis* var. *humilis*

  9 Stipules or their scars usually absent; leaf blade 2.5–5 cm long,
petiole 0.5–3 mm long . . . . *S. humilis* var. *microphylla*

S. alba L.*, White W.; infrequently planted and persisting in E TN. Eurasia.

S. babylonica L.*, Weeping W.; frequently planted for its graceful drooping
branches and persisting statewide. Asia. Plate 312.

S. caroliniana Michx., Carolina W.; streams and low areas; frequent; statewide.
Plate 313.

S. eriocephala Michx., Heart-Leaf W.; streambanks and low areas; infrequent;
W TN, WHR. Plate 314.

S. exigua Nutt., Sandbar W.; river bars, floodplains, margins of ponds and streams;
statewide except U, most common on the CP. *S. interior* Rowlee. Plate 315.

S. humilis Marshall var. *humilis*, Upland W.; open upland woods, balds, barrens,
damp fields; frequent; statewide. Plate 316.

S. humilis Marshall var. *microphylla* (Anderson) Fernald; open upland woods, low
fields; occasional; M and E TN. *S. tristis* Aiton. Plate 317.

S. nigra Marshall, Black W.; streambanks, low woods; common; statewide. Plate 318.

S. sericea Marshall, Silky W.; damp openings; occasional; M and E TN. Plate 319.

Caprifoliaceae   *Sambucus* (Elderberry)

Shrubs; flowers white, in terminal cymes (late spring–summer), fruit a berry
(summer–fall).

1 Pith nearly white; cyme flat or nearly so; fruit blue-
   black . . . . *S. canadensis*
1 Pith brownish in older wood; cyme paniculiform; fruit
   red . . . . *S. racemosa* ssp. *pubens*

S. canadensis L., Common E.; open, usually moist sites; common; statewide. Plate 320.

S. racemosa L. ssp. *pubens* (Michx.) House, Red E.; open woodlands at high elevations in the U, deep gorges on the CU; infrequent. *S. pubens* Michx. Plate 321.

Lauraceae   *Sassafras* (Sassafras)

S. albidum (Nutt.) Nees; aromatic, medium to large trees; dioecious; flowers yellowish, in panicles, the pistillate flowers slightly larger than the staminate (spring), fruit a blue-black drupe (fall); open woods and ruderal sites; common; statewide. The roots are used for tea. Plate 322.

Schisandraceae   *Schisandra* (Star Vine)

S. glabra (Bickn.) Rehd.; sprawling or twining vines; monoecious; flowers red, solitary, axillary (late spring), fruit an aggregate of berries (summer); rich bluffs and bottomland forests; infrequent; W TN. **Threatened.** *S. coccinea* Michx. Plate 323.

Solanaceae   *Solanum* (Bittersweet)

S. dulcamara L.*; suffrutescent, climbing or scrambling vines; flowers purplish, in panicles (late spring), fruit a berry (summer); riverbanks and ruderal sites; infrequent; CB, VR, U. Eurasia. Plate 324.

Rosaceae   *Sorbus* (Mountain Ash)

S. americana Marshall; shrubs or small trees; flowers white, in compound corymbs (summer), fruit a pome (fall); balds and open woods at high elevations; infrequent; U. *Pyrus americana* (Marshall) DC. Plate 325.

Rosaceae   *Spiraea* (Spiraea)

Shrubs; flowers white or pink, in branched or unbranched corymbs, umbels, or panicles (summer), fruit a capsule (fall). Numerous cultivars are planted and persist but do not successfully spread from cultivation in TN.

1   Inflorescence of simple, unbranched corymbs or umbels . . . . . S. prunifolia
1   Inflorescence of branched corymbs, umbels, or panicles 2
    2   Inflorescence paniculiform, at least longer than wide 3
        3   Lower leaf surface densely woolly . . . . S. tomentosa
        3   Lower leaf surface glabrous or slightly pubescent . . . . S. alba
    2   Inflorescence corymbiform, at least as wide as or wider than long 4
        4   Leaves entire at the base, but usually with a few teeth near the tip;
            petals white, rarely pink . . . . S. virginiana
        4   Leaves toothed along the entire margin; petals pink . . . . S. japonica

S. alba Du Roi, Meadowsweet; bogs and thickets; infrequent; upper U. **Endangered.** Plate 326.

S. japonica L.f.*, Japanese S.; streambanks and moist slopes; infrequent; M and E TN, but most common on the CU. Japan. Plate 327.

S. prunifolia Siebold & Zucc.*, Plum-Leaf S.; persistent around home sites and occasionally escaping; infrequent; statewide except the in the U. E. Asia. Plate 328.

S. tomentosa L.; Hardhack; wet meadows, marshes, streambanks; occasional; M TN, CU, U; most common on the CU. Plate 329.

S. virginiana Britton, Appalachian S.; damp margins of creeks and rivers; infrequent; CU, U. Often in competition with S. japonica. **Federally Threatened, Endangered.** Plate 330.

Staphyleaceae   Staphylea (Bladder Nut/Possum Cods)

S. trifolia L.; shrubs or small trees; flowers greenish white, in terminal panicles (spring), fruit an inflated capsule (summer); rich woods, especially along waterways; common; statewide. Plate 331.

Theaceae   Stewartia (Mountain Camellia)

S. ovata (Cav.) Weath.; shrubs or small trees; flowers white, solitary in leaf axils (summer), fruit a capsule (fall); woodlands and streambanks; infrequent; E TN. Plate 332.

Styracaceae   Styrax (Storax/Snowbell)

Shrubs or small trees; flowers white, axillary, and solitary or in few to several flowered racemes (spring); fruit a capsule (summer).

1 Leaves stellate beneath, mostly >5 cm wide; flowers
   >4/raceme .... *S. grandifolius*
1 Leaves glabrous beneath or with stellate hairs only on the main veins,
   mostly <5 cm wide; flowers solitary or <4/raceme .... *S. americanus*

S. americanus Lam., American S.; low woods and streambanks; infrequent, mostly
in W TN with a few populations in M TN and on the CU. Plate 333.

S. grandifolius Aiton, Big-Leaf S.; ravines and mesic slopes; infrequent; CP, M TN,
and CU. Plate 334.

Caprifoliaceae   *Symphoricarpos* (Coralberry)

S. orbiculatus Moench; shrubs; flowers greenish, in spikes (summer), fruit a berry
(fall–winter); margin of woods and fallow fields; frequent; statewide, but infrequent in W TN. Plate 335.

Symplocaceae   *Symplocos* (Horse Sugar/Sweet Leaf)

S. tinctoria (L.) L'Hér.; shrubs or small trees; flowers yellowish, in axillary clusters
(spring), fruit a drupe (fall); open woods and bottomlands; widely scattered across
the state (CP, CU, U). **Special Concern.** The young leaves are slightly sweet and
relished by browsing animals. Plate 336.

Tiliaceae   *Tilia* (Basswood/Linden)

Large trees; flowers white, in cymes, the peduncle slender and attached to a
leaflike bract (late spring), fruit nutlike (fall). These are valuable timber trees and
also are prized for honey made from their flowers. Root sprouts often are prolific
and may encircle the tree. Basswood is highly polymorphic in eastern North
America and has long been the subject of taxonomic debate. Various treatments
have recognized from one to 20 species. Hardin, J. W. 1990. Variation patterns
and recognition of *Tilia americana* s.l. Syst. Bot. 15:33–48.

1 Lower leaf surface puberulent with sparsely scattered stellate hairs;
   fruiting peduncles and pedicels glabrous or slightly
   puberulent .... *T. americana* var. *americana*
1 Lower leaf surface pale or whitish, densely stellate-tomentose,
   rarely puberulent; fruiting peduncles stellate-
   tomentose .... *T. americana* var. *heterophylla*

T. americana L. var. *americana;* rich woods and moist slopes; frequent; statewide.

T. americana L. var. *heterophylla* (Vent.) J.W.Loudon; rich woods and moist slopes; common; statewide. Plate 337.

Anacardiaceae   *Toxicodendron* (Poison Ivy/Oak/Sumac)

Vines or small shrubs; flowers greenish yellow, in axillary panicles (spring–early summer), fruit a creamy drupe (fall). All species contain contact poison that results in painful blisters and skin rashes. They are sometimes placed in the genus *Rhus* which has red fruits, terminal inflorescences, and lacks contact poison.

```
1  Leaflets 7–13; shrubs up to 5 m tall  . . . . . . . . . . . . . . . . . . . . . . T. vernix
1  Leaflets 3; vines climbing by aerial roots or shrubs up to 1 m  . . . . . . . . 2
   2  Stems climbing or trailing; fruit glabrous or nearly so  . . . . . T. radicans
   2  Stems mostly erect; fruit pubescent . . . . . . . . . . . . . . . . . T. pubescens
```

T. pubescens Mill., Poison O.; dry woodlands; infrequent; CP, EHR, CU, VR. *Rhus toxicodendron* L., *T. toxicarium* (Salisb.) Gillis. Plate 338.

T. radicans (L.) Kuntze, Poison I.; open woods and margins, fencerows, and disturbed sites; common; statewide. *Rhus radicans* L. Plate 339.

T. vernix (L.) Kuntze, Poison S.; open, wet thickets; infrequent; CU, U. *Rhus vernix* L. Plate 340.

Apocynaceae   *Trachelospermum* (Climbing Dogbane)

T. difforme (Walter) A.Gray; twining vines; flowers yellow-green (early summer), fruit a follicle (late summer); swamps, bottomlands, and wet thickets; infrequent; restricted to W TN except for Robertson (WHR) and Hamilton (VR) Counties. Plate 341.

Ulmaceae   *Ulmus* (Elm)

Medium to large trees; flowers small, in fascicles, racemes, or cymes (spring or fall—see key), fruit a flattened samara with a membranous wing surrounding the fruit body (spring or fall), the wing with marginal cilia in all of our species except *U. rubra.* Several species are commonly planted for shade and ornament. Dutch Elm Disease, resulting from a pathogenic fungus, has destroyed many specimens in both lawn and forest, especially American Elm.

1 Leaves subsessile, petiole ±3 mm long; leaf blade mostly <7 cm
   long 2
   2 Leaf apex rounded; leaf surface pubescent across the entire lower
      surface, rough above; fall flowering . . . . U. crassifolia
   2 Leaf apex acute; leaf surface pubescent beneath only on the major
      veins, smooth above; spring flowering . . . . U. alata
1 Leaves distinctly petiolate, petiole ±3 mm; leaf blade mostly
   >7 cm long 3
   3 Upper leaf surface scabrous, rough to touch (sandpaper-like); fruit
      margin not ciliate . . . . U. rubra
   3 Upper leaf surface glabrous or only slightly scabrous; fruit margin
      ciliate 4
      4 Branches without corky ridges or lines; flowers/fruits in
         fascicles <2.5 cm long; body of fruit glabrous; spring
         flowering . . . . U. americana
      4 Branches with corky ridges or lines, especially on younger trees;
         flowers/fruits in racemes up to 5 cm long; body of fruit pubescent;
         spring or fall flowering 5
         5 Bud scales pubescent; hairs on body of fruit of similar length to
            those of the fruit margin; spring flowering . . . . U. thomasii
         5 Bud scales glabrous; marginal cilia longer than hairs on body of
            fruit; fall flowering . . . . U. serotina

U. alata Michx., Winged E.; alluvial to dry ridge woods, old fields, thickets,
fencerows; common; statewide. Plate 342.

U. americana L., American E.; bottomlands, ravines, swamps, streambanks, old
fields, fencerows; common; statewide. Widely cultivated as a street and lawn tree.
Plate 343.

U. crassifolia Nutt., Cedar E.; riparian and mesic woods, bluffs; infrequent; MV,
CP. **Special Concern.** Plate 344.

U. rubra Muhl., Red/Slippery E.; bottomlands, ravines, mesic slopes, old fields and
fencerows; common; statewide. Sometimes planted. Plate 345.

U. serotina Sarg., September E.; limestone bluffs, rocky woods, bottomlands, riparian forests; occasional; M TN, with outliers in the MV and the VR. Plate 346.

U. thomasii Sarg., Rock E.; rocky woods, riparian sites, floodplains; occasional; M
TN, with a few reports from W TN and on the CU. This species has been credited to Tennessee since Gattinger (1901), including the FNA (1997) treatment.

Many of the specimens we used to document its distribution are sterile and as Shanks (1952) noted, are difficult to distinguish from *U. americana*. Further study is needed to document the distribution, and perhaps actual presence, of Rock Elm in Tennessee. Plate 347.

Ericaceae   *Vaccinium*

Deciduous or evergreen shrubs or small trees; flowers white to pink, in racemes or, rarely, solitary in leaf axils (late spring–summer), fruit a berry (summer–fall). A taxonomically difficult genus complicated by hybridization and polyploidy resulting in a plethora of named taxa. The highbush group presents the greatest problems and there is significant taxonomic disagreement among various authors. Vander Kloet, S.P. 1980. The taxonomy of the highbush blueberry, *Vaccinium corymbosum*. Canad. J. Bot. 58:1187–1201; Uttall, L. J. 1986. Updating the genus *Vaccinium* L. (Ericaceae) in West Virginia. Castanea 51:197–201; Uttall, L. J. 1987. The genus *Vaccinium* L. in Virginia. Castanea 52:231–261; Vander Kloet, S. P. 1988. The genus *Vaccinium* in North America. Publication 1828, Research Branch, Agriculture Canada, Ottawa.

1 Corolla lobes 4, the lobes longer than the tube; stamens 8 2
 2 Leaves deciduous, >2 cm long; erect shrubs . . . . *V. erythrocarpum*
 2 Leaves evergreen, <2 cm long; trailing to erect
    shrubs . . . . *V. macrocarpon*
1 Corolla lobes 5, the lobes shorter than the tube; stamens 10 3
 3 Corolla open-campanulate, broadest at the apex; anthers
    awned 4
   4 Stamens exserted at flowering; flowers subtended by a leafy bract;
      leaves deciduous . . . . *V. stamineum*
   4 Stamens not exserted at flowering; flowers not subtended by a
      leafy bract; leaves shiny, tardily deciduous . . . . *V. arboreum*
 3 Corolla urceolate, constricted at the apex; anthers not awned 5
   5 Corolla, fruit, and twigs densely pubescent . . . . *V. hirsutum*
   5 Corolla, fruit, and twigs glabrous or essentially so 6
     6 Leaf blades <3.5 cm long; corolla >2X as long as
        wide . . . . *V. elliottii*
     6 Leaf blades >3.5 cm long; corolla <2X as long as wide 7
       7 Plants 8 dm or less tall . . . . *V. pallidum*
       7 Plants >8 dm tall 8
         8 Leaves densely pubescent beneath . . . . *V. fuscatum*
         8 Leaves glabrous, often glaucous, or with a few hairs along
            the midrib . . . . *V. corymbosum*

V. arboreum Marshall, Farkleberry; dry woods and slopes; common; statewide. Plate 348.

V. corymbosum L., Highbush Blueberry; damp-dry woodlands and clearings; common in E TN, less frequent in W and M TN. *V. constablaei* A.Gray. Plate 349.

V. elliottii Chapm., Mayberry; pond margins; Coffee Co. (EHR). **Endangered.** Some recent authors place this taxon in synonymy with *V. corymbosum*. To us, it is distinct by its smaller leaves and primarily Coastal Plain distribution. Plate 350.

V. erythrocarpum Michx., Bearberry; mid to upper-elevation woods in the U and on moist sandstone ledges on the CU; infrequent. Plate 351.

V. fuscatum Aiton, Highbush Blueberry; damp woodlands; statewide, except in the CB, most frequent on the CU. Scarcely distinct from *V. corymbosum*. *A. atrococcum* (A.Gray) A.Heller. Plate 352.

V. hirsutum Buckley, Woolly Blueberry; dry pine ridges; infrequent; southern CU and U. This is our only blueberry with hairy, yet edible fruits. Plate 353.

V. macrocarpon Aiton, Cranberry; bogs and sandy, damp, open woodlands; infrequent; EHR, U. **Threatened.** With the exception of the Shady Valley Bog (Johnson Co.) sites, some of our populations may have been introduced to altered habitats. Plate 354.

V. pallidum Aiton, Lowbush Blueberry; dry woodlands; common; M and E TN. Perhaps the tastiest of our blueberries. *V. vacillans* Kalm *ex* Torr. Plate 355.

V. stamineum L., Deerberry; dry openings and margins of woodlands; common; statewide. Fruit often abundant, but scarcely edible. Plate 356.

Caprifoliaceae   *Viburnum*

Contributed by Tim J. Weckman

Shrubs or small trees; flowers white to creamy and sometimes tinged with pink, in compound cymes (spring–early summer), fruit a single seeded blue or black drupe (fall).

1   Buds naked; leaves broadly ovate, serrulate, often >10 cm wide; marginal
    flowers enlarged, sterile . . . . *V. lantanoides*
1   Buds valvate or imbricate; leaf shape and margins variable, but <10 cm
    wide; flowers of equal size, all fertile 2
      2   Leaves 3-lobed, palmately veined . . . . *V. acerifolium*

2 Leaves unlobed, pinnately veined 3
   3 Buds valvate; veins anastomosing before reaching leaf
      margin 4
      4 Lower leaf surface (at maturity) glabrous or essentially so
         or with rusty-brown upright, branched hairs; leaf margin
         serrulate; flowers/fruits on a common peduncle 5 mm or
         less long 5
         5 Buds and lower leaf surface glabrous or essentially so; leaves
           dull above; fruit calyx stipitate . . . . *V. prunifolium*
         5 Buds and lower leaf surface with upright, branched hairs
           (10X), tending to be denser near the midrib and primary
           lateral veins; leaves shiny above; fruit calyx
           sessile . . . . *V. rufidulum*
      4 Lower leaf surface (at maturity) with closely appressed rusty-
         brown scales (10X); leaf margin irregularly serrate, crenate,
         entire, or revolute; flowers/fruits on common peduncle
         5–50 mm long 6
         6 Leaves pubescent on and adjacent to the leaf margin; apex
           mostly rounded, rarely cuspidate . . . . *V. nudum*
         6 Leaves glabrous on and adjacent to the leaf margin; apex
           acuminate . . . . *V. cassinoides*
   3 Buds imbricate; veins ending in a marginal tooth 7
      7 Petioles and leaf bases with red glandular hairs; leaves cordate
         to subcordate; fruit elliptical; base of style glabrous 8
         8 Leaves lance-ovate, rounded to subcordate, marginal
           teeth <12/side; petioles absent or up to
           1 cm long . . . . *V. rafinesquianum*
         8 Leaves broadly ovate, cordate, marginal teeth >12/per side;
           petioles >1 cm long 9
           9 Lower ¼ of leaf lacking teeth; petioles 1.25 cm long;
              bark tight throughout; inflorescence
              bracteate . . . . *V. bracteatum*
           9 Leaves toothed throughout; petiole 1.5–2.5 cm long;
              bark exfoliating on mature stems; inflorescence not
              bracteate . . . . *V. molle*
      7 Petiole and leaf base pubescence variable, not red
         glandular; leaves elliptic-ovate; fruit rounded; base of style
         pubescent . . . . *V. dentatum*

V. acerifolium L., Maple-Leaf Viburnum; deciduous woodlands; frequent; M and E TN. Plate 357.

V. bracteatum Rehd., Limerock Arrow Wood; calcareous woods and ledges along streams; Franklin Co. (CU). **Endangered.** Plate 358.

V. cassinoides L., Withe Rod; streambanks in rich woods; occasional; E TN. Plate 359.

V. dentatum L., Arrow Wood; mesic woods, open meadows and streambanks; frequent; WHR, EHR, E TN. Plate 360.

V. lantanoides Michx., Witch Hobble; mesic woods, coves at middle to upper elevations; occasional; U. Plate 361.

V. molle Michx., Kentucky Viburnum; calcareous woods and ledges along streams; Smith Co. (EHR). Plate 362.

V. nudum L., Possum Haw; open meadows, streambanks, swamps; occasional; CP, EHR, CU. Plate 363.

V. prunifolium L., Plum-Leaf Viburnum; mesic woods, thickets; occasional in W and M TN, frequent in E TN. Plate 364.

V. rafinesquianum Schult., Rafinesque's Viburnum; calcareous woods; infrequent; VR, U. Plate 365.

V. rufidulum Raf., Rusty Blackhaw; woodlands and thickets; common; statewide. Plate 366.

Apocynaceae   *Vinca* (Periwinkle/Graveyard Plant)

Trailing or mat forming, only slightly woody evergreens; flowers blue to lavender (spring), axillary and usually solitary, fruit a follicle (summer).

1   Leaf margin ciliate, blades >2.5 cm wide . . . . *V. major*
1   Leaf margin without cilia, blades <2.5 cm wide . . . . *V. minor*

V. major L.*, Greater P.; escaping from cultivation and occasionally spreading to woodlands; infrequent; statewide. More robust than the following species. S Europe. Plate 367.

V. minor L.*, P.; escaping from cultivation and spreading to roadsides and woodlands; frequent; statewide. More often escaping than the preceding species. S Europe. Plate 368.

Vitaceae   *Vitis* (Grape)

Vines, climbing by tendrils; polygamous; flowers greenish white, in panicles (spring), fruit a pulpy berry (summer–fall). Hybridization is common and most species are morphologically variable in leaf shape and pubescence, often leading to taxonomic difficulty. Numerous cultivars of European and American species are grown for fruit and may persist around old home places and orchards. Moore, M. O. 1991. Classification and systematics of eastern North American *Vitis* L. (Vitaceae) north of Mexico. Sida 14:339–367.

1 Tendrils simple; bark tight, with prominent lenticels; pith continuous through nodes . . . . *V. rotundifolia*
1 Tendrils bifid to trifid; bark shredding, the lenticels inconspicuous; pith interrupted by nodal diaphragms 2
  2 Mature leaves glaucous beneath; nodes glaucous or not 3
    3 Mature berries <9 mm in diameter; mature leaves glabrous to glabrate beneath; nodes usually glaucous; nodal diaphragms usually <2 mm in diameter . . . . *V. aestivalis* var. *bicolor*
    3 Mature berries >9 mm in diameter; mature leaves slightly to heavily arachnoid pubescent beneath; nodes usually not glaucous; nodal diaphragms usually >2 mm in diameter . . . . *V. aestivalis* var. *aestivalis*
  2 Mature leaves not glaucous beneath; nodes not glaucous 4
    4 Tendrils or inflorescences present at 3-many consecutive nodes . . . . *V. labrusca*
    4 Tendrils or inflorescences present at only two consecutive nodes 5
      5 Leaves reniform, glabrous beneath at maturity; tendrils, when present, only opposite the uppermost nodes . . . . *V. rupestris*
      5 Leaves cordate to cordate-ovate, glabrous to pubescent beneath at maturity; tendrils present opposite most nodes 6
        6 Nodal diaphragms <1 mm wide, usually <0.5 mm wide; growing tips enveloped by enlarging, unfolding leaves . . . . *V. riparia*
        6 Nodal diaphragms >1 mm wide; growing tips not enveloped by enlarging, unfolding leaves 7
          7 Twigs more or less round, glabrous or arachnoid pubescent; mature berries usually >8 mm in diameter; nodes usually not banded with red pigmentation 8
            8 Nodal diaphragms >2.5 mm wide; leaf apex usually long acuminate; branchlets of the season purple-reddish . . . . *V. palmata*

8 Nodal diaphragms <2.5 mm wide; leaf apex usually acute to short acuminate; branchlets of the season gray to green or brown or with purplish pigmentation only on one side of the branchlet . . . . *V. vulpina*

7 Twigs angled, arachnoid and/or hirtellous pubescent, varying to glabrate; mature berries <8 mm in diameter; nodes frequently with red bands 9

9 Branchlets of the season sparsely to densely hairy, often with arachnoid pubescence as well; leaves more or less uniformly pubescent on veins below . . . . *V. cinerea* var. *cinerea*

9 Branchlets of the season without evident hairs (if present, then concealed by arachnoid pubescence); leaves usually without hairs below, or only with a few sparse hairs . . . . *V. cinerea* var. *baileyana*

V. aestivalis Michx. var. *aestivalis*, Summer/Pigeon G.; well-drained woodlands, thickets, fencerows; common; statewide. Plate 369.

V. aestivalis Michx. var. *bicolor* Deam; same habitats as the typical var.; frequent; U, with a few occurrences westward to southern WHR.

V. cinerea (Engelm. in A.Gray) Engelm. *ex* Millardet var. *baileyana* (Munson) Comeaux, Possum G.; moist soils, especially floodplains, low woods and streambanks; occasional; statewide. Plate 370.

V. cinerea (Engelm. in A.Gray) Engelm. *ex* Millardet var. *cinerea*, Downy/Sweet Winter G.; mostly low woods and thickets; infrequent; W TN, with a few occurrences on the WHR. Plate 371.

V. labrusca L., Fox G.; upland and lowland woods, fencerows, thickets; occasional; WHR, EHR, E TN. Plate 372.

V. palmata Vahl, Red/Cat G.; riverbanks and riparian forests; common; W TN, with a few occurrences on the WHR. Plate 373.

V. riparia Michx., Riverside/Frost G.; mostly in moist soils of streambanks and low woodlands, thickets and fencerows; infrequent; W TN, WHR, CB. Plate 374.

V. rotundifolia Michx., Muscadine; both lowland and upland sites, usually in woodlands; common; statewide, but mostly absent in the CB. Plate 375.

V. rupestris Scheele, Sand/Sugar G.; calcareous sites, usually in moist soils along riverbanks and in low woods; infrequent; CB. **Endangered—Possibly Extirpated.**

V. vulpina L., Frost G.; mostly well-drained upland sites, woodlands, thickets, fencerows; common; statewide. Plate 376.

Fabaceae   *Wisteria* (Wisteria)

Twining vines; flowers pinkish purple, blue, or white, in terminal, pendent racemes (spring); fruit a legume (summer–fall).

1   Ovary and fruit glabrous; pedicels mostly 5–10 mm long; corolla
    1.5–2 cm long; native species of low woodlands, thickets, and
    shorelines . . . . *W. frutescens*
1   Ovary and fruit villous; pedicels 15–20 mm long; corolla
    1.5–2.5 cm long; introduced species of urban sites 2
    2   Leaflets 7–11(–13); inflorescence 1–2 dm long, the flowers opening
        almost simultaneously . . . . *W. sinensis*
    2   Leaflets 13–19; inflorescence >2 dm long, the flowers opening from
        the base of the raceme to the apex . . . . *W. floribunda*

W. floribunda (Willd.) DC.*, Japanese W.; widely planted statewide and persisting around home sites. Japan. *Rehsonia floribunda* (Willd.) Stritch. Plate 377.

W. frutescens (L.) Poir., American W.; margins of wet woodlands, swamps, and waterways; frequent; statewide, but most common in W TN. *W. macrostachya* (Torr. & A.Gray) B.L.Rob. & Fernald. Plate 378.

W. sinensis (Sims) Sweet*, Chinese W.; infrequently planted statewide and its level of establishment is unknown. China. *Rehsonia sinensis* (Sims) Stritch.

Ranunculaceae   *Xanthorhiza* (Yellowroot)

X. simplicissima Marshall; low shrubs; flowers purple-brown, in axillary panicles (spring), fruit a follicle (summer–fall); streambanks and margins of moist woods; frequent in E TN, and a few disjunct populations in W and M TN. The yellow inner bark is used medicinally and for dye. Plate 379.

Rutaceae   *Zanthoxylum* (Toothache Tree/Prickly Ash)

Z. americanum Mill.; shrubs or small trees; dioecious; flowers yellow-green (spring), in axillary clusters, fruit a follicle (summer); cedar glade thickets and margins of woods; infrequent; WHR, CB, U. **Special Concern.** Most populations are persistent after cultivation. The bark is chewed as a folk remedy for toothache. Plate 380.

Appendix 1

Distribution of Tennessee's Native Woody Species and Lesser
Taxa within the Physiographic Provinces

Taxa	MV	CP	WHR	CB	EHR	CU	VR	U
Abies fraseri								+
Acer drummondii	+	+		+	+			
Acer negundo	+	+	+	+	+	+	+	+
Acer pensylvanicum						+	+	+
Acer rubrum	+	+	+	+	+	+	+	+
Acer saccharinum	+	+	+	+	+	+	+	+
Acer saccharum ssp. *floridanum*		+	+					
Acer saccharum ssp. *leucoderme*							+	+
Acer saccharum ssp. *nigrum*		+	+	+	+	+	+	+
Acer saccharum ssp. *saccharum*	+	+	+	+	+	+	+	+
Acer spicatum						+	+	+
Aesculus flava			+	+	+	+	+	+
Aesculus glabra			+	+	+		+	
Aesculus pavia	+	+	+		+	+	+	
Aesculus sylvatica						+	+	
Alnus serrulata	+	+	+	+	+	+	+	+
Alnus viridis ssp. *crispa*								+
Amelanchier arborea		+	+	+	+	+	+	+
Amelanchier canadensis						+	+	+
Amelanchier laevis		+	+	+	+	+	+	+

Taxa	MV	CP	WHR	CB	EHR	CU	VR	U
Amelanchier sanguinea						+	+	+
Amorpha fruticosa	+	+	+	+	+	+	+	+
Amorpha glabra			+	+	+	+	+	+
Ampelopsis arborea*	+	+						
Ampelopsis cordata	+	+	+	+	+	+	+	+
Aralia spinosa	+	+	+	+	+	+	+	+
Aristolochia macrophylla			+		+	+	+	+
Aristolochia tomentosa	+	+	+	+	+	+		
Aronia arbutifolia		+	+		+	+	+	+
Aronia melanocarpa		+	+		+	+	+	+
Arundinaria gigantea	+	+	+	+	+	+	+	+
Asimina triloba	+	+	+	+	+	+	+	+
Berberis canadensis						+	+	+
Berchemia scandens	+	+	+	+	+	+	+	
Betula alleghaniensis						+	+	+
Betula cordifolia								+
Betula lenta						+	+	+
Betula nigra	+	+	+	+	+	+	+	+
Bignonia capreolata	+	+	+	+	+	+	+	+
Brunnichia ovata	+	+	+					
Buckleya distichophylla								+
Bumelia lycioides	+	+	+	+	+	+	+	
Buxella brachycera						+		
Callicarpa americana*	+	+	+	+	+	+	+	+
Calycanthus floridus var. floridus			+		+	+	+	+
Calycanthus floridus var. glaucus						+	+	+
Calycocarpum lyonii	+	+	+	+	+	+	+	
Campsis radicans	+	+	+	+	+	+	+	+
Carpinus caroliniana	+	+	+	+	+	+	+	+
Carya aquatica	+	+						
Carya cordiformis	+	+	+	+	+	+	+	+
Carya glabra	+	+	+	+	+	+	+	+
Carya illinoinensis*	+	+	+					
Carya laciniosa	+	+	+	+	+	+	+	+
Carya ovata var. australis		+	+	+	+	+	+	+
Carya ovata var. ovata	+	+	+	+	+	+	+	+
Carya pallida		+	+		+	+	+	+
Carya tomentosa	+	+	+	+	+	+	+	+

Taxa	MV	CP	WHR	CB	EHR	CU	VR	U
Castanea dentata	+	+	+	+	+	+	+	+
Castanea pumila					+	+	+	+
Catalpa speciosa*	+	+						
Ceanothus americanus		+	+	+	+	+	+	+
Celastrus scandens	+	+	+	+	+	+	+	+
Celtis laevigata	+	+	+	+	+	+	+	+
Celtis occidentalis	+	+	+	+	+	+	+	+
Celtis tenuifolia	+	+	+	+	+	+	+	+
Cephalanthus occidentalis	+	+	+	+	+	+	+	+
Cercis canadensis	+	+	+	+	+	+	+	+
Chimaphila maculata		+	+	+	+	+	+	+
Chionanthus virginicus			+	+	+	+	+	+
Cladrastis kentukea*	+	+	+	+	+	+	+	+
Clematis catesbyana			+	+	+	+		
Clematis virginiana	+	+	+	+	+	+	+	+
Clethra acuminata						+		+
Clethra alnifolia					+			
Cocculus carolinus	+	+	+	+	+	+	+	+
Comptonia peregrina						+		
Conradina verticillata					+	+		
Cornus alternifolia			+	+	+	+	+	+
Cornus amomum		+	+	+	+	+	+	+
Cornus drummondii	+	+	+	+	+	+	+	
Cornus florida	+	+	+	+	+	+	+	+
Cornus foemina	+	+	+	+	+	+		
Corylus americana	+	+	+	+	+	+	+	+
Corylus cornuta						+	+	+
Cotinus obovatus						+		
Crataegus calpodendron			+	+	+	+	+	+
Crataegus collina			+	+		+	+	
Crataegus crus-galli	+	+	+	+	+	+	+	+
Crataegus harbisonii	+	+	+	+				
Crataegus intricata			+		+	+	+	+
Crataegus macrosperma		+	+		+	+	+	+
Crataegus marshallii		+	+				+	
Crataegus mollis			+	+				
Crataegus phaenopyrum		+	+	+		+		
Crataegus pruinosa		+	+	+	+	+	+	+

Taxa	MV	CP	WHR	CB	EHR	CU	VR	U
Crataegus punctata				+	+	+		+
Crataegus spathulata			+			+	+	
Crataegus uniflora			+		+		+	+
Crataegus viridis	+	+	+	+	+	+	+	+
Croton alabamensis					+			
Decodon verticillatus	+			+	+		+	
Decumaria barbara		+	+		+		+	+
Diervilla lonicera			+			+		+
Diervilla rivularis						+	+	+
Diervilla sessilifolia						+	+	+
Diospyros virginiana	+	+	+	+	+	+	+	+
Dirca palustris			+	+	+	+	+	+
Epigaea repens			+			+	+	+
Euonymus americanus	+	+	+	+	+	+	+	+
Euonymus atropurpureus	+	+	+	+	+	+	+	+
Euonymus obovatus			+		+		+	+
Fagus grandifolia	+	+	+	+	+	+	+	+
Forestiera acuminata	+	+	+	+		+		
Forestiera ligustrina			+	+	+	+		
Fothergilla major						+	+	+
Fraxinus americana	+	+	+	+	+	+	+	+
Fraxinus pennsylvanica	+	+	+	+	+	+	+	+
Fraxinus profunda	+	+						
Fraxinus quadrangulata			+	+	+	+	+	
Gaultheria procumbens					+	+	+	+
Gaylussacia baccata			+		+	+	+	+
Gaylussacia dumosa					+	+		
Gaylussacia ursina								+
Gelsemium sempervirens						+	+	
Gleditsia aquatica	+	+						
*Gleditsia triacanthos**	+	+	+	+	+	+	+	
*Gymnocladus dioicus**	+	+	+	+			+	
Halesia tetraptera			+			+	+	+
Hamamelis virginiana			+	+	+	+	+	+
Hydrangea arborescens	+	+	+	+	+	+	+	+
Hydrangea cinerea	+	+	+	+	+	+	+	+
*Hydrangea quercifolia**		+	+		+	+		
Hydrangea radiata								+

Taxa	MV	CP	WHR	CB	EHR	CU	VR	U
Hypericum crux-andreae			+		+	+	+	+
Hypericum densiflorum			+		+	+	+	+
Hypericum frondosum		+	+	+	+	+	+	+
Hypericum hypericoides	+	+	+	+	+	+	+	+
Hypericum lobocarpum		+			+			
Hypericum nudiflorum						+		
Hypericum prolificum		+	+	+	+	+	+	+
Hypericum stragulum		+	+	+	+	+	+	+
Ilex ambigua var. ambigua			+			+	+	+
Ilex ambigua var. montana					+	+	+	+
Ilex decidua var. decidua	+	+	+	+	+	+	+	
Ilex decidua var. longipes	+	+	+		+	+		
Ilex opaca*	+	+	+		+	+	+	+
Ilex verticillata		+	+		+	+	+	+
Itea virginica	+	+	+	+	+	+	+	+
Juglans cinerea	+	+	+	+	+	+	+	+
Juglans nigra	+	+	+	+	+	+	+	+
Juniperus virginiana	+	+	+	+	+	+	+	+
Kalmia angustifolia var. carolina								+
Kalmia latifolia			+	+	+	+	+	+
Leiophyllum buxifolium								+
Leucothoe fontanesiana						+	+	+
Leucothoe racemosa			+			+	+	
Leucothoe recurva								+
Lindera benzoin	+	+	+	+	+	+	+	+
Linnaea borealis								+
Liquidambar styraciflua	+	+	+	+	+	+	+	+
Liriodendron tulipifera	+	+	+	+	+	+	+	+
Lonicera canadensis								+
Lonicera dioica					+		+	+
Lonicera flava—see text								
Lonicera reticulata				+				
Lonicera sempervirens*	+	+	+	+	+	+	+	+
Lyonia ligustrina		+			+	+	+	+
Magnolia acuminata	+	+	+		+	+	+	+
Magnolia fraseri						+		+
Magnolia macrophylla			+	+	+	+	+	+
Magnolia tripetala			+		+	+	+	+

Taxa	MV	CP	WHR	CB	EHR	CU	VR	U
Magnolia virginiana		+						+
Malus angustifolia		+	+	+	+	+	+	+
Malus coronaria			+			+		+
Menispermum canadense	+	+	+	+	+	+	+	+
Menziesia pilosa								+
Mitchella repens	+	+	+	+	+	+	+	+
Morus rubra	+	+	+	+	+	+	+	+
Nemopanthus collinus								+
Nestronia umbellula					+			+
Neviusia alabamensis				+	+			
Nyssa aquatica	+	+	+		+	+		
Nyssa biflora	+	+	+		+			
Nyssa sylvatica	+	+	+	+	+	+	+	+
Opuntia humifusa		+	+	+	+	+	+	+
Ostrya virginiana	+	+	+	+	+	+	+	+
Oxydendrum arboreum		+	+	+	+	+	+	+
Parthenocissus quinquefolia	+	+	+	+	+	+	+	+
Paxistima canbyi							+	
Philadelphus hirsutus			+	+	+	+	+	+
Philadelphus inodorus		+	+	+	+	+		+
Philadelphus pubescens var. *intectus*			+	+	+		+	
Philadelphus pubescens var. *pubescens*			+	+	+		+	+
Phoradendron leucarpum	+	+	+	+	+	+	+	+
Physocarpus opulifolius			+	+	+	+	+	+
Picea rubens								+
Pieris floribunda								+
Pinus echinata			+	+	+	+	+	+
Pinus pungens						+	+	+
Pinus rigida							+	+
Pinus strobus			+			+	+	+
*Pinus taeda**		+	+	+	+	+	+	+
Pinus virginiana		+	+	+	+	+	+	+
Planera aquatica	+	+	+					
Platanus occidentalis	+	+	+	+	+	+	+	+
Polygonella americana						+		
Populus deltoides	+	+	+	+	+	+	+	
Populus grandidentata		+	+	+	+	+		+
Populus heterophylla	+	+	+		+			

Taxa	MV	CP	WHR	CB	EHR	CU	VR	U
Potentilla tridentata								+
Prunus americana	+	+	+	+	+	+	+	+
Prunus angustifolia	+	+	+	+	+	+	+	+
Prunus hortulana			+	+	+	+	+	+
Prunus mexicana	+	+	+	+	+	+		
Prunus munsoniana	+	+	+	+	+	+	+	+
Prunus pensylvanica								+
Prunus pumila					+			
Prunus serotina	+	+	+	+	+	+	+	+
Prunus umbellata			+	+	+	+		
Prunus virginiana								+
Ptelea trifoliata			+	+	+	+	+	+
Pyrularia pubera						+		+
Quercus alba	+	+	+	+	+	+	+	+
Quercus bicolor	+		+	+	+	+	+	
Quercus coccinea	+	+	+	+	+	+	+	+
Quercus falcata	+	+	+	+	+	+	+	+
Quercus imbricaria	+	+	+	+	+	+	+	+
Quercus lyrata	+	+	+	+	+	+	+	
Quercus macrocarpa	+	+	+	+				
Quercus marilandica		+	+	+	+	+	+	+
Quercus michauxii	+	+	+	+	+	+	+	+
Quercus montana		+	+	+	+	+	+	+
Quercus muhlenbergii	+	+	+	+	+	+	+	+
Quercus nigra	+	+	+		+	+	+	
Quercus pagoda	+	+	+			+	+	
Quercus palustris	+	+	+	+	+	+		
Quercus phellos	+	+	+	+	+	+	+	
Quercus prinoides				+			+	
Quercus rubra	+	+	+	+	+	+	+	+
Quercus shumardii	+	+	+	+	+	+	+	
Quercus stellata		+	+	+	+	+	+	+
Quercus texana	+	+						
Quercus velutina	+	+	+	+	+	+	+	+
Rhamnus alnifolia							+	
Rhamnus caroliniana		+	+	+	+	+	+	+
Rhamnus lanceolata				+			+	
Rhododendron alabamense			+	+	+	+		

Taxa	MV	CP	WHR	CB	EHR	CU	VR	U
Rhododendron arborescens					+	+	+	+
Rhododendron calendulaceum							+	+
Rhododendron canescens		+	+	+	+	+	+	+
Rhododendron catawbiense						+	+	+
Rhododendron cumberlandense					+	+	+	+
Rhododendron maximum			+		+	+	+	+
Rhododendron minus							+	+
Rhododendron periclymenoides		+	+	+	+	+	+	+
Rhododendron viscosum								+
Rhus aromatica		+	+	+	+	+	+	
Rhus copallinum	+	+	+	+	+	+	+	+
Rhus glabra	+	+	+	+	+	+	+	+
Rhus typhina				+	+	+	+	+
Ribes curvatum						+		
Ribes cynosbati						+	+	+
Ribes glandulosum								+
Ribes odoratum			+					
Ribes rotundifolium								+
*Robinia hispida**		+	+	+	+	+	+	+
*Robinia pseudoacacia**					+	+	+	+
Rosa carolina		+	+	+	+	+	+	+
Rosa palustris	+	+	+	+	+	+	+	+
Rosa setigera	+	+	+	+	+	+	+	+
Rosa virginiana			+	+				
Rubus allegheniensis		+				+	+	+
Rubus argutus	+	+	+	+	+	+	+	+
Rubus betulifolius			+	+	+	+	+	+
Rubus canadensis								+
Rubus flagellaris	+	+	+	+	+	+	+	+
Rubus hispidus				+	+	+	+	+
Rubus idaeus ssp. *strigosus*								+
Rubus occidentalis	+	+	+	+	+	+	+	+
Rubus odoratus						+	+	+
Rubus trivialis	+	+	+			+		+
Salix caroliniana	+	+	+	+	+	+	+	+
Salix eriocephala	+	+	+					
Salix exigua	+	+	+	+	+	+	+	
Salix humilis var. *humilis*		+	+	+	+	+	+	+

Taxs	MV	CP	WHR	CB	EHR	CU	VR	U
Salix humilis var. microphylla			+	+	+	+	+	+
Salix nigra	+	+	+	+	+	+	+	+
Salix sericea			+	+	+	+	+	+
Sambucus canadensis	+	+	+	+	+	+	+	+
Sambucus racemosa ssp. pubens						+		+
Sassafras albidum	+	+	+	+	+	+	+	
Schisandra glabra	+	+						
Smilax bona-nox	+	+	+	+	+	+	+	+
Smilax glauca	+	+	+	+	+	+	+	+
Smilax rotundifolia	+	+	+	+	+	+	+	+
Smilax tamnoides	+	+	+	+	+	+	+	+
Sorbus americana								+
Spiraea alba								+
Spiraea tomentosa			+	+	+	+		+
Spiraea virginiana						+		+
Staphylea trifolia	+	+	+	+	+	+	+	+
Stewartia ovata						+	+	+
Styrax americanus	+	+	+		+	+		
Styrax grandifolius		+	+	+	+	+		
Symphoricarpos orbiculatus			+	+	+	+	+	+
Symplocos tinctoria		+				+		+
Taxodium distichum	+	+	+					
Taxus canadensis						+		
Thuja occidentalis					+	+	+	+
Tilia americana var. americana	+	+	+	+		+	+	+
Tilia americana var. heterophylla	+	+	+	+	+	+	+	+
Toxicodendron pubescens		+			+	+	+	
Toxicodendron radicans	+	+	+	+	+	+	+	+
Toxicodendron vernix						+		+
Trachelospermum difforme	+	+	+				+	
Tsuga canadensis					+	+	+	+
Tsuga caroliniana								+
Ulmus alata	+	+	+	+	+	+	+	+
Ulmus americana	+	+	+	+	+	+	+	+
Ulmus crassifolia	+	+						
Ulmus rubra	+	+	+	+	+	+	+	+
Ulmus serotina	+		+	+	+		+	
Ulmus thomasii	+	+	+	+	+	+		

Taxa	MV	CP	WHR	CB	EHR	CU	VR	U
Vaccinium arboreum	+	+	+	+	+	+	+	+
Vaccinium corymbosum		+	+		+	+	+	+
Vaccinium elliottii					+			
Vaccinium erythrocarpum						+		+
Vaccinium fuscatum		+	+		+	+	+	+
Vaccinium hirsutum						+		+
Vaccinium macrocarpon					+			+
Vaccinium pallidum			+	+	+	+	+	+
Vaccinium stamineum		+	+	+	+	+	+	+
Viburnum acerifolium			+	+	+	+	+	+
Viburnum bracteatum						+		
Viburnum cassinoides						+	+	+
Viburnum dentatum			+		+	+	+	+
Viburnum lantanoides								+
Viburnum molle					+			
Viburnum nudum		+			+	+		
Viburnum prunifolium	+	+	+	+	+	+	+	+
Viburnum rafinesquianum							+	+
Viburnum rufidulum	+	+	+	+	+	+	+	+
Vitis aestivalis var. *aestivalis*	+	+	+	+	+	+	+	+
Vitis aestivalis var. *bicolor*			+		+	+	+	+
Vitis cinerea var. *baileyana*	+	+	+	+	+	+	+	+
Vitis cinerea var. *cinerea*	+	+	+					
Vitis labrusca			+		+	+	+	+
Vitis palmata	+	+	+					
Vitis riparia	+	+	+	+				
Vitis rotundifolia	+	+	+	+	+	+	+	+
Vitis rupestris				+				
Vitis vulpina	+	+	+	+	+	+	+	+
Wisteria frutescens	+	+	+	+	+	+	+	+
Xanthorhiza simplicissima		+	+		+	+	+	+
*Yucca filamentosa**			+	+		+		+
*Yucca flaccida**		+	+	+	+	+	+	+
*Zanthoxylum americanum**			+	+				
Totals (358)	153	200	244	200	238	271	251	261

*The distribution of these taxa within the state may have been influenced by cultivation and/or naturalization.

Appendix 2

Distribution of Tennessee's Native Woody Genera within the Physiographic Provinces

Taxa	No.	MV	CP	WHR	CB	EHR	CU	VR	U
Abies	1								+
Acer	10	+	+	+	+	+	+	+	+
Aesculus	4	+	+	+	+	+	+	+	+
Alnus	2	+	+	+	+	+	+	+	+
Amelanchier	4		+	+	+	+	+	+	+
Amorpha	2	+	+	+	+	+	+	+	+
*Ampelopsis**	2	+	+	+	+	+	+	+	+
Aralia	1	+	+	+	+	+	+	+	+
Aristolochia	2	+	+	+	+	+	+	+	+
Aronia	2		+	+		+	+	+	+
Arundinaria	1	+	+	+	+	+	+	+	+
Asimina	1	+	+	+	+	+	+	+	+
Berberis	1						+	+	+
Berchemia	1	+	+	+	+	+	+	+	
Betula	4	+	+	+	+	+	+	+	+
Bignonia	1	+	+	+	+	+	+	+	+
Brunnichia	1	+	+	+					
Buckleya	1								+

Taxa	No.	MV	CP	WHR	CB	EHR	CU	VR	U
Bumelia	1	+	+	+	+	+	+	+	
Buxella	1						+		
*Callicarpa**	1	+	+	+	+	+	+	+	+
Calycanthus	2			+		+	+	+	+
Calycocarpum	1	+	+	+	+	+	+	+	
Campsis	1	+	+	+	+	+	+	+	+
Carpinus	1	+	+	+	+	+	+	+	+
*Carya**	9	+	+	+	+	+	+	+	+
Castanea	2	+	+	+	+	+	+	+	+
*Catalpa**	1	+	+						
Ceanothus	1		+	+	+	+	+	+	+
Celastrus	1	+	+	+	+	+	+	+	+
Celtis	3	+	+	+	+	+	+	+	+
Cephalanthus	1	+	+	+	+	+	+	+	+
Cercis	1	+	+	+	+	+	+	+	+
Chimaphila	1		+	+	+	+	+	+	+
Chionanthus	1		+	+	+	+	+	+	+
*Cladrastis**	1	+	+	+	+	+	+	+	+
Clematis	2	+	+	+	+	+	+	+	+
Clethra	2					+	+		+
Cocculus	1	+	+	+	+	+	+	+	+
Comptonia	1						+		
Conradina	1					+	+		
Cornus	5	+	+	+	+	+	+	+	+
Corylus	2	+	+	+	+	+	+	+	+
Cotinus	1						+		
Crataegus	14	+	+	+	+	+	+	+	+
Croton	1					+			
Decodon	1	+			+	+		+	
Decumaria	1		+	+		+		+	+
Diervilla	3			+			+	+	+
Diospyros	1	+	+	+	+	+	+	+	+
Dirca	1			+	+	+	+	+	+
Epigaea	1			+			+	+	+
Euonymus	3	+	+	+	+	+	+	+	+
Fagus	1	+	+	+	+	+	+	+	+
Forestiera	2	+	+	+	+	+	+		
Fothergilla	1						+	+	+

Taxa	No.	MV	CP	WHR	CB	EHR	CU	VR	U
Fraxinus	4	+	+	+	+	+	+	+	+
Gaultheria	1					+	+	+	+
Gaylussacia	3			+		+	+	+	+
Gelsemium	1						+	+	
Gleditsia*	2	+	+	+	+	+	+	+	
Gymnocladus*	1	+	+	+	+			+	
Halesia	1			+			+	+	+
Hamamelis	1			+	+	+	+	+	+
Hydrangea*	4	+	+	+	+	+	+	+	+
Hypericum	8	+	+	+	+	+	+	+	+
Ilex*	6	+	+	+	+	+	+	+	+
Itea	1	+	+	+	+	+	+	+	+
Juglans	2	+	+	+	+	+	+	+	+
Juniperus	1	+	+	+	+	+	+	+	+
Kalmia	2			+	+	+	+	+	+
Leiophyllum	1								+
Leucothoe	3			+			+	+	+
Lindera	1	+	+	+	+	+	+	+	+
Linnaea	1								+
Liquidambar	1	+	+	+	+	+	+	+	+
Liriodendron	1	+	+	+	+	+	+	+	+
Lonicera*	5	+	+	+	+	+	+	+	+
Lyonia	1			+		+	+	+	+
Magnolia	5	+	+	+	+	+	+	+	+
Malus	2		+	+	+	+	+	+	+
Menispermum	1	+	+	+	+	+	+	+	+
Menziesia	1								+
Mitchella	1	+	+	+	+	+	+	+	+
Morus	1	+	+	+	+	+	+	+	+
Nemopanthus	1								+
Nestronia	1					+			+
Neviusia	1				+	+			
Nyssa	3	+	+	+	+	+	+	+	+
Opuntia	1		+	+	+	+	+	+	+
Ostrya	1	+	+	+	+	+	+	+	+
Oxydendrum	1		+	+	+	+	+	+	+
Parthenocissus	1	+	+	+	+	+	+	+	+
Paxistima	1							+	

Taxa	No.	MV	CP	WHR	CB	EHR	CU	VR	U
Philadelphus	4		+	+	+	+	+	+	+
Phoradendron	1	+	+	+	+	+	+	+	+
Physocarpus	1		+	+	+	+	+	+	+
Picea	1								+
Pieris	1								+
Pinus*	6		+	+	+	+	+	+	+
Planera	1	+	+	+					
Platanus	1	+	+	+	+	+	+	+	+
Polygonella	1						+		
Populus	3	+	+	+	+	+	+	+	+
Potentilla	1								+
Prunus	10	+	+	+	+	+	+	+	+
Ptelea	1			+	+	+	+	+	+
Pyrularia	1						+		+
Quercus	21	+	+	+	+	+	+	+	+
Rhamnus	3		+	+	+	+	+	+	+
Rhododendron	10		+	+	+	+	+	+	+
Rhus	4	+	+	+	+	+	+	+	+
Ribes	5		+				+	+	+
Robinia*	2		+	+	+	+	+	+	+
Rosa	4	+	+	+	+	+	+	+	+
Rubus	10	+	+	+	+	+	+	+	+
Salix	7	+	+	+	+	+	+	+	+
Sambucus	2	+	+	+	+	+	+	+	+
Sassafras	1	+	+	+	+	+	+	+	+
Schisandra	1	+	+						
Smilax	4	+	+	+	+	+	+	+	+
Sorbus	1								+
Spiraea	3			+	+	+	+		+
Staphylea	1	+	+	+	+	+	+	+	+
Stewartia	1						+	+	+
Styrax	2	+	+	+	+	+	+		
Symphoricarpos	1		+	+	+	+	+	+	+
Symplocos	1		+				+		+
Taxodium	1	+	+	+					
Taxus	1						+		
Thuja	1					+	+	+	+
Tilia	2	+	+	+	+	+	+	+	+

Taxa	No.	MV	CP	WHR	CB	EHR	CU	VR	U
Toxicodendron	3	+	+	+	+	+	+	+	+
Trachelospermum	1	+	+	+				+	
Tsuga	2					+	+	+	+
Ulmus	6	+	+	+	+	+	+	+	+
Vaccinium	9	+	+	+	+	+	+	+	+
Viburnum	10	+	+	+	+	+	+	+	+
Vitis	10	+	+	+	+	+	+	+	+
Wisteria	1	+	+	+	+	+	+	+	+
Xanthorhiza	1		+	+		+	+	+	+
Yucca*	2		+	+	+	+	+	+	+
Zanthoxylum*	1			+	+				
Totals (143)	358	81	98	109	97	108	119	112	118

*The distribution of these taxa (or at least one taxon in genera with more than one taxon) within the state may have been influenced by cultivation and/or naturalization.

Appendix 3

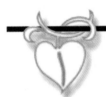

Cultivated and/or Persisting Taxa Represented by Collections but Not Included in Keys and Generic Treatments

Acer platanoides L., Norway Maple (Aceraceae)

Aesculus hippocastanum L., European Horse Chestnut (Hippocastanaceae)

Alnus glutinosa (L.) Gaertn., Black/European Alder (Betulaceae)

Buddleja davidii Franch., Butterfly Bush (Buddlejaceae)

Buxus sempervirens L., Boxwood (Buxaceae)

Callicarpa dichotoma (Lour.) K.Koch, Beautyberry (Verbenaceae)

Castanea mollissima Blume, Chinese Chestnut (Fagaceae)

Cotoneaster pyracantha (L.) Spach, Pyracantha (Rosaceae)

Cydonia japonica Pers., Flowering Quince (Rosaceae)

Cytisus scoparius (L.) Link, Scotch Broom (Fabaceae)

Euonymus fortunei (Turcz.) Hand.-Mazz., Wintercreeper (Celastraceae)

Fagus sylvatica L., European Beech (Fagaceae)

Ficus carica L., Fig (Moraceae)

Ginkgo biloba L., Maidenhair Tree (Ginkgoaceae)

Kerria japonica (L.) DC., Japanese Rose (Rosaceae)

Koelreuteria paniculata Laxm., Golden Rain Tree (Sapindaceae)

Lagerstroemia indica L., Crape Myrtle (Lythraceae)

Ligustrum amurense Carrière, Amur Privet (Oleaceae)

Ligustrum japonicum Thunb., Japanese Privet (Oleaceae)

Lonicera fragrantissima Lindl. & Paxton, Fragrant Honeysuckle (Caprifoliaceae)

Lonicera morrowii A.Gray, Morrow's Honeysuckle (Caprifoliaceae)

Nandina domestica Thunb., Nandina (Berberidaceae)

Philadelphus coronarius L., Mock Orange (Saxifragaceae)

Picea abies (L.) Karst., Norway Spruce (Pinaceae)

Populus x *canescens* (Aiton) J.E.Smith, Gray Poplar (Salicaceae)

Populus nigra L., Black Poplar (Salicaceae)

Prunus avium L., Sweet Cherry (Rosaceae)

Prunus cerasus L., Sour Cherry (Rosaceae)

Prunus institia L., Damson Plum (Rosaceae)

Prunus spinosa L., Blackthorn Plum (Rosaceae)

Pyrus calleryana Decne., Bradford Pear (Rosaceae)

Quercus acutissima Carruth., Sawtooth Oak (Fagaceae)

Quercus robur L., English Oak (Fagaceae)

Quercus virginiana Mill., Southern Live Oak (Fagaceae)

Rhamnus cathartica L., Common Buckthorn (Rhamnaceae)

Rhamnus frangula L., Alder Buckthorn (Rhamnaceae)

Ribes rubrum L., Red/Garden Currant (Saxifragaceae)

Rosa bracteata J.C.Wendl., Bracted Rose (Rosaceae)

Rosa canina L., Dog Rose (Rosaceae)

Rosa gallica L., French Rose (Rosaceae)

Rosa pimpinellifolia L., Scotch Rose (Rosaceae)

Rubus laciniatus Willd., Cut-Leaved Blackberry (Rosaceae)

Salix caprea L., Goat Willow (Salicaceae)

Salix cinerea L., Gray Willow (Salicaceae)

Salix fragilis L., Crack Willow (Salicaceae)

Spiraea douglasii Hook., Douglas' Spiraea (Rosaceae)

Spiraea salicifolia L., Willow-Leaf Spiraea (Rosaceae)

Syringa vulgaris L., Lilac (Oleaceae)

Tamarix parviflora DC., Tamarisk (Tamaricaceae)

Tilia cordata Mill., European Basswood (Tiliaceae)

Ulmus parviflora Jacq., Chinese/Lacebark Elm (Ulmaceae)

Ulmus procera Salisb., English Elm (Ulmaceae)

Ulmus pumila L., Siberian Elm (Ulmaceae)

Wigelia floribunda (Sieb. & Zucc.) C.A.Mey., Wigelia (Caprifoliaceae)

REFERENCES

Austin Peay State University Center for Field Biology and University of Tennessee Herbarium. Atlas of Tennessee Vascular Plants. <http://www.bio.utk.edu/botany/herbarium/vascular/atlas.html>.

Balmford, A., A. H. M. Jayasuriya, and J.B. Green. 1996. Using higher-taxon richness as a surrogate for species richness: II. Local applications. Proc. Royal Soc. Lond. B. 263:1571–1575.

Baskin, J. M., and C. C. Baskin. 1989. Cedar glade endemics in Tennessee, and a review of their autecology. J. Tennessee Acad. Sci. 64:63–74.

Baskin, J. M., and C. C. Baskin. 1999. Cedar glades of the southeastern United States. Pp. 206–219. *In:* Anderson, R. C., J. S. Fralish, and J. M. Baskin, eds. Savannas, barrens, and rock outcrop plant communities of North America. Cambridge Univ. Press, New York, NY.

Baskin, J. M., C. C. Baskin, and E. W. Chester. 1994. The Big Barrens Region of Kentucky and Tennessee: further observations and considerations. Castanea 59:226–254.

Baskin, J. M., C. C. Baskin, and E. W. Chester. 1999. The Big Barrens Region of Kentucky and Tennessee. Pp. 190–205. *In:* Anderson, R. C., J. S. Fralish, and J. M. Baskin, eds. Savannas, barrens, and rock outcrop plant communities of North America. Cambridge Univ. Press, New York, NY.

Baskin, J. M., E. W. Chester, and C. C. Baskin. 1997. Forest vegetation of the Kentucky Karst Plain (Kentucky and Tennessee): review and synthesis. J. Torrey Bot. Soc. 124:322–335.

Bates, V. 1985. The vascular plants of Tennessee: a taxonomic and geographic guide to the floristic literature. J. Tennessee Acad. Sci. 60:66–76.

Billings, W. D., S. A. Cain, and W. B. Drew. 1937. Winter key to trees of eastern Tennessee. Castanea 2:29–44.

Braun, E. L. 1950. Deciduous forests of eastern North America. The Blakiston Co., Inc., Philadelphia, PA.

Brummitt, R. K., and C. E. Powell, eds. 1992. Authors of plant names. Royal Botanic Gardens, Kew, Great Britian.

Bryant, W. S., W. C. McComb, and J. S. Fralish. 1993. Oak-Hickory Forests (Western Mesophytic/Oak-Hickory Forests). Pp. 143–201. In: Martin, W. H., S. G. Boyce, and A. C. Echternacht, eds. Biodiversity of the Southeastern United States: Upland terrestrial communities. John Wiley & Sons, Inc., New York, NY.

Caplenor, D. 1955. An annotated list of the vascular plants of the gorges of Fall Creek Falls State Park. J. Tennessee Acad. Sci. 30:93–108.

Caplenor, D. 1965. The vegetation of the gorges of the Fall Creek Falls State Park in Tennessee. J. Tennessee Acad. Sci. 40:27–39.

Carpenter, J. S., and E. W. Chester. 1987. Vascular flora of the Bear Creek Natural Area, Stewart County, Tennessee. Castanea 52:112–128.

Chester, E. W. 1993. Vascular flora of Land Between the Lakes, Kentucky and Tennessee: an updated checklist. J. Tennessee Acad. Sci. 68:1–14.

Chester, E. W., ed. 1989. The vegetation and flora of Tennessee. Proceedings of a symposium sponsored by the Austin Peay State University Center for Field Biology. J. Tennessee Acad. Sci. 64, No. 3.

Chester, E. W., R. J. Jensen, and J. Schibig. 1995. Forest communities of Montgomery and Stewart counties, northwestern Middle Tennessee. J. Tennessee Acad. Sci. 70:82–91.

Chester, E. W., B. E. Wofford, J. M. Baskin, and C. C. Baskin. 1997. A floristic study of barrens on the southwestern Pennyroyal Plain, Kentucky and Tennessee. Castanea 62:161–172.

Chester, E. W., B. E. Wofford, R. Kral, H. R. DeSelm, and A. M. Evans. 1993. Atlas of Tennessee vascular plants. Vol. 1. Pteridophytes, gymnosperms, angiosperms: monocots. Misc. Publ. No. 9. The Center for Field Biology, Austin Peay State University, Clarksville, TN.

Chester, E. W., B. E. Wofford, and R. Kral. 1997. Atlas of Tennessee vascular plants. Vol. 2. Angiosperms: dicots. Misc. Publ. No 13. The Center for Field Biology, Austin Peay State University, Clarksville, TN.

Clark, R. C. 1971. The woody plants of Alabama. Ann. Missouri Bot. Gard. 58:99–242.

Clark, R. C. 1974. Ilex collina, a second species of Nemopanthus in the southern Appalachians. J. Arnold Arb. 55:435–440.

Clebsch, E. E. C. 1989. Vegetation of the Appalachian Mountains of Tennessee east of the Great Valley. J. Tennessee Acad. Sci. 64:79–83.

Clements, R. K., and B. E. Wofford. 1991. The vascular flora of Wolf Cove, Franklin County, Tennessee. Castanea 56:268–285.

Crites, G. D., and E. E. C. Clebsch. 1986. Woody vegetation in the inner Nashville Basin: an example from the Cheek Bend area of the Central Duck River Valley. ASB Bull. 33:155–177.

DeSelm, H. R., and N. Murdock. 1993. Grass-dominated communities. Pp. 87–142. In: Martin, W. H., S. G. Boyce, and A. C. Echternacht, eds. Biodiversity of the Southeastern United States: Upland terrestrial communities. John Wiley & Sons, Inc., New York, NY.

Duncan, W. H. 1967. Woody vines of the southeastern states. Sida 3:1–76.

Duncan, W. H., and M.B. Duncan. 1988. Trees of the southeastern United States. Univ. of Georgia Press, Athens, GA.

Farmer, J. A., and J. L. Thomas. 1969. Disjunction and endemism in *Croton alabamensis*. Rhodora 71:94–103.

Fenneman, N. M. 1938. Physiography of Eastern United States. McGraw Hill Book Co., Inc., New York, NY.

Fernald, M. L. 1950. Gray's manual of botany. Eighth edition. Van Nostrand Reinhold Co., New York, NY.

Flora of North America Editorial Committee. 1993. Flora of North America north of Mexico. Vol. 2. Pteridophytes and Gymnosperms. Oxford Univ. Press, New York, NY.

Flora of North America Editorial Committee. 1997. Flora of North America north of Mexico. Vol. 3. Magnoliophyta: Magnoliidae and Hamamelidae. Oxford Univ. Press, New York, NY.

Foote, L. E., and S. B. Jones Jr. 1989. Native shrubs and woody vines of the southeast: landscaping uses and identification. Timber Press, Portland, OR.

Gattinger, A. 1901. The flora of Tennessee and a philosophy of botany. Gospel Advocate Publishing Co., Nashville, TN.

Ginzbarg, S. 1992. A new disjunct variety of *Croton alabamensis* (Euphorbiaceae) from Texas. Sida 15:41–52.

Gleason, H. A., and A. Cronquist. 1991. Manual of vascular plants of northeastern United States and adjacent Canada. Second edition. New York Botanical Garden, Bronx, New York, NY.

Godfrey, R. K. 1988. Trees, shrubs, and woody vines of northern Florida and adjacent Georgia and Alabama. Univ. of Georgia Press, Athens, GA.

Graham, A. 1965. Origin and evolution of the biota of southeastern North America: evidence from the fossil plant record. Evolution 18:571–585.

Graham, A. 1999. Late Cretaceous and Cenozoic history of North American vegetation north of Mexico. Oxford Univ. Press, New York, NY.

Griffith, G. E., J. M. Omernik, and S. H. Azevedo. 1997. Ecoregions of Tennessee. Western Ecology Division, National Health and Environmental Effects Research Laboratory, U.S. Environmental Protection Agency, Corvallis, OR.

Guthrie, M. 1989. A floristic and vegetational overview of Reelfoot Lake. J. Tennessee Acad. Sci. 64:113–116.

Hackney, C. T., S. M. Adams, and W. H. Martin. 1992. Biodiversity of the Southeastern United States: Aquatic communities. John Wiley & Sons, Inc., New York, NY.

Harlow, W. M., E. S. Harrar, J. W. Hardin, and F. M. White. 1996. Textbook of dendrology. Eighth edition. McGraw-Hill, Inc., New York, NY.

Heineke, T. E. 1989. Plant communities and flora of West Tennessee between the loess hills and the Tennessee River. J. Tennessee Acad. Sci. 64:117–119.

Henson, J. W. 1990. Aquatic and certain wetland vascular vegetation of Reelfoot Lake, 1920s–1980s. I. Floristic survey. J. Tennessee Acad. Sci. 65:63–68.

Hinkle, C. R. 1989. Forest communities on the Cumberland Plateau of Tennessee. J. Tennessee Acad. Sci. 64:123–129.

Hinkle, C. R., W. C. McComb, J. M. Safley Jr., and P. A. Schmalzer. 1993. Mixed Mesophytic Forests. Pp. 203–253. In: Martin, W. H., S. G. Boyce, and A. C. Echternacht, eds. Biodiversity of the Southeastern United States: Upland terrestrial communities. John Wiley & Sons, Inc., New York, NY.

Isely, D. 1990. Vascular flora of the southeastern United States. Volume 3, Part 2. Leguminosae (Fabaceae). Univ. of North Carolina Press, Chapel Hill, NC.

Isely, D. 1998. Native and naturalized Leguminosae (Fabaceae) of the United States (exclusive of Alaska and Hawaii). Brigham Young Univ., Provo, UT.

Jones, R. L. 1989. A floristic study of wetlands on the Cumberland Plateau of Tennessee. J. Tennessee Acad. Sci. 64:131–134.

Joyner, J. M., and E. W. Chester. 1994. The vascular flora of Cross Creeks National Wildlife Refuge, Stewart County, Tennessee. Castanea 59:117–145.

Kartesz, J. T. 1994. A synonymized checklist of the vascular flora of the United States, Canada, and Greenland. Vol. 1. Checklist. Timber Press, Portland, OR.

Kartesz, J. T., and C. A. Meacham. 1999. Synthesis of the North American flora, Version 1.0. North Carolina Botanical Garden, Chapel Hill, NC.

Küchler, A. W. 1964. Potential natural vegetation of the conterminous United States. Special Publ. 36. Amer. Geogr. Soc., Washington, DC.

Lance, R. W., and J. B. Phipps. 2000. Crataegus harbisonii Beadle rediscovered and amplified. Castanea 65:291–296.

Lewis, P. O., and E. T. Browne Jr. 1991. The vascular flora of Haywood County, Tennessee. J. Tennessee Acad. Sci. 66:37–44.

Little, E. L., Jr. 1971. Atlas of United States Trees. Vol. 1. Conifers and important hardwoods. Misc. Publ. No. 1146. U.S.D.A., Forest Service, Washington, DC.

Little, E. L., Jr. 1977. Atlas of United States Trees. Vol. 4. Minor eastern hardwoods. Misc. Publ. No. 1342. U.S.D.A., Forest Service, Washington, DC.

Little, E. L., Jr. 1981. Atlas of United States Trees. Vol. 6. Supplement. Misc. Publ. No. 1410. U.S.D.A., Forest Service, Washington, DC.

Luther, E. T. 1977. Our restless earth, the geologic regions of Tennessee. Univ. of Tennessee Press, Knoxville, TN.

Luteyn, J. L., W. S. Judd, S. P. Vander Kloet, L. L. Dorr, G. D. Wallace, K. A. Kron, P. F. Stevens, and S. E. Clemants. 1996. Ericaceae of the southeastern United States. Castanea 61:101–144.

Martin, W. H. 1989. Forest patterns in the Great Valley of Tennessee. J. Tennessee Acad. Sci. 64:137–143.

Martin, W. H., S. G. Boyce, and A. C. Echternacht. 1993a. Biodiversity of the Southeastern United States: Upland terrestrial communities. John Wiley & Sons, Inc., New York, NY.

Martin, W. H., S. G. Boyce, and A. C. Echternacht. 1993b. Biodiversity of the Southeastern United States: Lowland terrestrial communities. John Wiley & Sons, Inc., New York, NY.

McKinney, L. E. 1986. The vascular flora of Short Mountain (Cannon County) Tennessee. J. Tennessee Acad. Sci, 61:20–24.

McKinney, L. E. 1989. Vegetation of the Eastern Highland Rim of Tennessee. J. Tennessee Acad. Sci. 64:145–147.

McKinney, L. E., and T. E. Hemmerly. 1984. Preliminary study of deciduous forest within the inner Central Basin of Middle Tennessee. J. Tennessee Acad. Sci. 59:40–41.

Miller, R. A. 1974. The geologic history of Tennessee. Bulletin 74. Tennessee Division of Geology, Department of Environment and Conservation, Nashville, TN.

Miller, N. A., and J. Neiswender. 1989. A plant community study of the Third Chickasaw Bluff, Shelby County, Tennessee. J. Tennessee Acad. Sci. 64: 149–154.

Moore, M. O. 1991. Classification and systematics of eastern North American Vitis L. (Vitaceae) north of Mexico. Sida 14:339–367.

Murrell, S. E., and B. E. Wofford. 1987. Floristics and phytogeography of Big Frog Mountain, Polk County, Tennessee. Castanea 52:262–290.

Quarterman, E., B. H. Turner, and T. E. Hemmerly. 1972. Analysis of virgin mixed mesophytic forests in Savage Gulf, Tennessee. Bull. Torrey Bot. Club 99:228–232.

Quarterman, E., M. P. Burbanck, and D. J. Shure. 1993. Rock outcrop communities: limestone, sandstone, and granite. Pp. 35–86. In: Martin, W. H., S. G. Boyce, and A. C. Echternacht, eds. Biodiversity of the Southeastern United States: Upland terrestrial communities. John Wiley & Sons, Inc., New York, NY.

Radford, A. E., H. E. Ahles, and C. R. Bell. 1968. Manual of the vascular flora of the Carolinas. Univ. of North Carolina Press, Chapel Hill, NC.

Ramseur, G. S. 1989. Some changes in the vegetation of the Great Smoky Mountains. J. Tennessee Acad. Sci. 64:159–160.

Robertson, R. K. 1974. The genera of Rosaceae in the southeastern United States. J. Arnold Arb. 55:303–332, 344–401, 611–662.

Schibig, J. 1996. Twenty years of forest change at Radnor Lake Natural Area, Davidson County, Middle Tennessee. J. Tennessee Acad. Sci. 71:57–72.

Schmalzer, P. A. 1989. Vegetation and flora of the Obed River Gorge System, Cumberland Plateau, Tennessee. J. Tennessee Acad. Sci. 64:161–168.

Shanks, R. E. 1952a. Checklist of woody plants of Tennessee. J. Tennessee Acad. Sci. 27:27–50.

Shanks, R. E. 1952b. Notes on woody plant distribution in Tennessee. Castanea 17:90–96.

Shanks, R. E. 1953. Woody plants of Tennessee: first supplement. J. Tennessee Acad. Sci. 28:158–159.

Shanks, R. E. 1954. Woody plants of Tennessee: second supplement. J. Tennessee Acad. Sci. 29:234–237.

Shanks, R. E. 1958. Floristic regions of Tennessee. J. Tennessee Acad. Sci. 33:195–210.

Shanks, R. E., and A. J. Sharp. 1947. Summer key to the trees of eastern Tennessee. J. Tennessee Acad. Sci. 22:114–133.

Shanks, R. E., and A. J. Sharp. 1950. Summer key to Tennessee trees. New Series No. 124. Contribution from the Department of Botany, University of Tennessee, Knoxville, TN.

Shanks, R. E., and A. J. Sharp. 1963. Summer key to Tennessee trees. Univ. of Tennessee Press, Knoxville, TN.

Sharitz, R. R., and W. J. Mitsch. 1993. Southern floodplain forests. Pp. 311–372. In: Martin, W. H., S. G. Boyce, and A. C. Echternacht, eds. Biodiversity of the Southeastern United States: Lowland terrestrial communities. John Wiley & Sons, Inc., New York, NY.

Sharp, A. J. 1966. The origin and relationships of the Southern Appalachian flora. Phi Kappa Phi 11th Annual Faculty Lecture, 5 May, 1966. Typescript. Herbarium, Austin Peay State University, Clarksville, TN.

Sharp, A. J. 1970. Epilogue. Pp. 405–410. In: Holt, P. C., ed. The distributional history of the biota of the Southern Appalachians. Part II. Flora. Proceedings of a symposium sponsored by Virginia Polytechnic Institute and the Association of

Southeastern Biologists. Research Division Monograph 2. Virginia Polytechnic Institute, Blacksburg, VA.

Sharp, A. J., R. E. Shanks, J. K. Underwood, and E. McGilliard. 1956. A preliminary checklist of monocots of Tennessee. Mimeographed. Herbarium, University of Tennessee, Knoxville, TN.

Sharp, A. J., R. E. Shanks, H. L. Sherman, and D. H. Norris. 1960. A preliminary checklist of dicots of Tennessee. Mimeographed. Herbarium, University of Tennessee, Knoxville, TN.

Shaver, J. M. 1954. Ferns of Tennessee with the fern allies excluded. George Peabody College for Teachers, Nashville, TN.

Shaw, J. T. 2001. The woody plants of the Big South Fork National River and Recreation Area (BSFNRRA). M.S. thesis. University of Tennessee, Knoxville, TN.

Small, J. K. 1933. Manual of the southeastern flora. University of North Carolina Press, Chapel Hill, NC.

Smith, E. B. 1994. Keys to the flora of Arkansas. University of Arkansas Press, Fayetteville, AR.

Somers, P., ed. 1986. Proceedings of a symposium on the biota, ecology, and ecological history of cedar glades. ASB Bull. 33, No. 4.

Souza, K. H., and R. Kral. 1990. The vascular flora of Dickson County, Tennessee. J. Tennessee Acad. Sci. 65:91–100.

Stephenson, S. L., A. N. Ash, and D. F. Stauffer. 1993. Appalachian Oak Forests. Pp. 255–303. In: Martin, W. H., S. G. Boyce, and A. C. Echternacht, eds. Biodiversity of the Southeastern United States: Upland terrestrial communities. John Wiley & Sons, Inc., New York, NY.

Swanson, R. E. 1994. A field guide to the trees and shrubs of the southern Appalachians. Johns Hopkins University Press, Baltimore, MD.

Tennessee Natural Heritage Program. 2001. Tennessee rare plant list. Division of Natural Heritage, Tennessee Department of Environment and Conservation, Nashville, TN.

Thomas, R. D. 1989. The vegetation of Chilhowee Mountain, Tennessee. J. Tennessee Acad. Sci. 64:185–188.

Warden, J. C. 1989. Changes in the spruce-fir forest of Roan Mountain in Tennessee over the past fifty years as a result of logging. J. Tennessee Acad. Sci. 64:193–195.

Webb, D. E., and A.L. Bates. 1989. The aquatic vascular flora and plant communities along rivers and reservoirs of the Tennessee River system. J. Tennessee Acad. Sci. 64:197–203.

White, P. S. 1981. Looking for Linnaea. The Tennessee Conservationist 57(5): 14–16.

White, P. S. 1982. The flora of Great Smoky Mountains National Park: An annotated checklist of the vascular plants and a review of previous work. U.S.D.I., National Park Service, SE Region. Res./Resour. Manag. Dept. SER-55.

White, P. S., E. R. Buckner, J. D. Pittillo, and C. V. Cogbill. 1993. High-elevation forests: Spruce-Fir Forests, Northern Hardwoods Forests, and associated communities. Pp. 305–337. *In:* Martin, W. H., S. G. Boyce, and A. C. Echternacht, eds. Biodiversity of the Southeastern United States: Upland terrestrial communities. John Wiley & Sons, Inc., New York, NY.

Wofford, B. E. 1989a. Floristic elements of the Tennessee Blue Ridge. J. Tennessee Acad. Sci. 64:205–207.

Wofford, B. E. 1989b. Guide to the vascular plants of the Blue Ridge. University of Georgia Press, Athens, GA.

Wofford, B. E., and R. Kral. 1993. Checklist of the vascular plants of Tennessee. Sida, Botanical. Miscellany. No.10. Botanical Research Institute of Texas, Inc., Fort Worth, TX.

Wofford, B. E., T. S. Patrick, L. R. Phillippe, and D. H. Webb. 1979. The vascular flora of Savage Gulf, Tennessee. Sida 8:135–151.

Index to Families and Genera

Families are given alphabetically, followed by an alphabetical listing of included genera with page numbers.

INDEX TO SCIENTIFIC NAMES

This index includes all taxa found in the generic treatments; italics refer to synonyms. Families are not included here but are listed alphabetically with included genera in a separate index, as are common names. Excluded introductions are alphabetized by genus in Appendix 3.

INdEX to Common Names

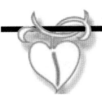

PLATES

Plate 1. *Abies fraseri*

Plate 2. *Juniperus virginiana*

Plate 3. *Picea rubens*

Plate 4. *Pinus echinata*

Plate 5. *Pinus pungens*

Plate 6. *Pinus rigida*

Plate 7. *Pinus strobus*

Plate 8. *Pinus taeda*

Plate 9. *Pinus virginiana*

Plate 10. *Taxodium distichum*

Plate 11. *Taxus canadensis*

Plate 12. *Thuja occidentalis*

Plate 13. *Tsuga canadensis*

Plate 14. *Tsuga caroliniana*

Plate 15. *Arundinaria gigantea*

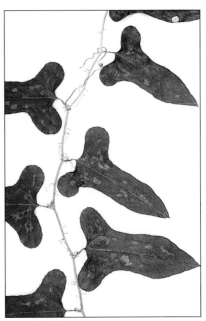

Plate 16. *Smilax bona-nox*

Plate 17. *Smilax glauca*   Plate 18. *Smilax rotundifolia*

Plate 19. *Smilax tamnoides*   Plate 20. *Yucca filamentosa*

Plate 21. *Acer drummondii*

Plate 22. *Acer negundo*

Plate 23. *Acer pensylvanicum*

Plate 24. *Acer rubrum*

Plate 25. *Acer saccharinum*

Plate 26. *Acer saccharum* ssp. *floridanum*

Plate 27. *Acer saccharum* ssp. *leucoderme*

Plate 28. *Acer saccharum* ssp. *nigrum*

Plate 29. *Acer saccharum* ssp. *saccharum*

Plate 30. *Acer spicatum*

Plate 31. *Aesculus flava*

Plate 32. *Aesculus glabra*

Plate 33. *Aesculus pavia*

Plate 34. *Aesculus sylvatica*

Plate 35. *Ailanthus altissima*

Plate 36. *Albizia julibrissin*

Plate 37. *Alnus serrulata*

Plate 38. *Alnus viridis* ssp. *crispa*

Plate 39. *Amelanchier arborea*

Plate 40. *Amelanchier canadensis*

Plate 41. *Amelanchier laevis*

Plate 42. *Amelanchier sanguinea*

Plate 43. *Amorpha fruticosa*

Plate 44. *Amorpha glabra*

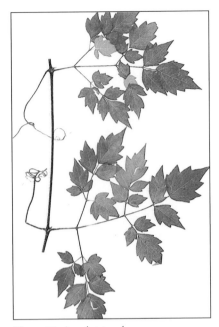

Plate 45. *Ampelopsis arborea*

Plate 46. *Ampelopsis cordata*

Plate 47. *Aralia spinosa*

Plate 48. *Aristolochia macrophylla*

Plate 49. *Aristolochia tomentosa*

Plate 50. *Aronia arbutifolia*

Plate 51. *Aronia melanocarpa*

Plate 52. *Asimina triloba*

Plate 53. *Berberis canadensis*

Plate 54. *Berberis thunbergii*

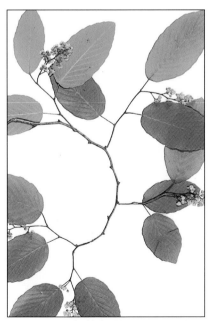

Plate 55. *Berchemia scandens*

Plate 56. *Betula alleghaniensis*

Plate 57. *Betula cordifolia*

Plate 58. *Betula lenta*

Plate 59. *Betula nigra*

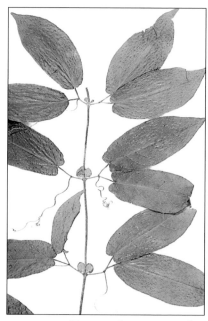

Plate 60. *Bignonia capreolata*

Plate 61. *Broussonetia papyrifera*

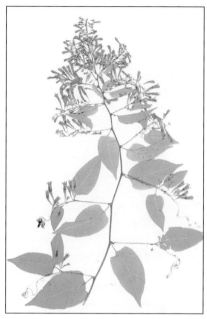

Plate 62. *Brunnichia ovata*

Plate 63. *Buckleya distichophylla*

Plate 64. *Bumelia lycioides*

Plate 65. *Buxella brachycera*

Plate 66. *Callicarpa americana*

Plate 67. *Calycanthus floridus* var. *floridus*

Plate 68. *Calycocarpum lyonii*

Plate 69. *Campsis radicans*

Plate 70. *Carpinus caroliniana*

Plate 71. *Carya aquatica* (80a)

Plate 72. *Carya cordiformis* (80b)

Plate 73. *Carya glabra* (80c)

Plate 74. *Carya illinoinensis* (80d)

Plate 75. *Carya laciniosa* (80e)

Plate 76. *Carya ovata* var. *australis* (80f)

Plate 77. *Carya ovata* var. *ovata* (80g)

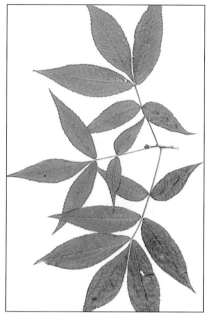

Plate 78. *Carya pallida* (80h)

Plate 79. *Carya tomentosa* (80i)

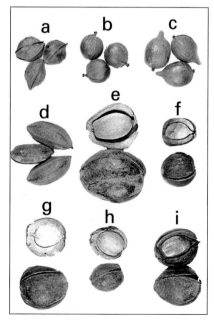

Plate 80. *Carya* fruits

Plate 81. *Castanea dentata*

Plate 82. *Castanea pumila*

Plate 83. *Catalpa bignonioides*

Plate 84. *Catalpa speciosa*

Plate 85. *Ceanothus americanus*

Plate 86. *Celastrus orbiculatus*

Plate 87. *Celastrus scandens*

Plate 88. *Celtis laevigata*

Plate 89. *Celtis occidentalis*

Plate 90. *Celtis tenuifolia*

Plate 91. *Cephalanthus occidentalis*

Plate 92. *Cercis canadensis*

Plate 93. *Chimaphila maculata*

Plate 94. *Chionanthus virginicus*

Plate 95. *Cladrastis kentukea*

Plate 96. *Clematis catesbyana*

Plate 97. *Clematis terniflora*

Plate 98. *Clematis virginiana*

Plate 99. *Clethra acuminata*

Plate 100. *Clethra alnifolia*

Plate 101. *Cocculus carolinus*

Plate 102. *Comptonia peregrina*

Plate 103. *Conradina verticillata*

Plate 104. *Cornus alternifolia*

Plate 105. *Cornus amomum*

Plate 106. *Cornus drummondii*

Plate 107. *Cornus florida*

Plate 108. *Cornus foemina*

Plate 109. *Corylus americana*

Plate 110. *Corylus cornuta*

Plate 111. *Cotinus obovatus*

Plate 112. *Crataegus calpodendron*

Plate 113. *Crataegus collina*

Plate 114. *Crataegus crus-galli*

Plate 115. *Crataegus harbisonii*

Plate 116. *Crataegus intricata*

Plate 117. *Crataegus macrosperma*

Plate 118. *Crataegus marshallii*

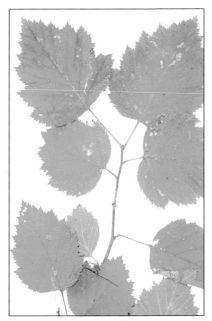

Plate 119. *Crataegus mollis*

Plate 120. *Crataegus phaenopyrum*

Plate 121. *Crataegus pruinosa*

Plate 122. *Crataegus punctata*

Plate 123. *Crataegus spathulata*

Plate 124. *Crataegus uniflora*

Plate 125. *Crataegus viridis*

Plate 126. *Croton alabamensis*

Plate 127. *Decodon verticillatus*

Plate 128. *Decumaria barbara*

Plate 129. *Diervilla lonicera*

Plate 130. *Diervilla sessilifolia*

Plate 131. *Diospyros virginiana*

Plate 132. *Dirca palustris*

Plate 133. *Elaeagnus pungens*

Plate 134. *Elaeagnus umbellata*

Plate 135. *Epigaea repens*

Plate 136. *Euonymus alatus*

Plate 137. *Euonymus americanus*

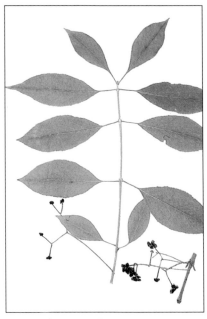

Plate 138. *Euonymus atropurpureus*

Plate 139. *Euonymus obovatus*

Plate 140. *Fagus grandifolia*

Plate 141. *Forestiera acuminata*

Plate 142. *Forestiera ligustrina*

Plate 143. *Forsythia viridissima*

Plate 144. *Fothergilla major*

Plate 145. *Fraxinus americana*

Plate 146. *Fraxinus pennsylvanica*

Plate 147. *Fraxinus profunda*

Plate 148. *Fraxinus quadrangulata*

Plate 149. *Gaultheria procumbens*

Plate 150. *Gaylussacia baccata*

Plate 151. *Gaylussacia dumosa*

Plate 152. *Gaylussacia ursina*

Plate 153. *Gelsemium sempervirens*

Plate 154. *Gleditsia aquatica*

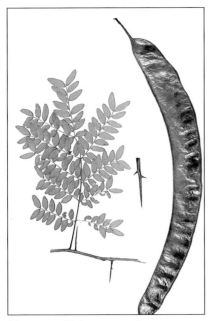

Plate 155. *Gleditsia triacanthos*

Plate 156. *Gymnocladus dioicus*

Plate 157. *Halesia tetraptera*

Plate 158. *Hamamelis virginiana*

Plate 159. *Hedera helix*

Plate 160. *Hibiscus syriacus*

Plate 161. *Hydrangea arborescens*

Plate 162. *Hydrangea cinerea*

Plate 163. *Hydrangea quercifolia*

Plate 164. *Hydrangea radiata*

Plate 165. *Hypericum crux-andreae*

Plate 166. *Hypericum densiflorum*

Plate 167. *Hypericum frondosum*

Plate 168. *Hypericum hypericoides*

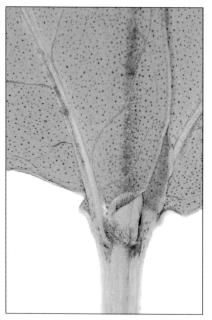

Plate 169. *Hypericum nudiflorum*

Plate 170. *Hypericum prolificum*

Plate 171. *Hypericum stragulum*

Plate 172. *Ilex ambigua* var. *ambigua*

Plate 173. *Ilex ambigua var. montana*

Plate 174. *Ilex decidua* var. *decidua*

Plate 175. *Ilex decidua* var. *longipes*

Plate 176. *Ilex opaca*

Plate 177. *Ilex verticillata*

Plate 178. *Itea virginica*

Plate 179. *Juglans cinerea*

Plate 180. *Juglans nigra*

Plate 181. *Kalmia angustifolia* var. *carolina*

Plate 182. *Kalmia latifolia*

Plate 183. *Leiophyllum buxifolium*

Plate 184. *Lespedeza bicolor*

Plate 185. *Leucothoe fontanesiana*

Plate 186. *Leucothoe racemosa*

Plate 187. *Leucothoe recurva*

Plate 188. *Ligustrum sinense*

Plate 189. *Lindera benzoin*

Plate 190. *Linnaea borealis*

Plate 191. *Liquidambar styraciflua*

Plate 192. *Liriodendron tulipifera*

Plate 193. *Lonicera canadensis*

Plate 194. *Lonicera dioica*

Plate 195. *Lonicera flava*

Plate 196. *Lonicera japonica*

Plate 197. *Lonicera maackii*

Plate 198. *Lonicera sempervirens*

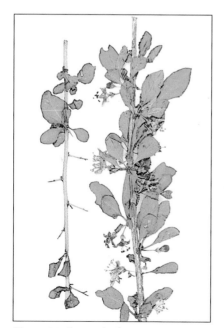

Plate 199. *Lycium barbarum*

Plate 200. *Lyonia ligustrina*

Plate 201. *Maclura pomifera*

Plate 202. *Magnolia acuminata*

Plate 203. *Magnolia fraseri*

Plate 204. *Magnolia grandiflora*

Plate 205. *Magnolia macrophylla*

Plate 206. *Magnolia tripetala*

Plate 207. *Magnolia virginiana*

Plate 208. *Malus angustifolia*

Plate 209. *Malus coronaria*

Plate 210. *Malus pumila*

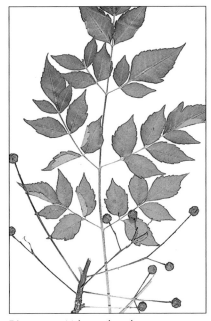

Plate 211. *Melia azedarach*

Plate 212. *Menispermum canadense*

Plate 213. *Menziesia pilosa*

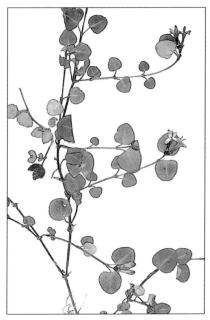

Plate 214. *Mitchella repens*

Plate 215. *Morus alba*

Plate 216. *Morus rubra*

Plate 217. *Nemopanthus collinus*

Plate 218. *Nestronia umbellula*

Plate 219. *Neviusia alabamensis*

Plate 220. *Nyssa aquatica*

Plate 221. *Nyssa biflora*

Plate 222. *Nyssa sylvatica*

Plate 223. *Opuntia humifusa*

Plate 224. *Ostrya virginiana*

Plate 225. *Oxydendrum arboreum*

Plate 226. *Parthenocissus quinquefolia*

Plate 227. *Paulownia tomentosa*

Plate 228. *Paxistima canbyi*

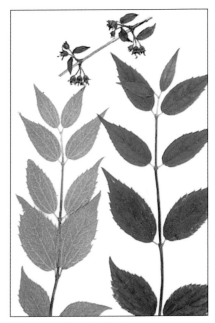

Plate 229. *Philadelphus hirsutus*

Plate 230. *Philadelphus inodorus*

Plate 231. *Philadelphus pubescens* var. *pubescens*

Plate 232. *Phoradendron leucarpum*

Plate 233. *Physocarpus opulifolius*

Plate 234. *Pieris floribunda*

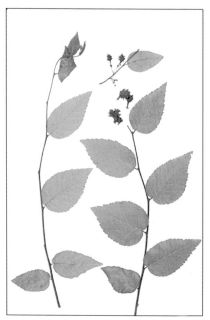

Plate 235. *Planera aquatica*

Plate 236. *Platanus occidentalis*

Plate 237. *Polygonella americana*

Plate 238. *Poncirus trifoliata*

Plate 239. *Populus alba*

Plate 240. *Populus deltoides*

Plate 241. *Populus grandidentata*

Plate 242. *Populus heterophylla*

Plate 243. *Populus x jackii*

Plate 244. *Potentilla tridentata*

Plate 245. *Prunus americana*

Plate 246. *Prunus angustifolia*

Plate 247. *Prunus hortulana*

Plate 248. *Prunus mahaleb*

Plate 249. *Prunus mexicana*

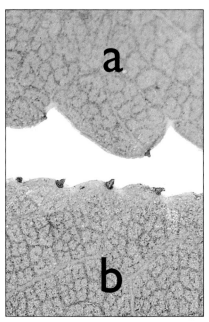

Plate 250. a. *Prunus hortuluana*
b. *P. munsoniana*

Plate 251. *Prunus pensylvanica*

Plate 252. *Prunus persica*

Plate 253. *Prunus pumila*

Plate 254. *Prunus serotina*

Plate 255. *Prunus umbellata*

Plate 256. *Prunus virginiana*

Plate 257. *Ptelea trifoliata*

Plate 258. *Pueraria montana var. lobata*

Plate 259. *Pyrularia pubera*

Plate 260. *Pyrus communis*

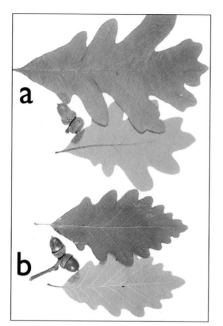

Plate 261. a. *Quercus alba;* b. *Q. bicolor*

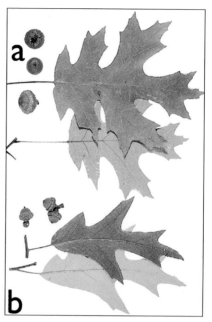

Plate 262. a. *Q. coccinea;* b. *Q. falcata*

Plate 263. a.*Q. imbricaria;* b.*Q. lyrata*

Plate 264. a. *Q. macrocarpa;* b. *Q. marilandica*

Plate 265. a.Q. *michauxii*; b. Q. *montana*

Plate 266. a. Q. *muhlenbergii*; b. Q. *nigra*

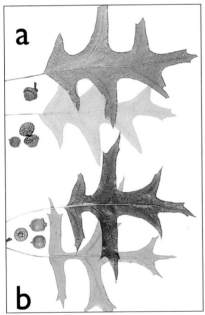

Plate 267. a. Q. *pagoda*; b. Q. *palustris*

Plate 268. a. Q. *phellos*; b. Q. *prinoides*;
c. Q. *rubra*

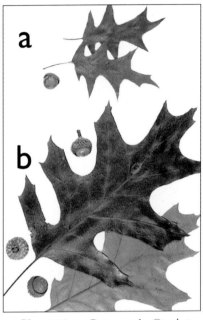

Plate 269. a. *Q. shumardii;* b. *Q. stellata*   Plate 270. a. *Q. texana;* b. *Q. velutina*

Plate 271. *Rhamnus alnifolia*   Plate 272. *Rhamnus caroliniana*

Plate 273. *Rhamnus lanceolata*

Plate 274. *Rhododendron alabamense*

Plate 275. *Rhododendron arborescens*

Plate 276. *Rhododendron calendulaceum*

Plate 277. *Rhododendron canescens*

Plate 278. *Rhododendron catawbiense*

Plate 279. *Rhododendron cumberlandense*

Plate 280. *Rhododendron maximum*

Plate 281. *Rhododendron minus*

Plate 282. *Rhododendron periclymenoides*

Plate 283. *Rhododendron viscosum*

Plate 284. *Rhus aromatica*

Plate 285. *Rhus copallinum*

Plate 286. *Rhus glabra*

Plate 287. *Rhus typhina*

Plate 288. *Ribes curvatum*

Plate 289. *Ribes cynosbati*

Plate 290. *Ribes glandulosum*

Plate 291. *Ribes odoratum*

Plate 292. *Ribes rotundifolium*

Plate 293. *Robinia hispida*

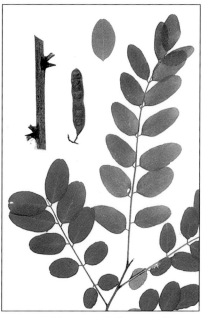

Plate 294. *Robinia pseudoacacia*

Plate 295. *Rosa carolina*

Plate 296. *Rosa eglanteria*

Plate 297. *Rosa multiflora*

Plate 298. *Rosa palustris*

Plate 299. *Rosa setigera*

Plate 300. *Rosa virginiana*

Plate 301. *Rubus allegheniensis*

Plate 302. *Rubus argutus*

Plate 303. *Rubus bifrons*

Plate 304. *Rubus canadensis*

Plate 305. *Rubus flagellaris*

Plate 306. *Rubus hispidus*

Plate 307. *Rubus idaeus* ssp. *strigosus*

Plate 308. *Rubus occidentalis*

Plate 309. *Rubus odoratus*

Plate 310. *Rubus phoenocalasius*

Plate 311. *Rubus trivialis*

Plate 312. *Salix babylonica*

Plate 313. *Salix caroliniana*

Plate 314. *Salix eriocephala*

Plate 315. *Salix exigua*

Plate 316. *Salix humilis* var. *humilis*

Plate 317. *Salix humilis* var. *microphyllus*

Plate 318. *Salix nigra*

Plate 319. *Salix sericea*

pith

Plate 320. *Sambucus canadensis*

Plate 321. *Sambucus racemosa* ssp. *pubens*

Plate 322. *Sassafras albidum*

Plate 323. *Schisandra glabra*

Plate 324. *Solanum dulcamara*

Plate 325. *Sorbus americana*

Plate 326. *Spiraea alba*

Plate 327. *Spiraea japonica*

Plate 328. *Spiraea prunifolia*

Plate 329. *Spiraea tomentosa*

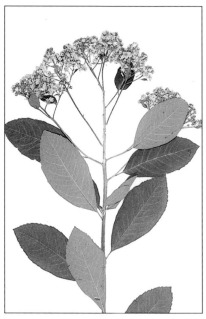

Plate 330. *Spiraea virginiana*

Plate 331. *Staphylea trifolia*

Plate 332. *Stewartia ovata*

Plate 333. *Styrax americana*

Plate 334. *Styrax grandifolia*

Plate 335. *Symphoricarpos orbiculatus*

Plate 336. *Symplocos tinctoria*

Plate 337. *Tilia americana* var. *heterophylla*

Plate 338. *Toxicodendron pubescens*

Plate 339. *Toxicodendron radicans*

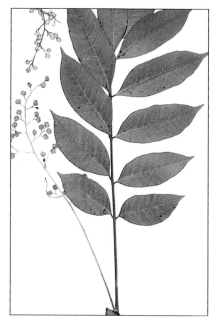

Plate 340. *Toxicodendron vernix*

Plate 341. *Trachelospermum difforme*

Plate 342. *Ulmus alata*

Plate 343. *Ulmus americana*

Plate 344. *Ulmus crassifolia*

Plate 345. *Ulmus rubra*

Plate 346. *Ulmus serotina*

Plate 347. *Ulmus thomasii*

Plate 348. *Vaccinium arboreum*

Plate 349. *Vaccinium corymbosum*

Plate 350. *Vaccinium elliottii*

Plate 351. *Vaccinium erythrocarpum*

Plate 352. *Vaccinium fuscatum*

Plate 353. *Vaccinium hirsutum*

Plate 354. *Vaccinium macrocarpon*

Plate 355. *Vaccinium pallidum*

Plate 356. *Vaccinium stamineum*

Plate 357. *Viburnum acerifolium*

Plate 358. *Viburnum bracteatum*

Plate 359. *Viburnum cassinoides*

Plate 360. *Viburnum dentatum*

Plate 361. *Viburnum lantanoides*

Plate 362. *Viburnum molle*

Plate 363. *Viburnum nudum*

Plate 364. *Viburnum prunifolium*

Plate 365. *Viburnum rafinesquianum*

Plate 366. *Viburnum rufidulum*

Plate 367. *Vinca major*

Plate 368. *Vinca minor*

Plate 369. *Vitis aestivalis* var. *aestivalis*

Plate 370. *Vitis cinerea* var. *baileyana*

Plate 371. *Vitis cinerea* var. *cinerea*

Plate 372. *Vitis labrusca*

Plate 373. *Vitis palmata*

Plate 374. *Vitis riparia*

Plate 375. *Vitis rotundifolia*

Plate 376. *Vitis vulpina*

Plate 377. *Wisteria floribunda*

Plate 378. *Wisteria frutescens*

Plate 379. *Xanthorhiza simplicissima*

Plate 380. *Zanthoxylum americanum*